Motion by Mean Curvature and Related Topics

Motion by Mean Curvature and Related Topics

Proceedings of the International Conference
held at Trento, July 20–24, 1992

Edited by

Giuseppe Buttazzo and Augusto Visintin

Walter de Gruyter · Berlin · New York 1994

Editors

Giuseppe Buttazzo
Istituto di Matematiche Applicate
Università di Pisa
I-56100 Pisa
Italy

Augusto Visintin
Dipartimento di Matematica
Università di Trento
I-38050 Povo (Trento)
Italy

1991 Mathematics Subject Classification: 49–06; 49Q 05

Key words: Minimal surfaces, geometric measure theory, mean curvature flow, Cahn-Hilliard equation, multi-phase systems

♾ Printed on acid-free paper which falls within the guidelines of the ANSI to ensure permanence and durability.

Library of Congress Cataloging-in-Publication Data

Motion by mean curvature and related topics : proceedings of the international conference held at Trento, July 20–24, 1992 / edited by Giuseppe Buttazzo and Augusto Visintin.
p. cm.
Includes bibliographical references.
ISBN 3-11-013881-6 (acid-free)
1. Geometric measure theory–Congresses. 2. Minimal surfaces–Congresses. 3. Curvature–Congresses.. I. Buttazzo, Giuseppe. II. Visintin, A. (Augusto)
QA312.M68 1994
516.3'62–dc20 94-680
CIP

Die Deutsche Bibliothek – Cataloging-in-Publication Data

Motion by mean curvature and related topics : proceedings of the international conference held at Trento, July 20–24, 1992 / ed. by Giuseppe Buttazzo and Augusto Visintin. – Berlin ; New York : de Gruyter, 1994
ISBN 3-11-013881-6
NE: Buttazzo, Giuseppe [Hrsg.]

Printed in Germany. Printing: Arthur Collignon GmbH, Berlin. Binding: Lüderitz & Bauer GmbH, Berlin.
Cover design: Thomas Bonnie, Hamburg.

Introduction

This volume collects contributions from participants at a meeting which was held in Trento in July 1992, in the framework of the activity of C.I.R.M. (Centro Internazionale per la Ricerca Matematica) of the Istituto Trentino di Cultura. This conference was under the auspices of the Dipartimento di Matematica dell'Università di Trento, and of a research group, supported by Italian C.N.R., on *Irregular Variational Problems.*

The modelling of surface tension has led to the important theory of *minimal surfaces* already several decades ago. In the last years also other topics related to minimal surfaces have been studied; for instance, problems of *pattern recognition* and of *image segmentation.* Evolution problems have also been investigated; several of them arise in the study of two-phase systems, when interface capillarity effects are not neglected. The *Stefan problem with surface tension,* and the *Cahn-Hilliard equation* are two examples.

Recently special attention has been devoted to the study of so called *motion by mean curvature,* namely, evolution of surfaces moving with velocity pointwise proportional to their mean curvature.

These developments motivated the organization of this meeting. Virtually all of the analysts active in this field were invited to present new results. Participation was fairly large (almost 90 people) and scientifically qualified. Speakers were then asked to contribute with a paper containing original results for these proceedings. Although not all of the lecturers accepted this proposal, we believe that this volume can offer a picture of a fully developing research body.

G. Buttazzo, A. Visintin

Table of Contents

A phase-field model with a double obstacle potential

J.F. Blowey† *and C.M. Elliott*‡

Abstract. A phase field model with a double obstacle is considered. Global existence in time is proved. Relationships with sharp interface models for phase transformations are discussed. Numerical computations of moving boundary problems are presented.

1. Introduction

In this paper we consider further the double obstacle phase field model introduced by Blowey & Elliott [BE3]. The phase-field equations are

$$\begin{aligned} c\theta_t + \frac{l}{2}u_t &= k\Delta\theta, \\ \tau u_t &= \gamma\Delta u - \psi'(u) + \alpha\theta, \end{aligned} \qquad (1.1\text{a, b})$$

where θ is the temperature and u is the order/phase parameter. The positive constants c, l and k denote respectively, the specific heat, latent heat and thermal conductivity. τ, γ and α are positive parameters which may be chosen to scale with a small parameter ϵ to yield free boundary limits as $\epsilon \to 0$. The Laplacian is denoted by Δ and the equations hold in some space-time domain $\Omega_T = \Omega \times (0,T)$. $\psi(\cdot)$ is a double well potential with equal minima at $u = \pm 1$, say. Then the two phases are characterized by u taking values close to $+1$ and -1 in each phase. Generally ψ has been taken to be a polynomial and in particular the quartic, Caginalp [Ca1,Ca2],

$$\psi(r) = \frac{1}{4}\left(r^2 - 1\right)^2 \qquad r \in \mathbb{R}. \qquad (1.2)$$

We consider

$$\psi(r) = \frac{1}{2}\left(1 - r^2\right) + I_{[-1,1]}(r) \qquad r \in \mathbb{R}, \qquad (1.3)$$

† Department of Mathematical Sciences, University of Durham, Durham. This work was supported by an SERC postdoctoral fellowship, grant number GR/F85659

‡ Mathematics Division, Sussex University, Brighton, BN1 9QH, U.K.

where $I_{[-1,1]}$ is the indicator function of the interval $[-1,1]$ so that $I_{[1,1]}$ vanishes for $|r|\leqslant 1$ and is $+\infty$ for $|r|>1$. It follows that (1.1b) becomes

$$-\tau u_t+\gamma\Delta u+u+\alpha\theta\in\partial I_{[-1,1]}(u), \tag{1.4}$$

or equivalently the complementarity problem

$$\begin{gathered}\big(\tau u_t-\gamma\Delta u-u-\alpha\theta\big)\big(|u|-1\big)=0,\\ \big(\tau u_t-\gamma\Delta u-u-\alpha\theta\big)\operatorname{sgn} u\leqslant 0,\quad |u|\leqslant 1,\end{gathered} \tag{1.5a, b}$$

which can also be written in variational form as

$$\big(\tau u_t-u-\alpha\theta,\eta-u\big)+\gamma\big(\nabla u,\nabla\eta-\nabla u\big)\geqslant 0\qquad\forall\,\eta\in K, \tag{1.6}$$

where $K=\{\eta\in H^1(\Omega):|\eta|\leqslant 1\}$ and $(\cdot,\cdot)$ denotes the L^2 inner product.

In section 2 global existence is proved for the phase field obstacle model (PFO) consisting of (1.1a, 1.5a,b) with homogeneous Neumann boundary conditions. Scaling α,τ,γ with a small parameter ϵ yields approximations to sharp interface models for phase transformation. These are discussed in section 3. The numerical solution of (PFO) is discussed in sections 4 and 5. Computations which approximate moving boundaries are described.

2. Global existence

Throughout this paper, we assume that Ω is a bounded domain in $\mathbb{R}^d$ ($d=1,2,3$); we denote the norm of $H^p(\Omega)$ ($p\geqslant 0$) by $\|\cdot\|_p$, the semi-norm $\|D^p\eta\|_0$ by $|\eta|_p$, the $L^2(\Omega)$ inner-product by $(\cdot,\cdot)$ and the duality pairing between $\big(H^1(\Omega)\big)'$ and $H^1(\Omega)$ by $\langle\cdot,\cdot\rangle$. For $d=2,3$, we assume that $\partial\Omega$ is Lipschitz and sufficiently smooth for any required elliptic regularity. Let $K:=\{\eta\in H^1(\Omega):|\eta|\leqslant 1\ a.e.\}$, we introduce the spaces $\boldsymbol{L}^2(\Omega)=\{L^2(\Omega)\}^2$ and $\boldsymbol{H}^1(\Omega)=\{H^1(\Omega)\}^2$.

Given $\{u_0,\theta_0\}$ and $T>0$, a weak formulation of (PFO) is:- find $\{u,\theta\}\in L^2(0,T;\boldsymbol{H}^1(\Omega))\cap H^1(0,T;(\boldsymbol{H}^1(\Omega))')\cap L^\infty(0,T;\boldsymbol{L}^2(\Omega))$ such that $a.e.\ t\in(0,T)$, $u(\cdot,t)\in K$

$$\begin{gathered}\Big\langle c\theta_t+\frac{l}{2}u_t,\eta\Big\rangle+\Big(k\nabla\theta,\nabla\eta\Big)=0\qquad\forall\,\eta\in H^1(\Omega),\\ \Big\langle\tau u_t,\xi-u\Big\rangle-\Big(u+\alpha\theta,\xi-u\Big)+\gamma\Big(\nabla u,\nabla\xi-\nabla u\Big)\geqslant 0\qquad\forall\,\xi\in K,\\ u(\boldsymbol{x},0)=u_0(\boldsymbol{x}),\qquad\theta(\boldsymbol{x},0)=\theta_0(\boldsymbol{x}).\end{gathered} \tag{2.1a, b, c}$$

We will find it useful to write (2.1a,b) as

$$\begin{gathered}\Big\langle e_t,\eta\Big\rangle+\frac{k}{c}\Big(\nabla e,\nabla\eta\Big)=\frac{kl}{2c}\Big(\nabla u,\nabla\eta\Big)\qquad\forall\,\eta\in H^1(\Omega),\\ \Big\langle\tau u_t,\xi-u\Big\rangle-\Big((1-\tfrac{\alpha l}{2c})u+\tfrac{\alpha}{c}e,\xi-u\Big)+\gamma\Big(\nabla u,\nabla\xi-\nabla u\Big)\geqslant 0\qquad\forall\,\xi\in K,\end{gathered} \tag{2.2a, b}$$

where $e := c\theta + \frac{l}{2}u$ is the 'enthalpy' or heat content. We observe that $\frac{d}{dt} \fint e = 0$ so that

$$\fint e = \fint e(0) = \frac{m}{|\Omega|}, \tag{2.3}$$

where $\fint \eta := (1, \eta)|\Omega|$.

It is convenient to consider a penalization of the above problem. For $\delta > 0$, define

$$\psi_\delta(r) := \frac{1}{\delta}\int_0^r \beta_\delta(s)ds + \tfrac{1}{2}\left(1 - r^2\right) =: \hat{\psi}_\delta(r) + \tfrac{1}{2}\left(1 - r^2\right),$$

where $\beta_\delta \in C^{1,1}(\mathbb{R})$ and is given by

$$\beta_\delta(r) := \begin{cases} r - (1 + \delta/2) & r \geqslant 1 + \delta, \\ (r-1)^2/(2\delta) & 1 < r < 1 + \delta, \\ 0 & |r| \leqslant 1, \\ (r+1)^2/(2\delta) & -(1+\delta) < r < -1, \\ r + 1 + \delta/2 & r \leqslant -(1 + \delta/2). \end{cases}$$

Note that the following properties hold: for $\delta < 1/4$ there exists $C_0 > 0$ (independent of δ) such that

$$\psi_\delta(r) \geqslant -C_0\delta \qquad \forall\, r \in \mathbb{R}, \tag{2.4}$$

and β_δ is a Lipschitz continuous function with

$$0 \leqslant \beta'_\delta(r) \leqslant 1 \text{ and } \beta_\delta(r)r \geqslant |\beta_\delta(r)| \qquad \forall\, r \in \mathbb{R}. \tag{2.5a, b}$$

Obviously $\hat{\psi}_\delta(r)$ satisfies $\hat{\psi}'_\delta(r) = \frac{1}{\delta}\beta_\delta(r)$, $0 \leqslant \hat{\psi}''_\delta(r) \leqslant 1/\delta$,

$$\hat{\psi}_\delta(r) \geqslant \tfrac{1}{2\delta}\left(\beta_\delta(r)\right)^2 \text{ and } \hat{\psi}_\delta(r) \geqslant \hat{\psi}_\delta(s) + \tfrac{1}{\delta}\beta_\delta(s)(r - s). \tag{2.6a, b}$$

Hence, rewriting (2.1a,b) as a penalized problem, we are led to consider the following weak formulation: find $\{u^\delta, \theta^\delta\}$ satisfying

$$\begin{aligned} \left\langle c\theta^\delta_t + \frac{l}{2}u^\delta_t, \eta\right\rangle + \left(k\nabla\theta^\delta, \nabla\eta\right) &= 0 \qquad \forall\, \eta \in H^1(\Omega), \\ \left\langle \tau u^\delta_t, \xi\right\rangle + \left(\psi'_\delta(u^\delta) - \alpha\theta^\delta, \xi\right) + \gamma\left(\nabla u^\delta, \nabla\xi\right) &= 0 \qquad \forall\, \xi \in H^1(\Omega). \end{aligned} \tag{2.7a, b}$$

We prove existence using the classical Galerkin method, c.f. [Li]. We introduce the orthogonal basis for $H^1(\Omega)$ consisting of eigenfunctions for

$$-\Delta z + z = \mu z; \qquad \frac{\partial z}{\partial \nu} = 0 \text{ on } \partial\Omega,$$

which we denote by $\{z_j\}_{j=1}^\infty$, with the property that $(z_i, z_j) = \delta_{ij}$ (the Kronecker delta function). Define $V^N = \operatorname*{Span}_{1 \leqslant i \leqslant N}\{z_i\} \subset H^1(\Omega)$, $\mathbf{V}^N := V^N \times V^N$ and consider a Galerkin approximation of (2.7) to be: find $\left\{U(t), \Theta(t)\right\} \in \mathbf{V}^N$ such that $U(0) = U_0 = P_0^N u_0$ and

$\Theta(0) = \Theta_0 = P_0^N \theta_0,$

$$U(t) = \sum_{i=1}^{N} f_i(t) z_i \qquad \text{and} \qquad \Theta(t) = \sum_{i=1}^{N} g_i(t) z_i,$$

where f_i and g_i $(i = 1, \cdots, N)$ are each continuously differentiable functions satisfying

$$\begin{aligned} c\frac{dg_i}{dt} + \frac{l}{2}\frac{df_i}{dt} &= -g_i(\mu_i - 1), \\ \tau\frac{df_i}{dt} + \frac{1}{\delta}\Big(\boldsymbol{H}^N(\boldsymbol{f})\Big)_i - f_i - \alpha g_i + \gamma(\mu_i - 1) f_i &= 0, \end{aligned} \tag{4.1a, b}$$

here $\Big(\boldsymbol{H}^N(\boldsymbol{f})\Big)_i := (\beta_\delta(U), z_i)$ is a globally Lipschitz continuous function. The L^2 projection into V^N is denoted by P_0^N, i.e. given $\eta \in L^2(\Omega)$

$$(P_0^N \eta, \chi) = (\eta, \chi) \qquad \forall\, \chi \in V^N.$$

So, for all $\chi \in V^N$,

$$\begin{aligned} (E_t, \chi) + k(\nabla\Theta, \nabla\chi) &= 0 \\ (\tau U_t + \frac{1}{\delta}\beta_\delta(U) - U + \alpha\Theta, \chi) + \gamma(\nabla U, \nabla\chi) &= 0, \end{aligned} \tag{2.9a, b}$$

where $E := c\Theta + \frac{l}{2}U$. Note that using the projection property of P_0^N and taking $\eta = 1$ in (2.9a),

$$⨍ E = ⨍ c\Theta_0 + \frac{l}{2}U_0 = m.$$

For $\tau > 0$ and $l > 0$, global existence and uniqueness of the vectors $\boldsymbol{f}$, $\boldsymbol{g}$ follows from ordinary differential equations theory. We now seek bounds which are independent of N and δ in order that we may pass to the limit. Setting $\chi = U$ in (2.9b) and $r = U$ in (2.5b), a calculation reveals

$$\frac{\tau}{2}\frac{d}{dt}|U|_0^2 + \gamma|U|_1^2 \leqslant \frac{\alpha}{c}\Big(E, U\Big) + \Big(1 - \frac{\alpha l}{2c}\Big)|U|_0^2 \leqslant \frac{\alpha\nu}{2c}\Big|E\Big|_0^2 + \Big(1 - \frac{\alpha l}{2c} + \frac{\alpha}{2c\nu}\Big)|U|_0^2, \tag{2.11}$$

for any $\nu > 0$. Similarly taking $\chi = E$ into (2.9a) yields

$$\frac{1}{2}\frac{d}{dt}\Big|E\Big|_0^2 + \frac{k}{c}\Big|E\Big|_1^2 = \frac{kl}{2c}(\nabla E, \nabla U) \leqslant \frac{k}{2c}\Big|E\Big|_1^2 + \frac{kl^2}{8c}\Big|U\Big|_1^2,$$

so that

$$\frac{d}{dt}\Big|E\Big|_0^2 + \frac{k}{c}\Big|E\Big|_1^2 \leqslant \frac{kl^2}{4c}\Big|U\Big|_1^2. \tag{2.12}$$

Taking a linear combination of (2.11) and (2.12) we find

$$\frac{d}{dt}\Big(\frac{\tau}{2}|U|_0^2 + \frac{2\gamma c}{kl^2}\Big|E\Big|_0^2\Big) + \frac{\gamma}{2}|U|_1^2 + \frac{2\gamma}{l^2}\Big|E\Big|_1^2 \leqslant \frac{\alpha\nu}{2c}\Big|E\Big|_0^2 + \Big(1 - \frac{\alpha l}{2c} + \frac{\alpha}{2c\nu}\Big)|U|_0^2. \tag{2.13}$$

Using a Grönwall inequality on (2.13) we find that Θ, U, E are bounded independently of N and δ in $L^\infty(0,T;L^2(\Omega)) \cap L^2(0,T;H^1(\Omega))$.

Setting $\chi = \beta_\delta(U)$ in (2.9b) we obtain, using (2.5a) and a kick-back argument,

$$\frac{\tau}{2}\frac{d}{dt}(\hat{\psi}_\delta(U),1) + \frac{1}{\delta}|\beta_\delta(U)|_0^2 \leqslant (U+\alpha\Theta, \beta_\delta(U)) \leqslant \delta(|U|_0^2 + \alpha^2|\Theta|_0^2) + \frac{1}{2\delta}|\beta_\delta(U)|_0^2,$$

$$\Longleftrightarrow \tau\frac{d}{dt}(\hat{\psi}_\delta(U),1) + \frac{1}{\delta}|\beta_\delta(U)|_0^2 \leqslant 2\delta(|U|_0^2 + \alpha^2|\Theta|_0^2). \tag{2.14}$$

Setting $s = P_0^N u_0$ in (2.6b), noting $|u_0| \leqslant 1$ *a.e.* and (2.5a)

$$\Big(\hat{\psi}_\delta(P_0^N u_0),1\Big) \leqslant \Big(\hat{\psi}_\delta(u_0),1\Big) + \tfrac{1}{\delta}\Big(\beta_\delta(P_0^N u_0) - \beta_\delta(u_0), P_0^N u_0 - u_0\Big) \leqslant \tfrac{1}{\delta}\Big|P_0^N u_0 - u_0\Big|_0^2 \to 0.$$

So integrating (2.14), noting the $L^2(\Omega_T)$ bounds on U and Θ are independent of δ and (2.6a)

$$\frac{\tau}{2\delta}|\beta_\delta(U)|_0^2 + \frac{1}{\delta}\int_0^t |\beta_\delta(U(t))|_0^2 dt \leqslant C(T)\delta$$

where C is independent of δ. Hence $\beta_\delta(U) \in L^\infty(0,T;L^2(\Omega))$ and from (2.9a,b) we deduce that

$$U_t, \Theta_t \in L^2(0,T;(H^1(\Omega))').$$

Furthermore, by an embedding theorem

$$\eta \in H^1(0,T;(H^1(\Omega))') \cap L^2(0,T;H^1(\Omega)) \Longrightarrow \eta \in C([0,T];L^2(\Omega)),$$

so that as $U(0) = P_0^N u_0$ and $\Theta(0) = P_0^N \theta_0$, it follows from convergence properties that $u^\delta(0) = u_0$. It also follows from compactness (c.f. [Li]) that $U \to u^\delta$ in $L^2(\Omega_T)$.

We can now pass to the limit as $N \to \infty$. Multiply (4.1a,b) respectively by $z_i\eta^N\zeta(t)$ and $z_i\xi^N\varrho(t)$ ($\eta^N, \xi^N \in V^N$ and $\zeta, \varrho \in L^2(0,T)$), sum over i and integrate over Ω_T, to obtain

$$\int_0^T \Big(\Big(c\Theta_t + \frac{l}{2}U_t, \eta^N\Big) + \Big(N\nabla\Theta, \nabla\eta^N\Big)\Big)\zeta(t)dt = 0 \qquad \forall\, \eta^N \in V^N,$$

$$\int_0^T \Big(\Big(\tau U_t, \xi^N\Big) + \Big(\psi_\delta'(U) - \alpha\Theta, \xi^N\Big) + \gamma\Big(\nabla U, \nabla\xi^N\Big)\Big)\varrho(t)dt = 0 \qquad \forall\, \xi^N \in V^N. \tag{2.15}$$

Using the estimates (2.13–2.14), we can extract convergent subsequences with the following properties

$$\Theta \rightharpoonup \theta^\delta \text{ in } L^2(0,T;H^1(\Omega)) \cap H^1(0,T;(H^1(\Omega))') \qquad \Theta \rightharpoonup^* \theta^\delta \text{ in } L^\infty(0,T;L^2(\Omega)),$$

$$U \rightharpoonup u^\delta \text{ in } L^2(0,T;H^1(\Omega)) \cap H^1(0,T;(H^1(\Omega))') \qquad U \rightharpoonup^* u^\delta \text{ in } L^\infty(0,T;L^2(\Omega)).$$

For arbitrary $\eta, \xi \in H^1(\Omega)$, set $\eta^N = P_1^N\eta$ and $\xi^N = P_1^N\xi$, where P_1^N is the $H^1(\Omega)$ projection into V^N. Since V^N is dense in $H^1(\Omega)$, it follows that $P^N\eta \to \eta$ in $H^1(\Omega)$. Using (2.5a) we obtain as $N \to \infty$

$$\left|\int_0^T \Big((\psi_\delta'(U) - \psi_\delta'(u^\delta), \xi)\varrho(t)\Big)dt\right| \leqslant (\tfrac{1}{\delta}+1)\|U - u^\delta\|_{L^2(\Omega_T)}\|\varrho\|_{L^2(0,T)}\|\xi\|_{L^2(\Omega)} \to 0.$$

Passing to the limit $N \to \infty$ we obtain the existence of $\{u^\delta, \theta^\delta\}$ satisfying (2.7a,b).

We wish to derive more estimates on $\{u^\delta, e^\delta\}$ which are independent of δ. It is convenient to derive them formally by using suitable test functions in (2.7a,b). However the calculations are essentially the same as those for the Galerkin approximation and we omit the details of passing to the limit as $N \to \infty$. First we remark that $|u^\delta|_0$ is uniformly bounded. Defining $\Omega_{1+\delta} := \{x \in \Omega : |u^\delta(x)| \leqslant 1+\delta\}$ and noting that outside this set $|u^\delta(x)| \leqslant |\beta_\delta(u^\delta(x))| + 1 + \delta/2$, gives

$$\int_\Omega (u^\delta)^2(x)dx = \int_{\Omega_{1+\delta}\cup\Omega\setminus\Omega_{1+\delta}} (u^\delta)^2(x)dx \leqslant |\Omega| \left[(1+\delta)^2 + 2\left(1+\tfrac{\delta}{2}\right)^2\right] + 2\left|\beta_\delta(u^\delta)\right|_0^2 \leqslant C_0, \tag{2.16}$$

where C_0 is bounded independent of N, δ and T. We recall the Poincaré inequality:- there exists a positive constant C_P such that for any $\eta \in H^1(\Omega)$ with $(\eta, 1) = 0$,

$$|\eta|_0 \leqslant C_P |\eta|_1. \tag{2.17}$$

Note that for $\eta \in H^1(\Omega)$

$$\left|\eta\right|_0^2 = \left|\eta - \frac{1}{|\Omega|}(\eta,1)\right|_0^2 + \frac{1}{|\Omega|}(\eta,1)^2 \leqslant C_P^2\left|\eta\right|_1^2 + \frac{1}{|\Omega|}(\eta,1)^2.$$

Setting $\nu = \frac{2c\gamma}{\alpha C_P^2 l^2}$ in (2.13), we obtain

$$\frac{d}{dt}\left(\frac{\tau}{2}|u^\delta|_0^2 + \frac{2\gamma c}{kl^2}\left|e^\delta\right|_0^2\right) + \frac{\gamma}{2}|u^\delta|_1^2 + \frac{\gamma}{l^2}\left|e^\delta\right|_1^2 \leqslant \frac{\gamma}{C_P^2 l^2}\frac{1}{|\Omega|}(e_0,1)^2 + \left(1 - \frac{\alpha l}{2c} + \frac{\alpha^2 C_P^2 l^2}{4c^2\gamma}\right)|u^\delta|_0^2. \tag{2.18}$$

Setting $D = \frac{1}{C_P^2}\min\left\{\frac{\gamma}{\tau}, \frac{k}{2c}\right\}$ and using the Poincaré and Cauchy Schwarz inequalities we obtain

$$\frac{d}{dt}\left(\frac{\tau}{2}|u^\delta|_0^2 + \frac{2\gamma c}{kl^2}\left|e^\delta\right|_0^2\right) + D\left(\frac{\tau}{2}|u^\delta|_0^2 + \frac{2\gamma c}{kl^2}\left|e^\delta\right|_0^2\right) \leqslant \left(\frac{2\gamma c D}{kl^2} + \frac{\gamma}{C_P^2 l^2}\right)\frac{1}{|\Omega|}(e_0,1)^2 + \left(1 - \frac{\alpha l}{2c} + \frac{\alpha^2 C_P^2 l^2}{4c^2\gamma} + \frac{\tau D}{2}\right)|u^\delta|_0^2.$$

Hence

$$\frac{\tau}{2}|u^\delta|_0^2 + \frac{2\gamma c}{kl^2}\left|e^\delta\right|_0^2 \leqslant C_1 + C_2 e^{-Dt},$$

where $C_1 = \left(1 - \frac{\alpha l}{2c} + \frac{\alpha^2 C_P^2 l^2}{4c^2\gamma} + \frac{\tau D}{2}\right)\frac{C_0}{D} + \left(\frac{2\gamma c D}{kl^2} + \frac{\gamma}{C_P^2 l^2}\right)\frac{|\Omega|}{D}m^2$

and $C_2 = \frac{\tau}{2}|u_0|_0^2 + \frac{2\gamma c}{kl^2}\left|c\theta_0 + \frac{l}{2}u_0\right|_0^2 - \left(\frac{2\gamma c D}{kl^2} + \frac{\gamma}{C_P^2 l^2}\right)\frac{|\Omega|}{D}m^2 - \left(1 - \frac{\alpha l}{2c} + \frac{\alpha^2 C_P^2 l^2}{4c^2\gamma} + \frac{\tau D}{2}\right)\frac{C_0}{D}.$

Notice C_1 and C_2 are bounded independently of δ and T. Hence, by taking $\rho_0 = \rho_0(m) = \sqrt{1+C_1}$ and $\mathcal{T}_0 = \frac{1}{D}\log_e(|C_2|)$, we deduce

$$\left\|\{u^\delta(t), e^\delta(t)\}\right\|_{L^2(\Omega)} \leqslant \rho_0 \qquad \text{for} \qquad t \geqslant \mathcal{T}_0\Big(\{u_0, \theta_0\}\Big). \tag{2.19}$$

Taking $\xi = -\Delta u^\delta$ in (2.7b) and using the fact that $\psi_\delta''(r) \geqslant -1$ we obtain

$$\frac{\tau}{2}\frac{d}{dt}\left|u^\delta\right|_1^2 + \gamma\left|\Delta u^\delta\right|_0^2 \leqslant \alpha(\theta^\delta, -\Delta u^\delta) + |u^\delta|_0^2 = \frac{\alpha l}{2c}(u^\delta, \Delta u) - \frac{\alpha}{c}(e^\delta, \Delta u^\delta) + |u^\delta|_0^2.$$

Taking $\eta = -\Delta e^\delta$ in (2.7a) yields

$$\frac{1}{2}\frac{d}{dt}\left|e^\delta\right|_1^2 + \frac{k}{2c}\left|\Delta e^\delta\right|_0^2 \leqslant \frac{l^2 k}{8c}\left|\Delta u^\delta\right|_0^2.$$

A linear combination of these inequalities implies

$$\frac{d}{dt}\left(\frac{\tau}{2}\left|u^\delta\right|_1^2 + \frac{c\gamma}{l^2 k}\left|e^\delta\right|_1^2\right) + \frac{\gamma}{4}\left|\Delta u^\delta\right|_0^2 + \frac{\gamma}{l^2}\left|\Delta e^\delta\right|_0^2 \leqslant \left(1 + \frac{\alpha^2 l^2}{4\gamma c^2}\right)\left|u^\delta\right|_0^2 + \frac{2\alpha^2}{\gamma c^2}\left|e^\delta\right|_0^2. \tag{2.20}$$

Thus for $t \geqslant \mathcal{T}(m)$, the inequality

$$\frac{d}{dt}\left(\left|u^\delta\right|_1^2 + \left|e^\delta\right|_1^2\right) \leqslant C\rho_0^2(m)$$

holds. Furthermore by (2.18) we have for $t \geqslant \mathcal{T}$

$$\int_t^{t+r}\left(\left|u^\delta\right|_1^2 + \left|e^\delta\right|_1^2\right) ds \leqslant C(m, r).$$

Applying the uniform Grönwall inequality [Te,p. 89] we have the existence of $\rho_1(m)$ and $\mathcal{T}_1(\{u_0, e_0\})$ such that

$$\|\{u^\delta(t), e^\delta(t)\}\|_{H^1(\Omega)} \leqslant \rho_1 \qquad \text{for all } t \geqslant \mathcal{T}_1. \tag{2.21}$$

Let us define

$$\mathcal{F}_\delta(u^\delta, \theta^\delta) = \int_\Omega \left(\frac{\gamma}{2}|\nabla u^\delta|^2 + \psi_\delta(u^\delta) + \frac{c\alpha}{l}(\theta^\delta)^2\right) dx. \tag{2.22}$$

Differentiating $\mathcal{F}_\delta(u^\delta, \theta^\delta)$ with respect to time, using (2.7b) with $\xi = u_t^\delta$ and (2.7a) with $\eta = \theta^\delta$ a calculation reveals

$$\begin{aligned}\frac{d}{dt}\mathcal{F}_\delta(u^\delta, \theta^\delta) &= \gamma(\nabla u^\delta, \nabla u_t^\delta) + (\psi_\delta'(u^\delta), u_t^\delta) + \tfrac{2c\alpha}{l}(\theta^\delta, \theta_t^\delta),\\ &= -\tau|u_t^\delta|_0^2 + \alpha(\theta^\delta, u_t^\delta) + \tfrac{2c\alpha}{l}(\theta^\delta, \theta_t^\delta) \;=\; -\tau|u_t^\delta|_0^2 - \tfrac{2\alpha k}{l}|\theta^\delta|_0^2.\end{aligned}$$

Hence, integration yields

$$\mathcal{F}_\delta(u^\delta(t), \theta^\delta(t)) + \int_{t_0}^t\left(\tau|u_t^\delta(s)|_0^2 + \frac{2\alpha k}{l}|\theta^\delta(s)|_0^2\right) ds = \mathcal{F}_\delta(u^\delta(t_0), \theta^\delta(t_0)). \tag{2.23}$$

Multiplying (2.20) by t and integrating yields

$$t\left(\frac{\tau}{2}\left|u^\delta(t)\right|_1^2+\frac{c\gamma}{l^2k}\left|e^\delta(t)\right|_1^2\right)+\int_0^t s\left(\frac{\gamma}{4}\left|\Delta u^\delta(s)\right|_0^2+\frac{\gamma}{l^2}\left|\Delta e^\delta(s)\right|_0^2\right)ds\leqslant$$
$$\int_0^t s\left(\left(1+\frac{\alpha^2l^2}{4\gamma c^2}\right)\left|u^\delta(s)\right|_0^2+\frac{2\alpha^2}{\gamma c^2}\left|e^\delta(s)\right|_0^2\right)ds+\int_0^t\left(\frac{\tau}{2}\left|u^\delta(s)\right|_1^2+\frac{c\gamma}{l^2k}\left|e^\delta(s)\right|_1^2\right)ds. \tag{2.24}$$

We are now in a position to pass to the limit with respect to δ, we repeat the arguments of [BE1]. Introduce $\beta:\mathbb{R}\to\mathbb{R}$ defined by

$$\beta(r):=\lim_{\delta\to 0}\beta_\delta(r)=\begin{cases} r-1 & r\geqslant 1,\\ 0 & |r|\leqslant 1,\\ r+1 & r\leqslant -1.\end{cases}$$

Notice that $\beta\in C^{0,1}$ and $\forall\ r,s\in\mathbb{R}$ satisfies the inequalities

$$\left|\beta(r)-\beta_\delta(r)\right|\leqslant\frac{\delta}{2}\text{ and }\left|\beta(r)-\beta(s)\right|\leqslant\left|r-s\right|. \tag{2.25a, b}$$

The stability bounds are independent of δ, we deduce the existence of subsequences of u^δ and θ^δ such that

$$\theta^\delta\rightharpoonup\theta\text{ in }L^2(0,T;H^1(\Omega))\cap H^1(0,T;(H^1(\Omega))')\qquad\theta^\delta\rightharpoonup^*\theta\text{ in }L^\infty(0,T;L^2(\Omega)),$$
$$u^\delta\rightharpoonup u\text{ in }L^2(0,T;H^1(\Omega))\cap H^1(0,T;(H^1(\Omega))')\qquad u^\delta\rightharpoonup^* u\text{ in }L^\infty(0,T;L^2(\Omega)).$$

Hence, using compactness (c.f. [Li]) and (2.5a) it is easy to show the weak convergence of $\beta_\delta(u^\delta)$ to $\beta(u)$ in $L^2(\Omega_T)$. However, $\beta_\delta(u_\delta)\to 0$ as $\delta\to 0$, thus $|u|\leqslant 1$. Now, restrict $\xi\in K$ in (2.15). Hence, existence of a solution to (2.1a,b) follows from the monotonicity of β_δ and the fact that $\beta_\delta(\xi)=0$. See [BE1] for detailed arguments when $c=\tau=0$.

We set

$$K_0=\{\eta\in L^2(\Omega):|\eta|\leqslant 1\},\ \boldsymbol{L}^2_m(\Omega)=\left\{\{\eta,\xi\}\in\boldsymbol{L}^2(\Omega):⨍\xi=m\right\}$$

and

$$\mathcal{F}(u,\theta)=\int_\Omega\left\{\frac{\gamma}{2}|\nabla u|^2+\psi(u)+\frac{c\alpha}{l}\theta^2\right\}dx.$$

Theorem 2.1

(i) For any $\{u_0,e_0\}\in\boldsymbol{L}^2_m(\Omega)$ and $u_0\in K_0$ (PFO) has a unique solution $\{u,e\}$ which satisfies

$$\{u,e\}\in L^\infty(0,T;\boldsymbol{L}^2_m(\Omega))\cap L^2(0,T;\boldsymbol{H}^1(\Omega))\cap H^1(0,T;(\boldsymbol{H}^1(\Omega))')$$

for all $T>0$ and

$$\{u,e\}\in C([0,\infty);\boldsymbol{L}^2(\Omega)).$$

(ii) Let $\{u^i, e^i\}$ solve (PFO) with initial data $\{u_0^i, e_0^i\} \in \boldsymbol{L}_m^2(\Omega)$ for $i = 1, 2$. Then there exists $C > 0$ depending on $\tau, \alpha, \gamma, c, k, l$ and Ω such that

$$\frac{\tau}{2}\left|u^1(t) - u^2(t)\right|_0^2 + \frac{2\gamma c}{kl^2}\left|e^1(t) - e^2(t)\right|_0^2 \leqslant e^{Ct}\left[\frac{\tau}{2}\left|u^1(0) - u^2(0)\right|_0^2 + \frac{2\gamma c}{kl^2}\left|e^1(0) - e^2(0)\right|_0^2\right] \tag{2.26}$$

(iii) For any m there exists $\rho = \rho(m) > 0$ and a $t^* = t^*(\|\{u_0, e_0\}\|_{\boldsymbol{L}^2(\Omega)})$ such that

$$\|\{u(t), e(t)\}\|_{\boldsymbol{H}^1(\Omega)} \leqslant \rho \qquad \text{for all } t > t^*$$

(iv) For any $t_0 > 0$ and $\{u_0, e_0\} \in \boldsymbol{L}_m^2(\Omega)$ with $u_0 \in K_0$ we have

$$\{u, e\} \in L^\infty(t_0, T; \boldsymbol{H}^1(\Omega)) \cap L^2(t_0, T; \boldsymbol{H}^2(\Omega)), \quad \{u_t, e_t\} \in L^2(t_0, T; \boldsymbol{L}^2(\Omega)),$$

and the energy inequality

$$\mathcal{F}(u(t), \theta(t)) + \int_{t_0}^t \left[\tau|u_t(s)|_0^2 + \frac{2k\alpha}{l}|\theta(s)|_1^2\right] ds \leqslant \mathcal{F}(u(t_0), \theta(t_0)), \tag{2.27}$$

holds for all $t > t_0 > 0$.

(v) Furthermore if $\{u_0, e_0\} \in \boldsymbol{H}^1(\Omega)$ then we can take $t_0 = 0$ in (2.27) and

$$\{u, e\} \in L^2(0, T; \boldsymbol{H}^1(\Omega)) \cap L^2(0, T; \boldsymbol{H}^2(\Omega)) \text{ and } \{u_t, e_t\} \in L^2(0, T; \boldsymbol{L}^2(\Omega)).$$

Proof. The statement of existence in *(i)* is an immediate consequence of the estimates and arguments. In order to prove inequality (2.26) and hence uniqueness, we note (2.2a,b) imply

$$\frac{1}{2}\frac{d}{dt}|e^1 - e^2|_0^2 + \frac{k}{c}|e^1 - e^2|_1^2 = \frac{l}{2c}(\nabla(u^1 - u^2), \nabla(e^1 - e^2))$$

and

$$\frac{\tau}{2}\frac{d}{dt}|u^1 - u^2|_0^2 + \gamma|u^1 - u^2|_1^2 \leqslant \frac{\alpha}{c}(e^1 - e^2, u^1 - u^2) + \left(1 - \frac{\alpha l}{2c}\right)|u^1 - u^2|_0^2,$$

so that taking a linear combination yields the inequality. The other statements of the theorem are straightforward consequences of (2.21), (2.24) and (2.20), apart from the energy inequality (2.27). The latter is a consequence of (2.23), the fact that $\limsup_{\delta \to 0}(\psi_\delta(u^\delta), 1) \geqslant (\psi_\delta(u), 1)$ and $u \in C([t_0, T]; \boldsymbol{H}^1(\Omega))$ for any $t_0 > 0$. See [BE1]. □

Remark It follows from (2.27) and by standard arguments that the ω-limit set for a particular initial data is contained in the set of equilibria. Stabilization to a particular equilibrium is difficult to prove because the steady states are not isolated in general, see [BE1] and [CE]. In the case of (1.5) with $\alpha = 0$, stabilization to an equilibrium was established for one space dimension in [CE]. The existence of a global attractor follows from the existence of an absorbing set in $\boldsymbol{H}^1(\Omega)$, see [BCH] and [Te].

3. Asymptotic behaviour in the scaling parameters

3.1 The classical Stefan problem

Let $1/\alpha = \epsilon$, $\tau/\alpha = \sigma_1(\epsilon)$ and $\gamma/\alpha = \sigma_2(\epsilon)$ where

$$\lim_{\epsilon\to 0} \sigma_1(\epsilon) = \lim_{\epsilon\to 0} \sigma_2(\epsilon) = 0.$$

We label the solutions of (PFO) by $\{u^\epsilon, \theta^\epsilon, e^\epsilon\}$ with obvious notation.

Suppose that the initial data for the phase parameter satisfies

$$u_0^\epsilon \in K \quad \text{and} \quad \limsup_{\epsilon\to 0} \sigma_2(\epsilon)|\nabla u_0^\epsilon|^2 \leqslant M.$$

The energy inequality (2.27) implies that

$$\frac{c}{l}|\theta^\epsilon|_0^2 + \frac{\sigma_2(\epsilon)}{2}|u^\epsilon|_1^2 + \int_0^t \left[\sigma_1(\epsilon)|u_t^\epsilon|_0^2 + \frac{2k}{l}|\theta^\epsilon|_1^2\right] ds \leqslant \frac{c}{l}|\theta_0^\epsilon|_0^2 + \frac{\sigma_2(\epsilon)}{2}|u_0^\epsilon|_1^2 + \frac{|\Omega|}{2}.$$

The variational inequality (2.1b) implies that for $\xi \in K$

$$(\theta^\epsilon, \xi - u^\epsilon) + \sigma_2(\epsilon)|u^\epsilon|_1^2 \leqslant (\sigma_1(\epsilon))^{1/2}\left(\sigma_1(\epsilon)^{1/2}|u_t^\epsilon|_0\right)|\xi - u^\epsilon|_0 + \epsilon|u^\epsilon|_0|\xi - u^\epsilon|_0$$
$$+ (\sigma_2(\epsilon))^{1/2}\left(\sigma_2(\epsilon)^{1/2}|u^\epsilon|_1\right)|\xi|_1.$$

Thus, in the limit $\epsilon \to 0$ we have

$$\theta^\epsilon \rightharpoonup \theta \text{ in } L^2(0,T;H^1(\Omega)) \qquad u^\epsilon \rightharpoonup^* u \text{ in } L^\infty(0,T;L^2(\Omega)),$$

where $|u| \leqslant 1$, (1.1a) holds in the sense of distributions and for *a.e.* $t \in (0,T)$

$$(\theta, \xi - u) \leqslant 0 \qquad \forall\, \xi \in K.$$

This last inequality implies that

$$u \in \text{sign}\ \theta = \begin{cases} 1 & \theta > 0, \\ \in [-1,1] & \theta = 0, \\ -1 & \theta < 0. \end{cases}$$

Thus $\{u, \theta\}$ forms a weak solution of the classical Stefan problem [EO].

3.2 The Stefan problem with Gibbs-Thomson and kinetic undercooling

Formal analysis suggests, [BE3], that with the scaling

$$\tau = \sigma_1\epsilon^2,\ \gamma = \epsilon^2 \text{ and } \alpha = \epsilon\sigma_2,$$

in the limit $\epsilon \to 0$, $\{u^\epsilon, \theta^\epsilon\} \to \{u, \theta\}$ with the following structure:- $\Omega = \Omega^+(t) \cap \Gamma(t) \cap \Omega^-(t)$, $\Omega^+(t) := \{\boldsymbol{x} \in \Omega : u(\boldsymbol{x}, t) = 1\}$, $\Omega^-(t) := \{\boldsymbol{x} \in \Omega : u(\boldsymbol{x}, t) = -1\}$, $\Gamma(t) = \partial\Omega^+(t) \cap \partial\Omega^-(t)$

and

$$\left.\begin{aligned} c\theta_t &= k\Delta\theta \quad \boldsymbol{x} \notin \Gamma(t), \\ k[\nabla\theta]_-^+ .\boldsymbol{n} &= -lV \\ \sigma_2\theta &= -\frac{\pi}{4}(\sigma_1 V + \kappa) \end{aligned}\right\} \quad \boldsymbol{x} \in \Gamma(t). \tag{3.1a, b, c}$$

Here $V\boldsymbol{n}$ is the normal velocity of $\Gamma(t)$ into $\Omega^+(t)$, $\boldsymbol{n}$ is the unit normal pointing pointing into $\Omega^+(t)$, $\left[\cdot\right]_-^+$ denotes the difference in the limits of the quantity in the brackets as $\boldsymbol{x}$ tends to $\Gamma(t)$ from each side and κ^m is the sum of the principal curvatures $\Gamma(t)$ (with the convention that κ^m is positive when $\Omega^-(t)$ is convex).

The appearance of κ^m in the relation (3.1b) for the solidification temperature is called the Gibbs-Thomson effect. If we take $\sigma_1 \neq 0$ then the expression of the interface temperature is known as kinetic undercooling. We are particularly interested in the moving boundary problem (3.1a,b,c) when $\theta < 0$ in $\Omega^+(t)$ and the solid region grows as the undercooled liquid in $\Omega^+(t)$ solidifies. Rigorous results are known in the radially symmetric case with smooth ψ, see [St].

4. Numerical Analysis

We consider a fully discrete implicit finite element approximation to the phase-field equations (2.1). We introduce some notation. Let $\mathcal{T}^h$ be a triangulation of Ω such that $\Omega = \bigcup_{\tau\in\mathcal{T}^h}\tau$, for ease of exposition we assume that Ω is polygonal. We define

$$\begin{aligned} S^h &:= \{\chi \in C(\overline{\Omega}) : \chi|_\tau \text{ is linear, } \tau \in \mathcal{T}^h\}, \\ K^h &:= \{\chi \in S^h : |\chi| \leqslant 1\}. \end{aligned}$$

The approximation we consider is: find $\{\Theta^n, U^n\}_{n\geqslant 0} \in S^h \times K^h$ such that

$$\begin{aligned} c(\partial\Theta^n, \chi)^h + \frac{l}{2}(\partial U^n, \chi)^h + k(\nabla\Theta^n, \nabla\chi) = 0 \qquad \forall\, \chi \in S^h, \\ \tau(\partial U^n, \chi - U^n)^h + \gamma(\nabla U^n, \nabla\chi - \nabla U^n) - (U^n, \chi - U^n)^h \geqslant \alpha(\Theta^n, \chi - U^n)^h \;\forall\, \chi \in K^h. \end{aligned} \tag{4.1a, b}$$

We denote by $(\cdot,\cdot)^h$ an approximation to the L^2 inner product, that may be easily evaluated according to a vertex quadrature rule. We define $\|\chi\|_h := \sqrt{(\chi,\chi)^h}$ and assume it is an equivalent norm to the L^2 norm with constants bounded independently of h. Also, we assume that a discrete Poincaré inequality holds. In what follows, it will be useful to define the mass and stiffness matrices

$$M_{ij} = (\phi_i, \phi_j)^h \qquad \text{and} \qquad K_{ij} = (\nabla\phi_i, \nabla\phi_j),$$

where $\{\phi_i\}_{i=1}^D$ is the cardinal basis for the nodes $\{\boldsymbol{x}_i\}_{i=1}^D$ of the triangulation $\mathcal{T}^h$.

As $c>0$, given $f\in C^0(\overline{\Omega})$ there exists a unique solution $\mathcal{G}^h f\in S^h$ to the following equation

$$k\Delta t(\nabla\mathcal{G}^h f,\nabla\xi)+c(\mathcal{G}^h f,\xi)^h=\frac{l}{2}(f,\xi)^h \qquad \forall\,\xi\in S^h.$$

We can write the linear operator $\mathcal{G}^h$ in terms of a matrix $G^h:=\frac{l}{2}\left(k\Delta tK+cM\right)^{-1}M$.

It is interesting to note that if $c=0$ the operator $\mathcal{G}^h$ is defined by imposing the constraints $(\mathcal{G}^h f,1)^h=(f,1)^h=0$. Furthermore, as we shall see, uniqueness is obtained for the discrete problem with $c=0$ when $\tau=0$.

Returning to the case $c>0$, using the operator $\mathcal{G}^h$ we can rewrite (4.1a,b) in terms of a single unknown U^n, since

$$\Theta^n=-\mathcal{G}^h U^n+\mathcal{G}^h\big(U^{n-1}+\frac{2c}{l}\Theta^{n-1}\big), \tag{4.2}$$

Thus (4.1a,b) becomes:- find $U^n\in K^h$ such that for all $\xi\in K^h$

$$\tau(\partial U^n,\xi-U^n)^h+\gamma(\nabla U^n,\nabla\xi-\nabla U^n)-(U^n,\xi-U^n)^h+\alpha(\mathcal{G}^h U^n,\xi-U^n)^h\geqslant \alpha(\mathcal{G}^h\big(U^{n-1}+\frac{2c}{l}\Theta^{n-1}\big),\xi-U^n)^h. \tag{4.3}$$

Notice that if we can find U^n satisfying (4.3), then defining Θ^n by (4.2) it is automatic that

$$c(\Theta^n,1)^h+\frac{l}{2}(U^n,1)^h=c(\Theta^{n-1},1)^h+\frac{l}{2}(U^{n-1},1)^h,$$

and (4.1a,b) holds.

Theorem 4.2 *Suppose either $\Delta t\leqslant\tau$ or $\tau<\Delta t$ and*

$$\frac{k\Delta t}{\gamma}\left(1-\frac{\tau}{\Delta t}\right)+c<\frac{\alpha l}{1-\frac{\tau}{\Delta t}}\left(1+\sqrt{1-\frac{2c}{\alpha l}\left(1-\frac{\tau}{\Delta t}\right)}\right),$$

with $c<\frac{\alpha l}{2\left(1-\frac{\tau}{\Delta t}\right)}$. Then there exists a unique solution to (4.1).

Proof. Notice that (4.3) is the Euler-Lagrange equation of the minimization problem: find $U^n\in K^h$ which minimizes on K^h the functional

$$\frac{\tau}{2\Delta t}|\chi-U^{n-1}|_h^2+\frac{\gamma}{2}|\chi|_1^2+\frac{\alpha}{2}(G^h\chi,\chi)^h-\frac{1}{2}|\chi|_0^2-\alpha(\mathcal{G}^h\big(U^{n-1}+\frac{2c}{l}\Theta^{n-1}\big),\chi)^h. \tag{4.4}$$

Existence is immediate. We now prove uniqueness. Let U^1 and U^2 be two different solutions to (4.3) and define $\Xi:=U_1-U_2$. It follows by setting $\xi=U^2$ in (4.3) when U_1 is the solution (and interchanging 1 and 2) that by addition we are led to the inequality

$$\frac{\tau}{\Delta t}|\Xi|_h^2+\gamma|\Xi|_1^2+\alpha(\mathcal{G}^h\Xi,\Xi)^h-|\Xi|_h^2\leqslant 0, \tag{4.5}$$

so that uniqueness follows immediately for $\Delta t\leqslant\tau$ (notice that $(\mathcal{G}^h\cdot,\cdot)$ defines a norm). For the case $\Delta t>\tau$ we must do some extra work. Let us analyze $|\Xi|_h^2$ in terms of a vector, where $\xi_i = \Xi(x_i)$, G^h and use the generalized Cauchy Schwarz inequality ($\eta > 0$)

$$\begin{aligned}|\Xi|_h^2 &= \boldsymbol{\xi}^T M \boldsymbol{\xi} = \boldsymbol{\xi}^T (k\Delta t K + cM)^{1/2} M^{-1/2} M M^{-1/2} (k\Delta t K + cM)^{-1/2} M \boldsymbol{\xi},\\ &\leqslant \frac{1}{4\eta}(k\Delta t |\Xi|_1^2 + c|\Xi|_h^2) + \eta \boldsymbol{\xi}^T M (k\Delta t K + cM)^{-1} M \boldsymbol{\xi},\\ &= \frac{1}{4\eta}(k\Delta t |\Xi|_1^2 + c|\Xi|_h^2) + \frac{2}{l}\eta(\Xi, \mathcal{G}^h \Xi)^h,\end{aligned}$$

so that rearranging, assuming that $4\eta > c$,

$$|\Xi|_h^2 \leqslant \frac{k\Delta t}{4\eta - c}|\Xi|_1^2 + \frac{8\eta^2}{l(4\eta - c)}(\Xi, \mathcal{G}^h \Xi)^h,$$

and substituting into (4.5) yields

$$\left(\gamma - \frac{k\Delta t}{4\eta - c}\left(1 - \frac{\tau}{\Delta t}\right)\right)|\Xi|_1^2 + \left(\alpha - \frac{8\eta^2}{l(4\eta - c)}\left(1 - \frac{\tau}{\Delta t}\right)\right)(\Xi, \mathcal{G}^h \Xi)^h \leqslant 0.$$

For uniqueness, it is enough to show that

$$\gamma - \frac{k\Delta t}{4\eta - c}\left(1 - \frac{\tau}{\Delta t}\right) \geqslant 0 \qquad \text{and} \qquad \alpha - \frac{8\eta^2}{l(4\eta - c)}\left(1 - \frac{\tau}{\Delta t}\right) > 0.$$

A calculation then yields the result. □

Remark If $\tau = 0$ then for uniqueness, we must take $c < \alpha l/2$. If $\alpha = O(\epsilon)$ where ϵ is small, then it is clear this inequality will only hold for all ϵ if $c = 0$, in which case uniqueness is obtained for $\Delta t < 2\gamma\alpha l/k$. In the case where $\tau = c = 0$, (2.1) is the deep quench limit of the Cahn-Hilliard equation studied in [BE1] and [BE2] where $\alpha = k = l/2 = 1$.

We now prove that the discrete scheme has the same Lyapunov functional as the continuous version.

Theorem 4.3 *For $\Delta t < \tau$ the sequence $\{U^n, \Theta^n\}$ has a Lyapunov functional given by*

$$\mathcal{F}^h(U, \Theta) := \frac{\gamma}{2}|U|_1^2 + \frac{1}{2}|\Omega| - \frac{1}{2}|U|_h^2 + \frac{\alpha c}{l}|\Theta|_0^2,$$

that is

$$\mathcal{F}^h(U^n, \Theta^n) \leqslant \mathcal{F}^h(U^{n-1}, \Theta^{n-1}).$$

Proof. Setting $\chi = U^{n-1}$ in (4.1b) and using $2a(b - a) = a^2 + (a - b)^2 - b^2$ yields

$$\frac{\gamma}{2}|U^n|_1^2 - \frac{1}{2}|U^n|_h^2 - \frac{\gamma}{2}|U^{n-1}|_1^2 + \frac{1}{2}|U^{n-1}|_h^2 + \left(\frac{\tau}{\Delta t} - 1\right)\frac{1}{2}|U^n - U^{n-1}|_h^2 + \frac{\gamma}{2}|U^n - U^{n-1}|_1^2 \leqslant \alpha(\Theta^n, U^n - U^{n-1}),$$

similarly setting $\chi = \Theta^{n-1}$ in (4.1a) and multiplying by $2\alpha\Delta t/l$ gives

$$\frac{2\alpha c}{l}(\Theta^n - \Theta^{n-1}, \Theta^n) + \alpha(U^n - U^{n-1}, \Theta^n) + \frac{2\alpha k\Delta t}{l}|\Theta^n|_1^2 = 0,$$

so that adding the inequalities yields the result. □

5. Numerical Computations

In this section we describe numerical computations to (2.1). The scaling parameters are chosen, together with the mesh size, to ensure that the computations simulate the sharp interface moving boundary problem (3.1). The spatial domain in general is taken to be a two dimensional square region. The numerical technique used to solve (4.3) for n fixed is as follows: let $(\phi_i)_{i=1}^{D}$ be the piecewise linear cardinal basis for $\mathcal{T}^h$, where D is the number of vertices of the triangulation. Using the mass and stiffness matrices, we can rewrite (4.3) in an equivalent vector formulation as: $\forall\, \boldsymbol{\xi} \in \mathbb{R}^D$, $|\xi_i|\leqslant 1$ $(1\leqslant i\leqslant D)$ find $\boldsymbol{U} \in \mathbb{R}^D$, $|U_i|\leqslant 1$ satisfying

$$(\boldsymbol{\xi}-\boldsymbol{U})^T\left(M\left(\frac{\tau}{\Delta t}(\boldsymbol{U}-\boldsymbol{U}^{n-1})\right)+\gamma K\boldsymbol{U}-M\boldsymbol{U}+\alpha G^h\boldsymbol{U}-\alpha G^h\left(\boldsymbol{U}^{n-1}+\frac{2c}{l}\boldsymbol{\Theta}^{n-1}\right)\right)\geqslant 0. \tag{5.1}$$

For convenience, we consider a uniform triangulation of the square domain Ω and define the inner product $(\cdot,\cdot)^h$ by

$$(\chi,\phi)^h=\int_\Omega \Pi^h\left(\chi(x)\phi(x)\right)dx,$$

where $\Pi^h\cdot$ is the piecewise bilinear functional on each subsquare of Ω. The vector variational inequality (5.1) is solved by a splitting algorithm for the sum of two nonlinear operators (see [LM]). The choice of this inner product, makes the resulting algorithm easy to solve using a fast cosine transform. Computations, based on this methodology, for approximating motion by mean curvature are described in [BE3] and [PV].

5.1 Gibbs-Thomson with kinetic undercooling

In this subsection we take the scaling $\tau=\sigma_1\epsilon^2$, $\alpha=\sigma_2\epsilon$ and $\gamma=\epsilon^2$. Unless we explicitly state otherwise, in all experiments the following data was taken: $\Omega=(0,0{\cdot}5)^2$, $c=k=l=1$, $\Delta t=0{\cdot}25\epsilon^2$.

In these experiments, we focus on the discrete level set

$$\Gamma^h(n\Delta t):=\{(x,y)\in\Omega : U^n(x,y)=0\},$$

which formally represents the discretization of the free boundary $\Gamma(t)$ solving the Stefan problem with Gibbs-Thomson and kinetic undercooling, see §3.2. We will also find it useful to define $\Omega_h^+(n\Delta t)=\{(x,y)\in\Omega : U^n(x,y)>0\}$ and $\Omega_h^-(n\Delta t)$ similarly.

The initial data U^0 was constructed in the following way: let (x_i,y_i) be the nodes of the triangulation and $d_i=d((x_i,y_i),\Gamma(0))$ be the value of the signed distance from (x_i,y_i) to $\Gamma(0)$ (d_i takes positive and negative values in $\Omega^+(0)$ and $\Omega^-(0)$ respectively), then we define

$$U^0(x_i)=\begin{cases}1 & \text{if } d_i>\pi\epsilon/2,\\ \sin(d_i/\epsilon) & \text{if } |d_i|\leqslant\pi\epsilon/2,\\ -1 & \text{if } d_i<-\pi\epsilon/2.\end{cases}$$

A planar interface

If $\Omega = \{(x,y) \in (0,0{\cdot}5)\times\mathbb{R}\}$ and $\Gamma(0) = \{(x,y) \in \Omega : y = 0.05\}$, then an exact solution to (3.1a,b,c) is given by

$$\theta(x,y,t) = \begin{cases} e^{-V(y-0\cdot05-Vt)} - \left(1 + \dfrac{\pi}{4}\dfrac{\sigma_1}{\sigma_2}V\right) & (x,y) \in \Omega^+(t) \\ -\dfrac{\pi}{4}\dfrac{\sigma_1}{\sigma_2}V & (x,y) \in \Omega^-(t), \end{cases}$$

where $\Omega^+(t) = \{(x,y) \in \Omega : y > Vt + 0.05\}$, $\Omega^-(t) = \{(x,y) \in \Omega : y < Vt + 0.05\}$ and $\Gamma(t) = \{(x,y) \in \Omega : y = 0.05 + Vt\}$ is moving with constant velocity $V\boldsymbol{n}$ into $\Omega^+(t)$.

To reduce the computational effort, we notice that θ is independent of x and so reduce the problem to that of one dimension. We performed the simulation on an interval $[0,2]$ with homogeneous Neumann boundary conditions although these are inappropriate for θ to be a solution of (3.1). However, we expect θ to be close to the solution over a short interval of time. Setting $V = 1$, $\sigma_1 = 1$, $\sigma_2 = 256 \times 9 \times \pi/70$, $T = 0{\cdot}3$ and taking the numerical parameters $\Delta t = \epsilon^2/4$, $h = 2/768$, $\epsilon = \left(\frac{7h}{\pi}\right)^{1/1.1}$ (see [BPV]) we found that $\mathrm{dist}(\Gamma(n\Delta t), \Gamma^h(n\Delta t)) < 3.58 \times 10^{-3}$ and $\max_n |\theta(x,y,n\Delta t) - \Theta^n(x,y)| < 2.24 \times 10^{-3}$ $((x,y) \in \Gamma^h(n\Delta t))$. We also found that the number of points in the region $|U^n(x_i,y_i)| < 1$ remained approximately constant throughout each computation.

A perturbed planar interface

Define $s : [-0.5, 0.5] \to \mathbb{R}$

$$s(x) = \begin{cases} 0.05 & \text{if } |x| \leqslant 3/8, \\ \dfrac{1}{16}\cos(8\pi x) + \dfrac{9}{80} & \text{if } |x| > 3/8, \end{cases}$$

$\Omega^+(0) = \{(x,y) \in \Omega : y > s(x)\}$, $\Omega^-(0) = \{(x,y) \in \Omega : y < s(x)\}$ and $\Gamma(0) = \{(x,y) \in \Omega : y = s(x)\}$. Hence $\Gamma(0)$ is a large perturbation of a flat interface. We define Θ^0 as follows: for $(x,y) \in \Omega$

$$\Theta^0(x,y) = \begin{cases} 0 & \text{if } U^0(x,y) = -1, \\ -1 & \text{if } U^0(x,y) = 1, \\ \dfrac{U^0(x,y) - 1}{2} & \text{if } |U^0(x,y)| < 1. \end{cases}$$

This is taken to be the definition of Θ^0 in the remainder of this subsection. Outside the interfacial region, θ is a uniform temperature. We take $\Omega = (-0{\cdot}5, 0{\cdot}5)\times(0,1)$ with homogeneous Neumann boundary conditions. All of the numerical parameters were identical to the previous experiment except $\sigma_1 = 0\cdot5$, $\sigma_2 = 9/(10\epsilon)$, $h = 1/512$ and $\Delta t = \epsilon^2/4$. For y sufficiently large $\Theta^0(x,y) \equiv -1$. In figure 5.1a, $\Gamma^h(n\Delta t)$ is plotted when $n = 0, 20, \cdots, 520$ on the square $(0,0{\cdot}5)\times(0,0{\cdot}5)$. The free boundary $\Gamma^h(0)$ moves into $\Omega_h^+(n\Delta t)$. In figures 5.1b,c,d we plot $\Gamma^h(n\Delta t)$ on the square $(-0{\cdot}5,0{\cdot}5)\times(0,1)$ when $n = 780, 820, \cdots, 1220$, $n = 1260, 1300, \cdots, 1700$ and $n = 1720, 1760, \cdots, 2180$ respectively. It is interesting to note that between $n = 1220$ and 1260, two parts of $\Gamma^h(n\Delta t)$ collide and develop a singularity, leaving $\Omega_h^+(n\Delta t)$ disconnected for several time steps. The experiment was stopped when

$\Theta^n(x,1)$ increased above the value -1. At the end of the simulation, the symmetry of $\Gamma^h(n\Delta t)$ had been lost.

A perturbed circle

In this experiment we attempt to replicate a simulation of [SS]. The initial shape of the free boundary is a symmetric perturbation of a circle, centred on the origin. To reduce the amount of computation, we use the symmetry and only compute on one quarter of the free boundary. Let $\Omega = (0,1)^2$ and

$$\Gamma(0) := \{(x,y) \in \Omega : x = (0.15 + 0.08\sin(4 \times 2\pi s))\cos(2\pi s) \text{ and } y = (0.15 + 0.05\sin(4 \times 2\pi s))\sin(2\pi s) \text{ where } s \in [0,1/4]\}\,.$$

We take $\Omega^-(0)$ to be the interior of $\Gamma(0)$ and $\Omega^+(0)$ to be the exterior. All other parameters were taken to be the same as in the previous experiment. We plot $\Gamma^h(n\Delta t)$ on the square $(0,0{\cdot}5)^2$ in figure 5.2a when $(n = 0, 20, \cdots, 280)$ and on $(0,1)^2$ in figures 5.2b,c,d where $(n = 300, 320, \cdots, 600)$, $(n = 620, 640, \cdots, 920)$ and $(n = 940, 960, \cdots, 1260)$ respectively. Again, we see that when $n = 1260$, $\Gamma^h(n\Delta t)$ is not symmetric.

Another perturbed circle

This experiment is identical to the one previously described in every respect, except we chose $\sigma_1 = 1$ and accordingly set $\Delta t = \epsilon^2/8$. In figure 5.3a $\Gamma^h(n\Delta t)$ is plotted on the square $(0,0{\cdot}5)^2$ when $(n = 0, 50, \cdots, 550)$ and on $(0,1)^2$ in figures 5.3b,c,d when $(n = 600, 650, \cdots, 950)$, $(n = 1000, 1050, \cdots, 1450)$, $(n = 1500, \cdots, 1950)$. Notice that $\Gamma^h(n\Delta t)$ moves on a slightly slower time scale than the previous experiment (Δt is half the value and we take more time steps). In the later stages, the morphology of $\Gamma^h(n\Delta t)$ is different to the last computation, the tip splitting which ensues tends to fan out whereas in the previous computation the tips appear to close in on each other.

5.2 Zero specific heat and zero kinetic undercooling

Taking $\tau = 0, \alpha = \frac{\pi}{4}\epsilon, \gamma = \epsilon^2, c = 0, k = 1$ and $l = 2$ in the phase field equations leads to a scaling of the Cahn-Hilliard equation of phase separation. The sharp interface limit is then a moving boundary problem for Laplace's equation. It has the property that $|\Omega^+(t)| = |\Omega^+(0)|$ for all t and $d|\Gamma(t)|/dt < 0$; [Ch]. It is clear that in two dimensions a steady state is a circle of prescribed area and θ is constant in Ω with value given by the curvature of the circle. Another steady state is a planar interface with $\theta = 0$. Depending on the ratio $|\Omega^+(0)|/|\Omega|$ the interface with minimal length is either a line or a quarter circle. Chen, [Ch], has shown that for small perturbations a circular interface is stable. We describe some experiments where the numerical parameters are the same as in the previous subsection, except we take $\Delta t = \epsilon^3$. (From Theorem 4.2 we have existence and uniqueness for our discretization when $\Delta t < \pi\epsilon^3$.)

A circular interface

Computations were performed with $\Gamma(0) = \{(x,y) \in \Omega : \max\{x,y\} = 0\cdot 25\}$. The discrete steady state was computed by time stepping and found to be such that the discrete level set Γ^h_∞ was an approximation to a quarter circle with a near radius 0.143. This is in good agreement with the minimal interface radius of $1/\sqrt{4\pi}$. The average of the discrete values of Θ^h was 3.511 with a maximum deviation from the mean of 3.4×10^{-2}. This is in reasonable agreement with $\sqrt{4\pi}$.

The two dimensional dumbbell

In this experiment we wish to take the initial free boundary to be that of a dumbbell. Making the dumbbell have a symmetric profile, we can reduce the computations by only computing on one quarter of its profile. For $x \in [0.05, 0.5]$ define

$$s(x) = \begin{cases} 0 \cdot 475\sqrt{1 - \left(\dfrac{x - 0.1}{0.05}\right)^2} & \text{if } x \leqslant 0 \cdot 1 + 0 \cdot 05\sqrt{1 - \left(\dfrac{4}{95}\right)^2} \\ 0.02 & \text{if } x > 0 \cdot 1 + 0 \cdot 05\sqrt{1 - \left(\dfrac{4}{95}\right)^2} \end{cases},$$

and $\Gamma(0) = \{(x, y) \in \Omega : y = s(x)\}$. We set $h = 0.5/196$ and $\epsilon = 7/(2 \times 196\pi)$, all other parameters remain unchanged. We performed a second experiment with different values for h and ϵ, the graphical output was the same. In figure 5.4, $\Gamma^h(n\Delta t)$ is displayed for various times. Eventually the handle pinches off and the final stationary profile is that of a quarter circle, corresponding to a minimal interface.

Two adjacent blobs

In this experiment we take initially two large segments of semi-circles separated by a small gap. As time evolves these coalesce and evolve into a circle. To save on computer time, we use the inherent symmetry and only compute on one quarter of the domain. We take

$$\Gamma(0) := \{(x, y) \in \Omega : x = 0.5 - 0.4\cos(\pi s/2), y = 0.5 - 0.4\sin(\pi s/2) \\ \text{where } s \in [0, 2\cos^{-1}(2.5)/\pi] \text{ and } x = 0.4, y \in [0, 0.5 - 0.4\sin(\cos^{-1}(2.5))]\},$$

for one of the blobs. We set $h = 0.5/64$ and $\epsilon = 7/(128\pi)$ and took the same parameters as in the previous experiment. Plotting $\Gamma^h(n\Delta t)$ at various times in figure 5.5, one sees the evolution of half a blob and eventual coalescence.

A cigar shape

For our initial shape we take a long strip with semi-circles at each end and again use the inherent symmetry in the computation. All parameters were taken to have the previously stated values except $h = 0.5/128$, $\epsilon = 7/(256\pi)$ and

$$\Gamma(0) := \{(x, y) \in \Omega : x = 0.0625 - 0.015\cos(\pi s/2), \\ y = 0.5 - 0.4\sin(\pi s/2) \text{ where } s \in [0, 1] \text{ and } y = 0.0625, x \in [0.0625, 0.5]\}.$$

Figure 5.6 shows $\Gamma^h(n\Delta t)$ plotted at various times. Notice that $\Omega_h^+(0)$ loses its convexity as the computations proceed. This contrasts with the preservation of convexity in motion by mean curvature. Also $\Gamma^h(n\Delta t)$ eventually takes the minimal interface shape of a quarter circle.

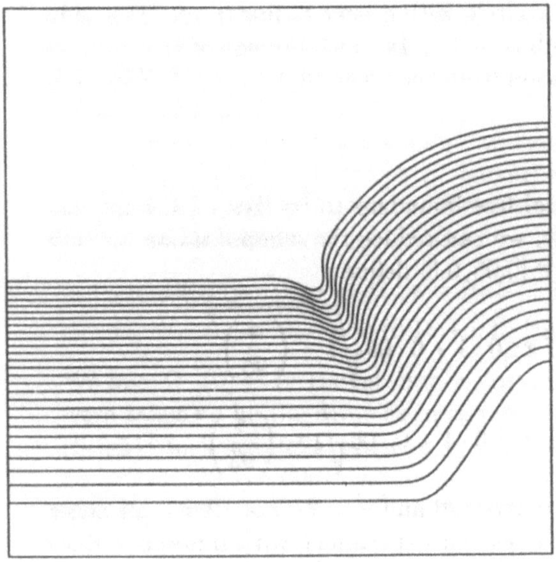

Figure 5.1a

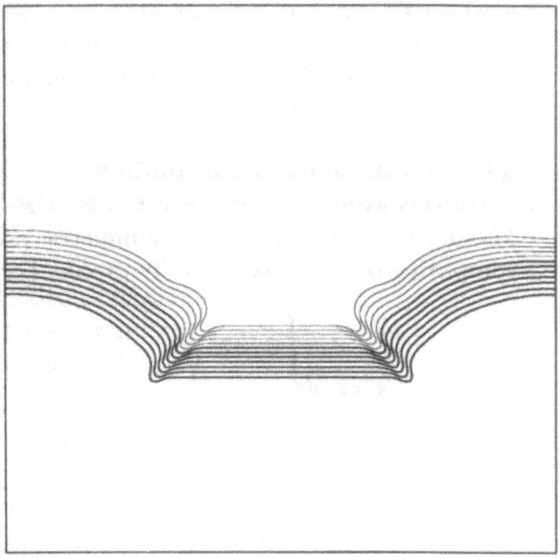

Figure 5.1b

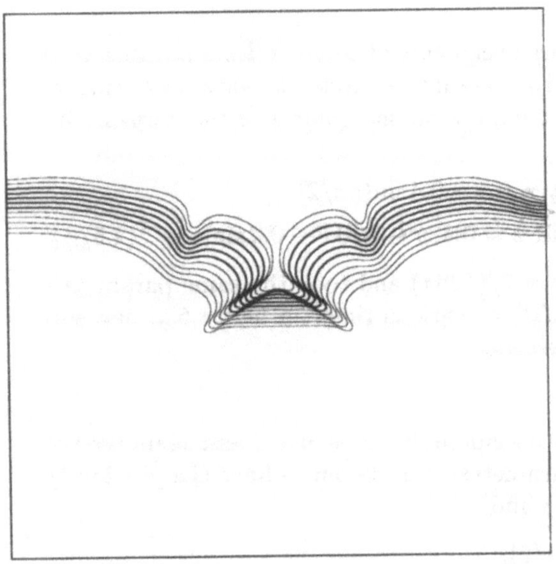

Figure 5.1c

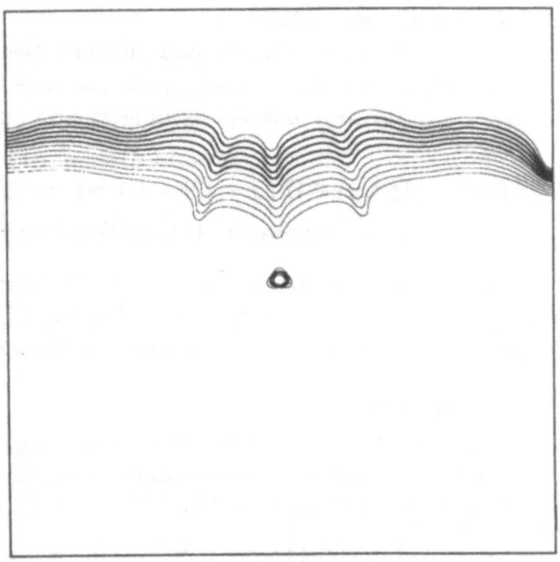

Figure 5.1d

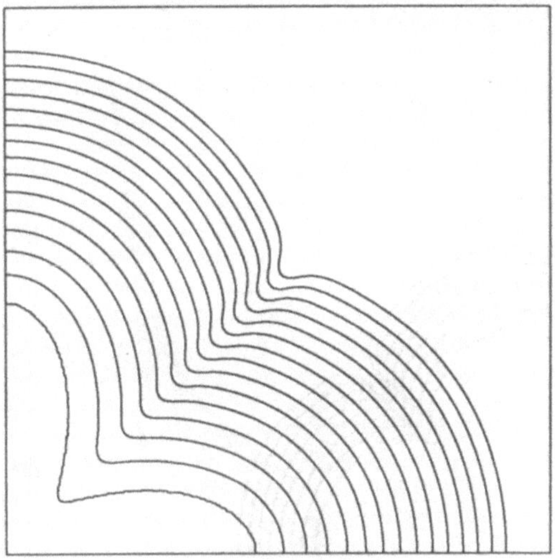

Figure 5.2a

Figure 5.2b

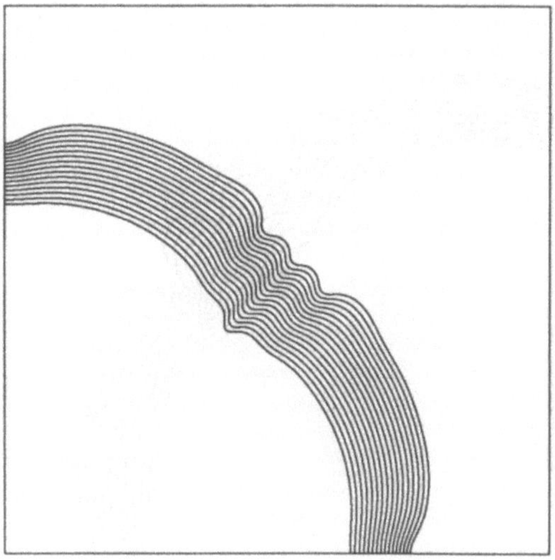

Figure 5.2c

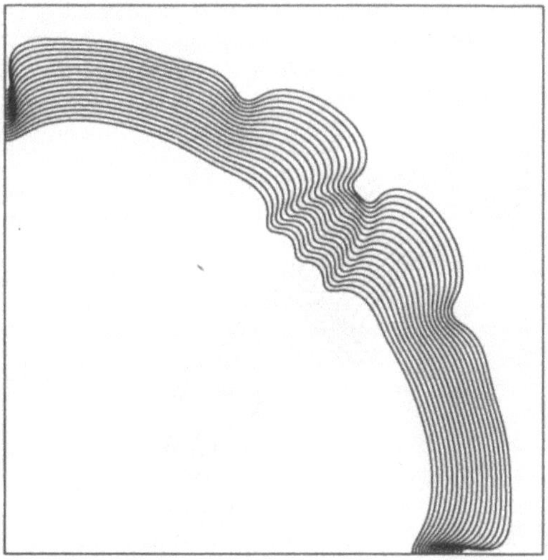

Figure 5.2d

Figure 5.3a

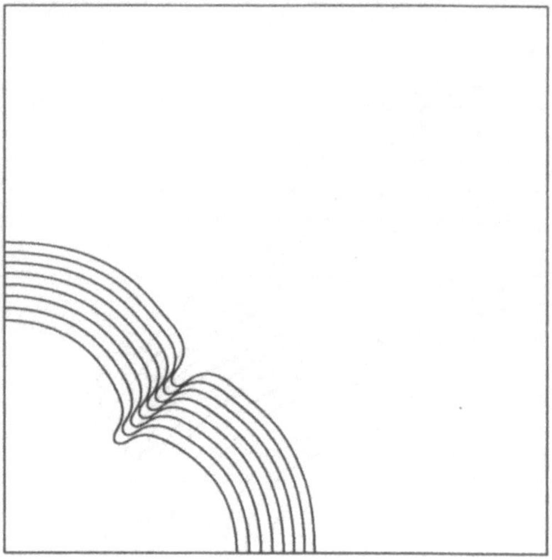

Figure 5.3b

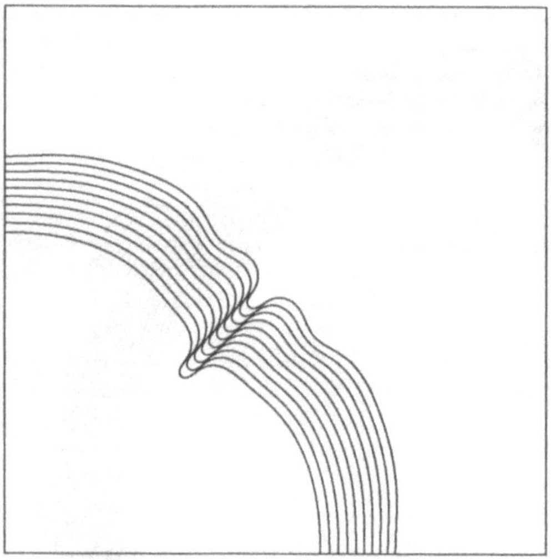

Figure 5.3c

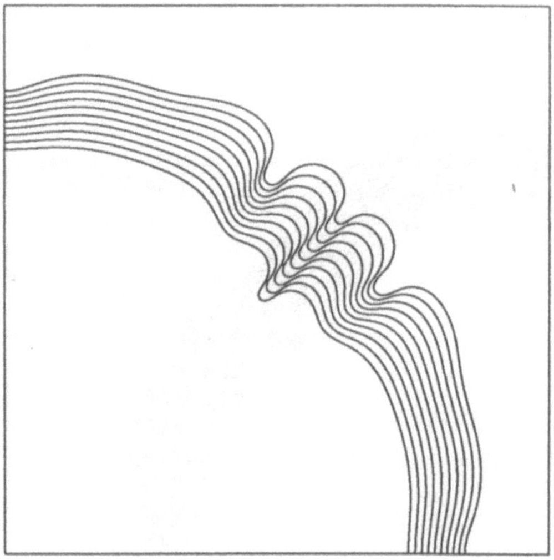

Figure 5.3d

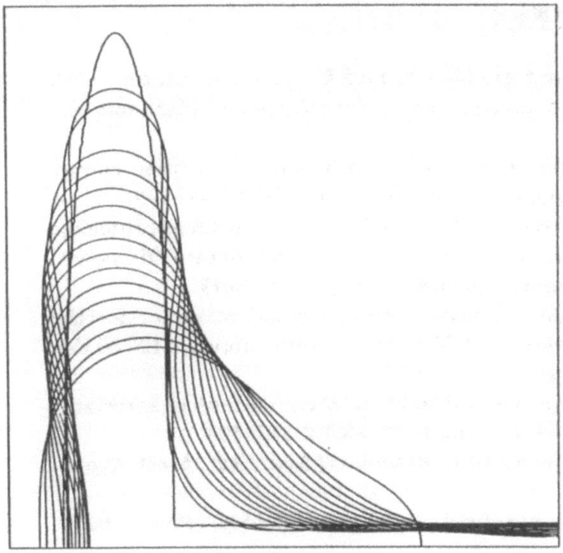

Figure 5.4

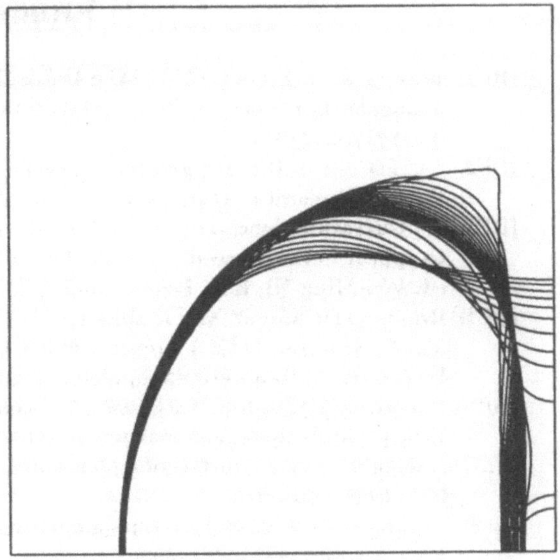

Figure 5.5

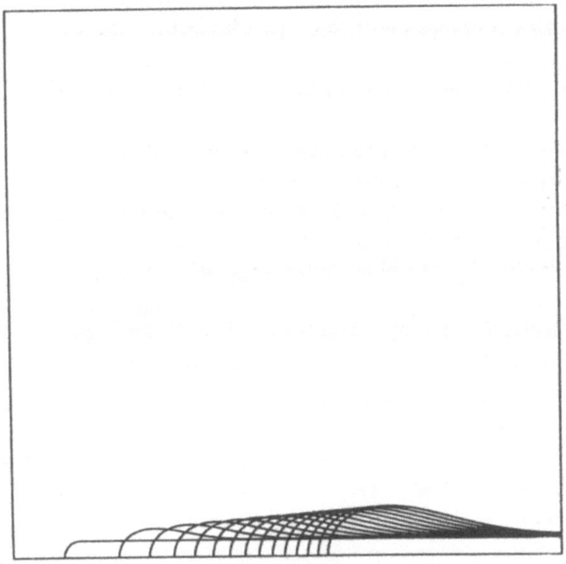

Figure 5.6

References

[BE1] Blowey, J.F., Elliott, C.M., The Cahn-Hilliard gradient theory for phase separation with non-smooth free energy Part I: Mathematical analysis, Euro. Jnl. of Applied Mathematics, **2** (1991), 233–279.

[BE2] —, The Cahn-Hilliard gradient theory for phase separation with non-smooth free energy Part II: Numerical analysis, Euro. Jnl. of Applied Mathematics, **3** (1992), 147–179.

[BE3] —, Curvature dependent phase boundary motion and parabolic double obstacle problems, to appear in the proceedings of the IMA (Minneapolis) workshop:- in 'Degenerate Diffusions' ed. Wei-Ming Ni, L.A. Peletier and J.L. Vazquez. Springer Verlag, New York.

[BCH] Brochet, D., Chen, Xinfu, Hilhorst, D., Finite dimensional exponential attractor for the phase field model, IMA preprint #858 University of Minnesota, Minneapolis, MN 55455, USA (1991). To appear in Applicable Analysis.

[BPV] Bellettini G., Paolini M., Verdi C., Γ-convergence of discrete approximations to interfaces with prescribed mean curvature, Rend. Atti. Naz. Lincei, **1**, (1990), 317–328

[Ca1] Caginalp, G., An analysis of a phase field model of a free boundary, Arch. Rat. Mech. Anal., **92**, (1986), 205–245.

[Ca2] —, Stefan and Hele-Shaw type models as asymptotic limits of phase field equations, Phys. Rev. A., **39** (1989), 887–896.

[Ch] Chen Xinfu, Hele-Shaw problem and area preserved curve shortening motion, Preprint

[CE] Chen Xinfu, Elliott C.M., Asymptotics for a parabolic double obstacle problem, Preprint submitted for publication

[EO] Elliott C.M., Ockendon J.R., Weak & variational methods for free and moving boundary problems, Pitman, London (1982).

[Li] Lions J.L., Quelques méthods de résolution des probèmes aux limit non linéaires, Dunod, Paris (1969)

[LM] Lions P.L., Mercier B., Splitting algorithms for the sum of two nonlinear operators, SIAM J. Numer. Anal., **16**(6) (1979), 964–979.

[PV] Paolini M., Verdi C., Asymptotic and numerical analyses of the mean curvature flow with space-dependent relaxation parameter, to appear in Asymptotic Analysis.

[St] Stoth, B.E.E., The coupled Stefan problem with the Gibbs-Thomson law as singular limit of the phase-field equations in the radial case, submitted to E.J.A.M.

[SS] Sethian, J.A., Strain, J., Crystal growth and dendritic solidification, to appear in J. Comp. Phys.

[Te] Témam, R., Infinite-Dimensional Dynamical Systems in Mechanics and Physics, Springer-Verlag (1988).

Cahn and Hilliard fluid on an oscillating boundary

G. Bouchitté and P. Seppecher

Abstract. A Cahn and Hilliard fluid is in equilibrium in a solid container. The rugosity of the boundaries is taken into account through the assumption of an oscillating boundary whose period and amplitude are of the same order of magnitude as the thickness of the interface. We study the Γ-limit of this problem when this length tends to zero. We obtain a homogenized boundary energy and we show that rugosity may modify the wetting property of the wall until it is completely wet.

1. Introduction

In 1959 Cahn and Hilliard [1] wrote a continuum model for two-phase fluids. The postulated energy was of the form:

$$E(u) = \int_{\Omega} W(u)\, dx + \lambda \int_{\Omega} |Du|^2\, dx + \int_{\partial\Omega} \sigma(u)\, dH^{N-1}$$

u is the density of the fluid. u is positive, $\int_{\Omega} u\, dx = m$. $W(u)$ is the volumic free energy for an homogeneous fluid of density u. It is a non convex function (for example the energy associated with the Van der Waals pressure). See figure 1 which defines α_1 and α_2, values of u in the phases. λ is a physical parameter which may be deduced from experimental surface tension as far as the function W is known. $\sigma(u)$ is a surfacic energy which characterizes the interactions between a fluid of density u and the wall ($\sigma(u)$ is positive). This model has its own length L^c :

$$L^c = \sqrt{\lambda}\frac{\alpha_2 - \alpha_1}{2\int_{\alpha_1}^{\alpha_2} \sqrt{W(u)}\, du}$$

which is characteristic of the thickness of the transition layer between the two phases $u = \alpha_1$ and $u = \alpha_2$.

As L^c is very small it is natural to study the asymptotic behaviour of the model as L^c tends to zero. This procedure is purely mathematical and the model does not include assumptions on the behaviour of the other physical quantities W, σ, m and Ω itself with respect to L^c. We emphasize that the postulated behaviour for these quantities are primordial for the resulting model.

If W, σ, m, Ω are constant, we lose every surface tension effect. For that the usual asymptotic problem is concerned with the limit as ε tends to zero of the *rescaled energy* :

$$E^{\varepsilon}(u) = \int_{\Omega} \frac{W_0(u)}{\varepsilon}\, dx + \varepsilon \int_{\Omega} |Du|^2\, dx + \int_{\partial\Omega} \sigma(u)\, dH^{N-1}$$

where $W_0 \geq 0, W_0(\alpha_1) = W_0(\alpha_2) = 0$ ($W_0(u) = W(u) - l(u)$ see figure 1)

This problem was completely solved in 1987 by L. Modica [2]. The resulting energy is given by:

$$E_0(u) = c\, H^{N-1}(\partial^* A \cap \Omega) - \widehat{c}\, H^{N-1}(\partial^* A \cap \partial\Omega)$$
$$\text{if } u = \alpha_1\, 1_A + \alpha_2\, 1_{\Omega\setminus A} \text{ and } per_{\Omega}(A) < +\infty$$
$$E_0(u) = +\infty \text{ otherwise,}$$

where

$$c = 2\int_{\alpha_1}^{\alpha_2} \sqrt{W_0(s)}\, ds$$
$$\widehat{c} = \widehat{\sigma}(\alpha_2) - \widehat{\sigma}(\alpha_1)$$
$$\widehat{\sigma}(t) = inf\{\sigma(s) + 2 \mid \int_s^t \sqrt{W_0(u)}\, du \mid, s \in \mathrm{R}\}$$

c is the surface tension. Note that $|\, \widehat{c}\, | \leq c$ and that the ratio $\widehat{c}/c = \cos\theta$ gives the contact angle θ between the interface and the wall, following Young's law.

Let us now list some other dependances of W, σ, m and Ω with respect to ε, which lead to different models.

The form chosen by L. Modica for W ($W = W_0/\varepsilon$) ensure that c is a finite non vanishing quantity, but it avoids any compressibility effect (the values of u in the phases are prescribed). A way to include some compressibility effects is to consider a family of functions W_{ε} so that, as ε tends to zero, $\int_{\alpha_1}^{\alpha_2} \sqrt{W_{\varepsilon}(u)}\, du$ tends to infinity and $d^2W_{\varepsilon}/du^2(\alpha_1)$ remains finite. That was done by Buttazo and Al.[3] who considered the family $W_{\varepsilon}(u) = W_0(u) + 1/\varepsilon^3\, \psi(u/\varepsilon)$ where W_0 is of the type described in figure 1 and β is a positive function with $supp(\psi) \subset\subset [\alpha_1, \alpha_2]$.

When a liquid film lies on a wall, the vicinity of the interface and the wall leads to strengths which may stabilize or not the film [4]. The dependence of the total energy with respect to the thickness of the film can be obtained by letting m depend on ε in such a way that it converges to $\int_{\Omega} \alpha_2\, dx$.

In this paper we concentrate on the effects due to the rugosity of the wall already pointed out for their relationship with friction and hysteresis phenomenon [5] (i.e. the difference between the receeding and the advancing contact angle).

We will then assume that the boundary of Ω is oscillating with period d_ε and amplitude h_ε tending to zero with ε: Ω depends upon ε.

In the second section we will give a precise mathematical setting of the problem and state the main theorem of convergence (Γ-convergence of associated functionals). We are led to a relaxed boundary contact energy $\widehat{\widehat{\sigma}}$ which is related to a local problem.

In section 3 we study the influence of the surface parameters (*rugosity* parameters) upon this energy $\widehat{\widehat{\sigma}}$. We show that, for a given fluid and a given material for the wall, the rugosity parameters can increase the contact angle continuously from its value on a flat wall to π (complete wetting).

Section 4 is devoted to proofs.

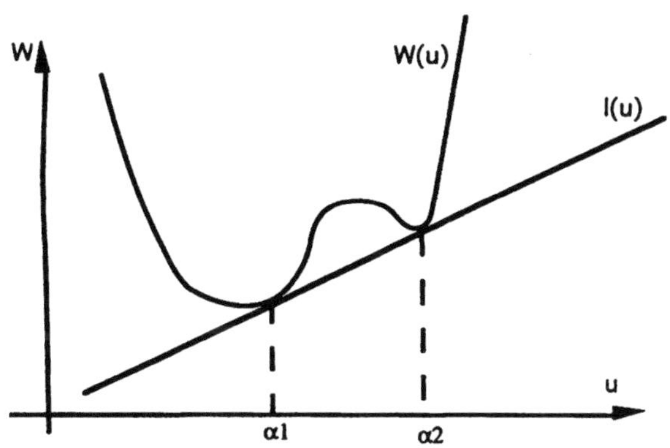

figure 1

2.The main theorem

Let Ω be a bounded open subset R^N with a smooth boundary. An element $x \in \mathrm{R}^N$ is described by its coordinates in an orthogonal basis. As the last coordinate x_N plays a special role in our problem we shall write $x = (x', x_N)$ where $x' = (x_1, x_2, ..., x_{N-1})$.

The oscillating boundary is expressed in terms of a function $f : \mathrm{R}^{N-1} \longrightarrow [0,1]$ which is assumed to be C^1 and Y-periodic where $Y =]-1/2, 1/2[^{N-1}$. Define for every positive d and h:

$$\Delta(d,h) = \{x_N > -hf(x'/d)\}$$
$$\Lambda(d,h) = \{x_N = -hf(x'/d)\}$$

Let d_ε and h_ε be two sequences of positive parameters tending to zero as $\varepsilon \longrightarrow 0$. We consider the following subsets of $\mathbf{R}^N$ (see figure 2):

$$\begin{array}{rclcrcl} \Omega_\varepsilon & = & \Delta(d_\varepsilon, h_\varepsilon) \cap \Omega & ; & \Omega_0 & = & \Delta(0,1) \cap \Omega \\ \Gamma_\varepsilon & = & \Lambda(d_\varepsilon, h_\varepsilon) \cap \Omega & ; & \Gamma_0 & = & \Lambda(0,1) \cap \Omega \\ \partial\Omega_\varepsilon & = & \Gamma_\varepsilon \cup \Gamma' & ; & \partial\Omega_0 & = & \Gamma_0 \cup \Gamma' \end{array}$$

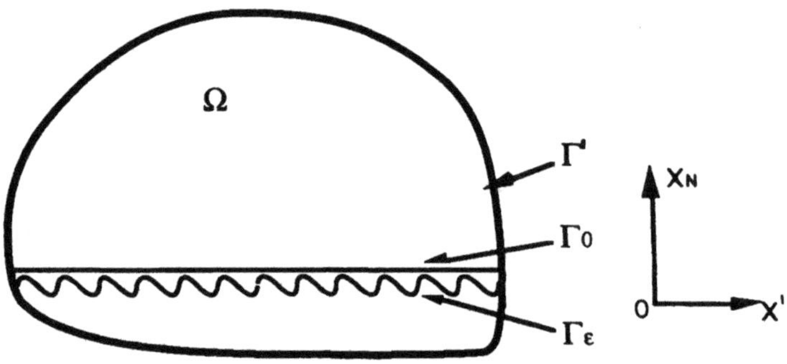

figure 2

A zooming of a part of Γ_ε brings us to consider, λ and α being two real parameters, the subset: $B_\lambda = \lambda Y \times \mathbf{R}$ (see figure 3) and the following subsets of L^2_{loc}:

$$\mathcal{A}^L_\lambda(\alpha) = \{u \in H^1_{loc}(\Delta(1,1));\ u\ \lambda Y \text{ periodic in } x';\ u = \alpha \text{ for } x_N > L\}$$
$$\mathcal{B}^L_\lambda(\alpha) = \{u \in \mathcal{A}^L_\lambda(\alpha);\ u = \alpha \text{ on } \partial B_\lambda \cap \{x_N > 0\}\}$$

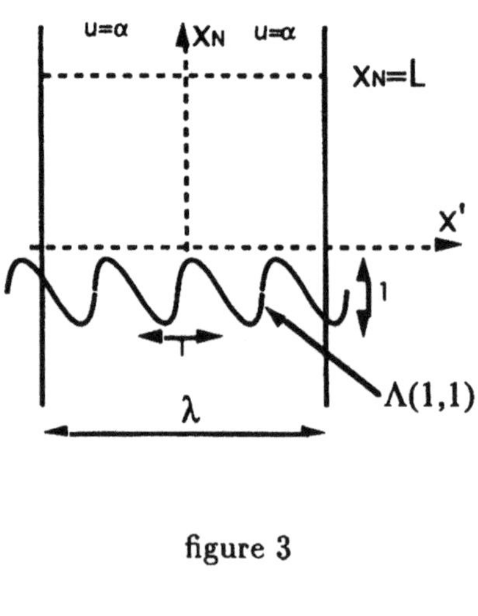

figure 3

The multiphase problem with the rough boundary Γ_ε reads as:

$$(\mathcal{P}^\varepsilon) \qquad inf\ \{E^\varepsilon(u),\ u \in L^2(\Omega),\ \int_\Omega u = m\} \qquad \text{where :}$$

$$E^\varepsilon(u) = \int_{\Omega_\varepsilon} [\varepsilon \mid Du \mid^2 + \frac{1}{\varepsilon} W(u)] dx + \int_{\Gamma_\varepsilon} \sigma(u)\ dH^{N-1}$$

if $u \in H^1(\Omega_\varepsilon)$ and $u = 0$ on $\Omega \backslash \overline{\Omega_\varepsilon}$,

$E^\varepsilon(u) = +\infty$ otherwise.

We will make the following assumptions:

$(H1)$ $\quad \lim \dfrac{h_\varepsilon}{d_\varepsilon} = \delta;\ \lim \dfrac{\varepsilon}{d_\varepsilon} = \gamma; \quad \delta, \gamma \in]0, +\infty[$

$(H2a)$ $\quad W$ is C^2; $W \geq 0$ and satisfies :

$(\ b)$ $\quad W(u) = 0 \Longleftrightarrow u \in \{\alpha_1, \alpha_2\}$

$(\ c)$ $\quad W - W^{**} \leq M$ for a suitable constant M

$(\ d)$ $\quad W(u) \geq \lambda_0 \mid u \mid^2 - \mu_0$ where $\lambda_0 > 0$

$(\ e)$ $\quad W''(\alpha_i) > 0$ (finite compressibility of the two phases)

$(H3)$ $\quad \sigma(u)$ is continuous; $\sigma(\mathrm{u}) \geq 0$ and $\sigma - \sigma^{**} \leq M$

Here W^{**} and σ^{**} denote the convexification of W and σ.

Our main result states that solutions of $\mathcal{P}^\varepsilon$ converge in $L^2(\Omega)$ to the solutions of a problem $\mathcal{P}^0$. To express this limit problem $\mathcal{P}^0$ we need the space $BV(\Omega_0)$ of functions $u \in L^1(\Omega_0)$ such that $\mid Du \mid (\Omega_0) = \sup\{\int_{\Omega_0} u\ div\ g\ dx\ ;\ g \in C_0^1(\Omega; \mathrm{R^N}), |g| \leq 1\} < +\infty$.

We also use the following surface energies:

$$c = 2 \int_{\alpha_1}^{\alpha_2} \sqrt{W(s)}\ ds \tag{2.1}$$

$$\widehat{c} = \widehat{\sigma}(\alpha_2) - \widehat{\sigma}(\alpha_1) \tag{2.2}$$

$$\widehat{\sigma}(\alpha) = inf\{\sigma(s) + 2 \mid \int_s^\alpha \sqrt{W(u)}\ du \mid, s \in \mathrm{R}\} \tag{2.3}$$

$$\widehat{\widehat{c}} = \widehat{\widehat{\sigma}}(\alpha_2) - \widehat{\widehat{\sigma}}(\alpha_1) \tag{2.4}$$

$$\widehat{\widehat{\sigma}}(\alpha) = \inf_{L>0} \inf_{u \in \mathcal{A}_1^L(\alpha)} \{ \int_{B\cap\Delta(1,1)} \{\frac{\delta}{\gamma} W(u) + \frac{\gamma}{\delta} \mid \frac{\partial u}{\partial x_N} \mid^2 + \gamma\delta \mid \frac{\partial u}{\partial x'} \mid^2 \} \, dx + \\ + \int_{B\cap\Lambda(1,1)} a_\delta(x') \, \sigma(u) dH^{N-1} \} \tag{2.5}$$

where a_δ is a distortion factor defined as:

$$a_\delta = \frac{(1 + \delta^2 \mid \nabla f \mid^2)^{1/2}}{(1 + \mid \nabla f \mid^2)^{1/2}} \tag{2.6}$$

Now $\mathcal{P}^0$ reads as:

$$(\mathcal{P}^0) \qquad \inf \{E^0(u), \; u \in L^2(\Omega), \int_\Omega u = m\} \qquad \text{where :}$$

$$E^0(u) = \frac{c}{\alpha_2 - \alpha_1} \int_{\Omega_0} \mid Du \mid \, dx + \int_{\Gamma'} \widehat{\sigma}(u) \, dH^{N-1} + \int_{\Gamma_0} \widehat{\widehat{\sigma}}(u) \, dH^{N-1}$$

if $u \in BV(\Omega)$, $u(x) \in \{\alpha_1, \alpha_2\}$ a.e. on Ω_0, $u(x) = 0$ a.e. on $\Omega\backslash\Omega_0$,

$E^0(u) = +\infty$ otherwise.

Theorem 2.1. *Under (H1), (H2), (H3), the sequence E^ε Γ-converges to E^0 in $L^2(\Omega)$, that is:*

(i) *For every sequence (u_ε) converging to $u \in L^2(\Omega)$, one has :* $\liminf_{\varepsilon \to 0} E^\varepsilon(u_\varepsilon) \geq E^0(u)$

(ii) *For every $u \in L^2(\Omega)$, there exists a sequence (u_ε) such that :* $u_\varepsilon \longrightarrow u$ in $L^2(\Omega)$, $\limsup_{\varepsilon \to 0} E^\varepsilon(u_\varepsilon) \leq E^0(u)$

Moreover we can choose u_ε so that $\int_\Omega u_\varepsilon \, dx = \int_\Omega u \, dx$ holds for every ε.

We notice that for $E^0(u) < +\infty$, u takes the form $u = \alpha_1 1_A + \alpha_2 1_{\Omega_0\backslash A}$ where $A \subset \Omega_0$ is a measurable subset with a finite perimeter in Ω_0 that is: $\int_{\Omega_0} |D1_A| < +\infty$. Denoting by $\partial^* A$ the reduced boundary of A (see the book by Giusti [6] for all related concepts), we find that $(\mathcal{P}^0)$ reduces to a purely geometrical problem with respect to A (liquid drop problem):

$$(\widetilde{\mathcal{P}}^0) \qquad \inf \{c \, H^{N-1}(\partial^* A \cap \Omega_0) - \widehat{c} \, H^{N-1}(\partial^* A \cap \Gamma') + \\ - \widehat{\widehat{c}} \, H^{N-1}(\partial^* A \cap \Gamma_0) \; ; \; A \subset \Omega_0, \mid A \mid = m_1\}$$

where $m_1 = \dfrac{|\Omega| \, \alpha_2 - m}{\alpha_2 - \alpha_1}$, $m_1 \in [0, |\Omega|]$.

As a consequence of theorem 2.1 we get the convergence of $(\mathcal{P}^\varepsilon)$:

Theorem 2.2. *Fix $m \in]\alpha_1|\Omega|, \alpha_2|\Omega|[$ (existence of two phases). Let (u_ε) be a sequence in $L^2(\Omega)$ such that:*

$$E^\varepsilon(u_\varepsilon) - \inf \mathcal{P}^\varepsilon = o(\varepsilon) \quad , \quad \int_\Omega u_\varepsilon \, dx = m$$

Then (u_ε) is relatively compact and every cluster point u is a solution of $(\mathcal{P}^0)$ that is $u = \alpha_1 \, 1_A + \alpha_2 \, 1_{\Omega_0 \setminus A}$ where A is solution of the liquid drop problem $(\widetilde{\mathcal{P}}^0)$.

Comment: The wetting properties of the rough wall Γ_0 will be deduced from the ratio $\widehat{\widehat{c}}/c$ and compared with that of the flat wall Γ' characterized by $\widehat{c}/c$. These ratios will be shown to be in [-1,1] (see Sec. 3) and have a precise geometrical meaning: they are the cosine of the contact angle between the fluid phase α_1 and the wall in case this contact occurs on Γ_0 or on Γ'.

3. Estimates on the local problem and dependence of the homogenized boundary energy with respect to the rugosity parameters

In what follows, α is assigned to take value α_1 or α_2. Given δ, $\gamma \in]0, +\infty[$, it is convenient to define for every $u \in H^1_{loc}$ and B Borel subset of R^N:

$$G(u, B) = \int_{\Delta(1,1)\cap B} \left(\frac{\delta}{\gamma} W(u) + \frac{\gamma}{\delta} \left| \frac{\partial u}{\partial x_N} \right|^2 + \delta\gamma \left| \frac{\partial u}{\partial x'} \right|^2\right) dx + \int_{\Lambda(1,1)\cap B} \sigma(u^+) \, dH^{N-1} \tag{3.1}$$

u^+ denoting the trace from above of u on $\Lambda(1,1)$.

We will use (also in sec. 4), the following local problems:

$$\sigma^L_\lambda(\alpha) = \inf \left\{\frac{G(u, B_\lambda)}{\lambda^{N-1}} \; ; \; u \in \mathcal{A}^L_\lambda(\alpha)\right\} \tag{3.2}$$

$$\tau^L_\lambda(\alpha) = \inf \left\{\frac{G(u, B_\lambda)}{\lambda^{N-1}} \; ; \; u \in \mathcal{B}^L_\lambda(\alpha)\right\} \tag{3.3}$$

where λ, L are positive reals.

As $\widehat{\widehat{\sigma}}$ (see (2.5)) and the last expressions depend on δ and γ, we will sometimes write $\widehat{\widehat{\sigma}}(\gamma, \delta)$, $\sigma^L_\lambda(\alpha, \gamma, \delta)$, $\tau^L_\lambda(\alpha, \gamma, \delta)$ and $G_{\gamma,\delta}(u, B)$.

It is easy to check that $\sigma_\lambda^L(\alpha)$ and $\tau_\lambda^L(\alpha)$ are decreasing functions of L. Moreover:

$$\widehat{\widehat{\sigma}}(\alpha) = \lim_{L \to \infty} \searrow \sigma_1^L(\alpha) \tag{3.4}$$

$$\sigma_\lambda^L(\alpha) \le \tau_\lambda^L(\alpha) \qquad \forall\lambda,\ \forall L. \tag{3.5}$$

The first estimate (Proposition 3.1) will be crucial for the proof of theorem 2.1.

Proposition 3.1
i) For every δ, $\gamma, L \in]0,+\infty[$ and every integer λ, *one has:*

$$\sigma_\lambda^L(\alpha) = \sigma_1^L(\alpha) \tag{3.6}$$

ii) Let I a compact interval in $]0,+\infty[$. Then the following inequalities hold uniformly for $\gamma, \delta \in I$ and for suitable C, $C'(L)$:

$$\sigma_1^L(\alpha) + \frac{C}{\lambda^{N-1}} \le \tau_\lambda^L(\alpha) \le \sigma_1^L(\alpha) + \frac{C'(L)}{\lambda^{(N-1)/2}} \tag{3.7}$$

$$\sigma_1^L(\alpha) \le \widehat{\widehat{\sigma}}(\alpha) \le \sigma_1^L(\alpha) + o(L) \tag{3.8}$$

Comments:

a) Let us stress the fact that the equality (3.6) is not trivial since the functional involved G is non-convex. In the homogenization theory, several non-convex examples exhibit a gap between the average energy on periodic cells of length $\lambda \in \mathrm{N}$ and the minimal energy taken on a unit cell (see S. Müller [7]). Fortunately this gap does not appear in our problem.

b) The first inequality in (3.7) is straightforward. The procedure to obtain the second one consists in taking the solution u^L associated with $\sigma_1^L(\alpha)$ and in using cutt-off functions on $B_\lambda \backslash B_{\lambda'}$ (where $\lambda' \in \mathrm{N}$) and $\lambda' \le \lambda - 1$) in order to fit the boundary condition $u = \alpha$ on $\partial B_\lambda \cap \{x_N > 0\}$ (cf. the definition of $\mathcal{B}_\lambda^L(\alpha)$ in sec. 2). The parameter $C'(L)$ depends on the norm of $(u^L - \alpha)$ in $L^2(B_1)$.

To prove prosition 3.1 we need the following lemma:

Lemma 3.2 *The variational problem associated with $\sigma_\lambda^L(\alpha)$ has at least one solution. Moreover:*
i) If $\lambda = k$ is an integer and u is a solution, so is $u(x' - i, x_N)$ for every $i \in \{1, \dots, k-1\}^{N-1}$
ii) If u and v are two solutions, so are $u \wedge v$ and $u \vee v$.

Proof. The existence of a solution is obtained classically by the weak compactness of minimizing sequences in the closed subset $\mathcal{A}_\lambda^L(\alpha)$ of H^1_{loc} and by the lowersemicontinuity of $G(., B_\lambda)$ for this topology.

Assertion (i) is obvious since $u(x' - i, x_N)$ is still kY periodic in x' and agrees with α for $x_N > L$.

Let us prove that $w = u \wedge v$ and $w' = u \vee v$ are also solutions associated with $\sigma_\lambda^L(\alpha)$. Let:

$$A = \{x \in B_\lambda \cap \Delta(1,1)\ ;\ u(x) < v(x)\} \cup \{x \in B_\lambda \cap \Lambda(1,1)\ ;\ u^+(x) < v^+(x)\}$$

(u^+, v^+ denote the traces of u, v on $\Lambda(1,1)$ from above)
We have:

$$Dw = 1_A\ Du + 1_{A^c}\ Dv \quad \text{a.e. on } B_\lambda \cap \Delta(1,1)$$
$$w^+ = 1_A\ u^+ + 1_{A^c}\ v^+ \quad H^{N-1} \text{ a.e. on } B_\lambda \cap \Lambda(1,1)$$

Hence:

$$G(w, B\lambda) = G(u, B_\lambda \cap E) + G(v, B_\lambda \cap A^c)$$
$$G(w', B\lambda) = G(v, B_\lambda \cap E) + G(u, B_\lambda \cap A^c)$$

By adding these two equalities, one gets:

$$G(w, B_\lambda) + G(w', B_\lambda) = G(u, B_\lambda) + G(v, B_\lambda) = 2\ \sigma_\lambda^L(\alpha)$$

Since w and w' are also in the admissible set $\mathcal{A}_\lambda^L(\alpha)$, one has:

$$G(w, B_\lambda) = G(w', B_\lambda) = \sigma_\lambda^L(\alpha)$$

□

Proof of Proposition 3.1

$$\text{Define}: \widehat{u}(x', x_N) = \inf_{i \in \{1,2,\dots,k-1\}^{N-1}} u(x' - i, x_N)$$

From lemma 3.2 we deduce that $\widehat{u}$ is a solution associated with $\sigma_\lambda^L(\alpha)$ ($\lambda \in$ N)

Obviously $\widehat{u}$ satisfies $\widehat{u} = \alpha$ for $x_N > L$ and $\widehat{u}(., x_N)$ is Y-periodic. Hence $\widehat{u} \in \mathcal{A}_1^L(\alpha)$ and:

$$\sigma_\lambda^L(\alpha) = \frac{G(\widehat{u}, B_\lambda)}{\lambda^{N-1}} = G(\widehat{u}, B_1) = \sigma_1^L(\alpha)$$

Let us just outline the proof of (ii). Let δ_L, γ_L in I. As I is compact, we can assume that, as $L \longrightarrow \infty$, $\delta_L \longrightarrow \delta$, $\gamma_L \longrightarrow \gamma$ and that u_L, a solution associated with $\sigma_1^L(\alpha, \gamma_L, \delta_L)$ converges weakly to some u in H^1_{loc}. Using the weak lower semicontinuity in H^1_{loc} of $G(., B_1)$, one gets:

$$\widehat{\widehat{\sigma}}(\alpha) = \lim_{L \longrightarrow +\infty} \sigma_1^L(\alpha) = \lim_{L \longrightarrow +\infty} G(u_L, B_1) \geq G(u, B_1) \tag{3.9}$$

Let us define:

$$\psi(s) = 2 \int_\alpha^s \sqrt{W(t)}dt\ ,\ v_L = \psi(u_L)\ ,\ v = \psi(u)$$

Since, for suitable $C_0 > 0$:

$$G(u_L, B_1) \geq C_0 \int_{B_1 \cap \Delta(1,1)} |\ Dv_L\ |\ dx$$

we obtain:

$$\int_{B_1\cap\Delta(1,1)} | Dv | \; dx \leq \liminf_{L\longrightarrow+\infty} \int_{B_1\cap\Delta(1,1)} | Dv_L | \; dx < +\infty$$

From assumptions (H2c), (H2d) and (H2e), we can show that $\psi(s) \geq C_1 \; | s - \alpha |^2$ for some suitable $C_1 > 0$. Then:

$$\int_Y | u - \alpha |^2 (x', t) \; dx' \leq \frac{1}{C_1} \int_{Y\times\{x_N>t\}} | Dv |$$

Integrating this inequality between $L-1$ and L gives:

$$\beta_L = \int_{Y\times[L-1,L]} | u - \alpha |^2 \; dx \longrightarrow 0 \text{ as } L \longrightarrow +\infty$$

We conclude by multiplying u by a suitable cutt-off function $\varphi(x_N)$ on $[L-1, L]$ (see the proof of lemma 4.2 below for a similar construction). That yields:

$$\sigma_1^L(\alpha) \leq G(u, B_1) + o(\beta_L)$$

Combined with (3.9), we obtain (3.7). □

Proposition 3.3

i) $\widehat{\widehat{\sigma}}(\alpha,\gamma,\delta)$ *is continuous with respect to* δ, $\gamma \in]0,+\infty[$.

ii) For every γ, $\delta \in]0,+\infty[$, *we have:*

$$|\widehat{\widehat{c}}(\gamma,\delta)| = |\widehat{\widehat{\sigma}}(\alpha_2,\gamma,\delta) - \widehat{\widehat{\sigma}}(\alpha_1,\gamma,\delta)| \leq c$$

Proof. The continuity of $\sigma_1^L(\alpha,.,.)$ for every $L > 0$ is straightforward. Then (i) is deduced by using estimate (3.8). For (ii) we prove the inequality $\widehat{\widehat{\sigma}}(\alpha_i) \leq \widehat{\widehat{\sigma}}(\alpha_j) + c$ for $i \neq j$ by extending a solution associated with the definition of $\sigma_1^L(\alpha_j)$ (L being fixed) by a function φ depending only on x_N such that $\varphi(L) = \alpha_j$ and $\varphi(+\infty) = \alpha_2$. □

Now we have to pay attention to the ratio $\widehat{\widehat{c}}/\widehat{c}$ which determines the contact angle on the rough surface associated with γ and δ. This ratio is ruled by the propositions 3.4 and 3.5 below, for which we leave out the proofs in order to be concise.

Proposition 3.4

i) Let us fix $\gamma \in]0, +\infty[$. *Then:*

$$\lim_{\delta \longrightarrow 0+} \widehat{\widehat{\sigma}}(\alpha, \gamma, \delta) = \widehat{\sigma}(\alpha) \ , \quad \lim_{\delta \longrightarrow 0+} \widehat{\widehat{c}}(\gamma, \delta) = \widehat{c}$$

ii) Assume that σ *does not reach a minimum on* $]\alpha_1, \alpha_2[$. *Then:*

$$\lim_{\delta \longrightarrow +\infty} \widehat{\widehat{c}}(\gamma, \delta) = c$$

Proof. Assertion (i) is obtained by approximating the infimum associated with $\sigma_1^L(\alpha, \gamma, \delta)$ by a function depending only on x_N and constant for $x_N < 0$. □

Comment: By the continuity of $\widehat{\widehat{c}}(\gamma, .)$, we see that $\widehat{\widehat{c}}(\gamma, .)$ ranges onto the interval $[c, \widehat{c}]$. In other words, every situation between the case of a flat surface ($\widehat{\widehat{c}} = \widehat{c}$) and the perfectly wetting case ($\widehat{\widehat{c}} = c$) is reached by increasing the slope factor δ from 0 to $+\infty$.

When the scale of rugosity is large with respect to the thickness of the phase transition ($\gamma << 1$), we are led to a local anisotropic Plateau problem:

Recalling that a_δ is given by (2.6), we define, for every fixed $\delta \in]0, +\infty[$, the quantity R_δ and the positively 1-homogeneous convex function $h_\delta(p)$ on R^{N} by:

$$h_\delta(p) = ({p_N}^2 + \delta^2 \, p'^2)^{1/2} \ , \quad R_\delta = \int_Y (1 + \delta^2 |\nabla f|^2)^{1/2} \, dx'$$

Proposition 3.5:

i) $$\lim_{\gamma \longrightarrow 0+} \widehat{\widehat{\sigma}}(\alpha, \gamma, \delta) = \inf_{u \in \mathcal{C}(\alpha)} \{ \frac{c}{\alpha_2 - \alpha_1} \int_{B_1 \cap \Delta(1,1)} h_\delta(Du) \, dx + \int_{B_1 \cap \Lambda(1,1)} a_\delta \, \widehat{\sigma}(u) dH^{\mathrm{N}-1} \}$$

where

$$\mathcal{C}(\alpha) = \{ u \in BV_{loc}(\Delta(1,1)); \ u(x) \in \{\alpha_1, \alpha_2\} \text{ a.e.}; \ u = \alpha \text{ for } x_N > 0 \}$$

ii) $\widehat{\widehat{c}}(+\infty, \delta)$ *is continuously increasing with respect to* δ. *Moreover:*

$$\widehat{\widehat{c}}(+\infty, \delta) = R_\delta \, \widehat{c} \quad \text{for } o \le \delta \le \frac{[(\frac{\widehat{c}}{c})^2 - 1]^{1/2}}{Lip(f)}$$

$$\lim_{\delta \longrightarrow +\infty} \widehat{\widehat{c}}(+\infty, \delta) = c$$

Proof. For assertion (i), we refer to Bouchitte [9] where limits of phase transition models with general anisotropic perturbations are described. □

4. Proofs of theorems 2.1 and 2.2

It will be convenient to localize the energy associated with E^ε as follows; let us define for every $u \in L^2(\Omega)$ and every Borel subset $B \subset \overline{\Omega}$:

$$F_{\varepsilon,d,h}(u,B) = \int_{B\cap\Delta(d,h)} [\varepsilon|Du|^2 + \frac{1}{\varepsilon}W(u)]\, dx + \int_{B\cap(\Lambda(d,h)\cup\Gamma')} \sigma(u^+)\, dH^{N-1}$$

if $u \in H^1(B \cap \Delta(d,h))$,

$F_{\varepsilon,d,h}(u,B) = +\infty$ otherwise.

For simplicity $F_{\varepsilon,d_\varepsilon,h_\varepsilon}$ will be denoted F_ε so that, if $u = 0$ on $\Omega\backslash\Omega_\varepsilon$, one has $E^\varepsilon(u) = F_\varepsilon(u,\overline{\Omega\varepsilon})$.

We will use the following lemmas:

Lemma 4.1 (scaling) *Let ε, h, d > 0, real parameters. Then, for every $\lambda > 0$ and every v in $L^1(Q)$, one has:*

$$F_{\lambda\varepsilon,\lambda h,\lambda d}(v,Q_\lambda) = \lambda^{N-1}\; F_{\varepsilon,h,d}(v_\lambda,Q)$$

where $v_\lambda(x) = v(\lambda x)$.

Proof. We use the change of variable $y = x/\lambda$ so that $\nabla v(x) = \lambda^{-1}\, \nabla v_\lambda(y)$. □

Lemma 4.2 *Let $u_0 = \alpha$ for $x_N > 0$ and $u_0 = 0$ for $x_N < 0$ with $\alpha \in \{\alpha_1,\alpha_2\}$ and assume that h/ε remains bounded. Then for every sequence $u_\varepsilon \in L^1(Q)$ such that $u_\varepsilon \longrightarrow u_0$ there exists $v_\varepsilon \in L^1(Q)$ such that:*

(*i*) $v_\varepsilon \in H^1(Q\cap\{x_N>0\})$; $v_\varepsilon = \alpha$ on $\partial Q \cap \{x_N > 0\}$

(*ii*) $\liminf_{\varepsilon\longrightarrow 0} F^\varepsilon(v_\varepsilon,Q) \le \liminf_{\varepsilon\longrightarrow 0} F^\varepsilon(u_\varepsilon,Q)$

(*iii*) $v_\varepsilon \longrightarrow u_0$ in $L^1(Q)$

Proof. First, possibly by extracting a subsequence, we can assume:

$$\liminf_{\varepsilon\longrightarrow 0} F^\varepsilon(v_\varepsilon,Q) = \lim_{\varepsilon\longrightarrow 0} F^\varepsilon(u_\varepsilon,Q) = \beta < +\infty \tag{4.1}$$

Hence

$$\lim_{\varepsilon\longrightarrow 0} \int_Q W(u_\varepsilon)\, dx = 0$$

and from growth condition (H2c), the sequence $| u_\varepsilon |^2$ is uniformly integrable. Thus:

$$u_\varepsilon \longrightarrow u_0 \text{ in } L^2(Q) \tag{4.2}$$

Let h_ε, k_ε be sequences such that:

$$k_\varepsilon \in \mathrm{N}, \quad \varepsilon k_\varepsilon \longrightarrow 1 \quad (\text{hence } k_\varepsilon \longrightarrow +\infty) \tag{4.3}$$

$$h_\varepsilon = o(|u_\varepsilon - u_0|)_{L^2(Q)} \tag{4.4}$$

Then we define a boundary layer on Q by setting:

$$T_\varepsilon = Q \backslash (1 - h_\varepsilon) Q \tag{4.5}$$

We divide T_ε into k_ε slices of width $h_\varepsilon / k_\varepsilon$.

$$S_\varepsilon^i = Q_\varepsilon^i \backslash Q_\varepsilon^{i-1}, \quad Q_\varepsilon^i = (1 - i\, h_\varepsilon / k_\varepsilon) Q, \quad i \in \{0, 1, ..., k_\varepsilon\} \tag{4.6}$$

so that: $T_\varepsilon = \cup_{i=1}^{k_\varepsilon} S_\varepsilon^i$.

Let φ_ε^i be a smooth function such that:

$$0 \leq \varphi_\varepsilon^i \leq 1, \quad \varphi_\varepsilon^i = 1 \text{ on } Q_\varepsilon^i, \quad \varphi_\varepsilon^i = 0 \text{ on } Q \backslash Q_\varepsilon^{i-1}, | D\varphi_\varepsilon^i | \leq \frac{h_\varepsilon}{k_\varepsilon} \tag{4.7}$$

For a suitable i (we shall choose later), let us consider $v_\varepsilon = u_\varepsilon \varphi_\varepsilon^i + \alpha\, (1 - \varphi_\varepsilon^i)$. We have $v_\varepsilon = \alpha$ on $\partial Q \cap \{x_N > 0\}$, and owing to assumptions (H2c) and (H3), the following inequalities hold:

$$W(v_\varepsilon) \leq W(u_\varepsilon) + C, \quad (W(\alpha) = 0) \tag{4.8}$$

$$\sigma(v_\varepsilon) \leq \sigma(u_\varepsilon) + C, \quad \text{on } \Gamma_\varepsilon \cap Q \tag{4.9}$$

On the other hand from (4.7) $v_\varepsilon = u_\varepsilon$ on $Q_\varepsilon^i, v_\varepsilon = \alpha$ on $Q \backslash Q_\varepsilon^{i-1}$ and $Dv_\varepsilon = \varphi_\varepsilon^i Du_\varepsilon + (u_\varepsilon - \alpha)\, D\varphi_\varepsilon^i$, so we have:

$$| Dv_\varepsilon |^2 \leq 2\, | Du_\varepsilon |^2 + | u_\varepsilon - \alpha |^2 \;\; (k_\varepsilon / h_\varepsilon)^2 \quad \text{on } S_\varepsilon^i \tag{4.10}$$

$$\begin{aligned} F^\varepsilon(v_\varepsilon, Q) &= F^\varepsilon(u_\varepsilon, Q_\varepsilon^i) + F^\varepsilon(v_\varepsilon, S_\varepsilon^i) + F^\varepsilon(\alpha, Q \backslash Q_\varepsilon^{i-1}) \\ &\leq F^\varepsilon(u_\varepsilon, Q) + R_\varepsilon^i \end{aligned} \tag{4.11}$$

where

$$R_\varepsilon^i = 2\,\varepsilon \int_{S_\varepsilon^i \cap \Delta_\varepsilon} | Du_\varepsilon |^2 + \frac{2\,\varepsilon k_\varepsilon^2}{h_\varepsilon^2} \int_{S_\varepsilon^i \cap \Delta_\varepsilon} | u_\varepsilon - \alpha |^2 + \frac{c}{\varepsilon} | S_\varepsilon^i | + c\, H^{N-1}(S_\varepsilon^i \cap \Gamma_\varepsilon) \tag{4.12}$$

Choose i such that $R_\varepsilon^i \leq \sum_{j=1}^{k_\varepsilon} R_\varepsilon^j / k_\varepsilon$. One gets from (4.12):

$$\begin{aligned} R_\varepsilon^i \leq \frac{2\,\varepsilon}{k_\varepsilon} \int_{T_\varepsilon \cap \Delta_\varepsilon} | Du_\varepsilon |^2 + \frac{2\,\varepsilon k_\varepsilon^2}{h_\varepsilon^2} \int_{T_\varepsilon \cap \Delta_\varepsilon} | u_\varepsilon - u_0 |^2 \\ + \frac{c\, h_\varepsilon^{N-1}}{\varepsilon k_\varepsilon} + \frac{c}{k_\varepsilon} H^{N-1}(T_\varepsilon \cap \Gamma_\varepsilon) \end{aligned}$$

Noticing that: $\varepsilon \int_{T_\varepsilon \cap \Delta_\varepsilon} | Du_\varepsilon |^2 \leq F^\varepsilon(u_\varepsilon, Q)$ and $H^{N-1}(T_\varepsilon \cap \Gamma_\varepsilon) \leq H^{N-1}(Q \cap \Gamma_\varepsilon) \leq \sup\ (1 + h_\varepsilon/d_\varepsilon \mid Df \mid^2)^{1/2}$ we obtain owing to (4.1),(4.3),(4.4) and to the assumption that $h_\varepsilon/d_\varepsilon$ is bounded:

$$\limsup_{\varepsilon \longrightarrow 0} R_\varepsilon^i = \lim_{\varepsilon \longrightarrow 0} R_\varepsilon^i = O$$

which yields by (4.11) to the inequality (ii). The assertion (iii) is trivial since $| v_\varepsilon - u_0 | \leq | u_\varepsilon - u_0 |$. □

Lemma 4.3 *Let* $Q =]-1/2, 1/2[^{N-1}$, $\alpha \in \{\alpha_1, \alpha_2\}$ *and* $(\varepsilon,\ d_\varepsilon,\ h_\varepsilon)$ *a sequence such that (H1) holds. Define* $u_0(x) = \alpha$ *if* $x_N > 0$, $u_0(x) = 0$ *if* $x_N < 0$. *Then:*

i) For every sequence (u_ε) *such that* $u_\varepsilon \longrightarrow u_0$ *in* $L^2(Q)$, *one has:*

$$\liminf_{\varepsilon \longrightarrow 0} F_\varepsilon(u_\varepsilon, Q) \geq \hat{\hat{\sigma}}(\alpha, \gamma, \delta)$$

ii) There exists (u_ε) *such that* $u_\varepsilon = \alpha$ *on* $\partial Q \cap \{x_N > 0\}$ *and:*

$$u_\varepsilon \longrightarrow u_0 \text{ in } L^2(Q)\ ,\quad \limsup_{\varepsilon \longrightarrow 0} F_\varepsilon(u_\varepsilon, Q) \leq \hat{\hat{\sigma}}(\alpha, \gamma, \delta)$$

Proof. By lemma 4.2, one can assume that $u_\varepsilon = \alpha$ on $\partial Q \cap \{x_N > 0\}$. Apply the lemma 4.1 with $\lambda = 1/d_\varepsilon$; one gets:

$$F_{\varepsilon, d_\varepsilon, h_\varepsilon}(u_\varepsilon, Q) = d_\varepsilon^{\ N-1}\ F_{\varepsilon/d_\varepsilon, 1, h_\varepsilon/d_\varepsilon}(v_\varepsilon, Q_{1/d_\varepsilon})$$

where $v_\varepsilon(x) = u_\varepsilon(x/d_\varepsilon)$.

Noticing that v_ε belongs to $\mathcal{B}_{L_\varepsilon}^{L_\varepsilon}(\alpha)$ with $L_\varepsilon = 1/d_\varepsilon$, we deduce from proposition 3.1 (cf. (3.7)):

$$F_{\varepsilon, d_\varepsilon, h_\varepsilon}(u_\varepsilon, Q) \geq \tau_{L_\varepsilon}^{L_\varepsilon}(\alpha, \varepsilon/d_\varepsilon, h_\varepsilon/d_\varepsilon) \geq$$
$$\geq \hat{\hat{\sigma}}(\alpha, \varepsilon/d_\varepsilon, h_\varepsilon/d_\varepsilon) - \frac{C}{L_\varepsilon^{\ N-1}}$$

The conclusion (i) follows by letting ε tend to 0 and using the continuity of $\hat{\hat{\sigma}}(\alpha, ., .)$ at (γ, δ) proved in proposition 3.2.

Now, let us prove assertion (ii): Let $w \in \mathcal{B}_1^1(\alpha)$ be the solution associated with the definition of $\tau_1^1(\alpha, \gamma, \delta)$. Define u_ε by:

$$\begin{aligned} u_\varepsilon(x) &= w(x'/d_\varepsilon, x_N/h_\varepsilon) \quad \text{if } x \in \Delta(d_\varepsilon, h_\varepsilon) \\ &= 0 \quad \text{otherwise} \end{aligned}$$

Through the $d_\varepsilon Y$-periodicity of u_ε with respect to x', we discover:

$$\int_{|x_N| \leq h_\varepsilon} |u_\varepsilon|^2\ dx = o(d_\varepsilon)$$

Since $u_\varepsilon = u_0$ for $|u_\varepsilon| > h\varepsilon$, one gets $u_\varepsilon \longrightarrow u_0$ in $L^2(Q)$. Now using the scaling lemma 4.1 with $\lambda = 1/d_\varepsilon$:

$$F_{\varepsilon,d_\varepsilon,h_\varepsilon}(u_\varepsilon, Q) = d_\varepsilon{}^{N-1}\ F_{\varepsilon/d_\varepsilon,1,h_\varepsilon/d_\varepsilon}(v_\varepsilon, Q_{1/d_\varepsilon})$$

where

$$v_\varepsilon(x) = w(x', \frac{d_\varepsilon}{h_\varepsilon} x_N)$$

Since $w = \alpha$ for $x_N > h_\varepsilon/d_\varepsilon$, we have $v_\varepsilon \in B_1^L(\alpha)$ for every $L > \sup(h_\varepsilon/d_\varepsilon)$. According to proposition 3.1, we have:

$$\begin{aligned} F_{\varepsilon,d_\varepsilon,h_\varepsilon}(u_\varepsilon, Q) &\leq \tau^L_{1/d_\varepsilon}(\alpha, \gamma_\varepsilon, \delta_\varepsilon) \\ &\leq \sigma_1^L(\alpha, \gamma_\varepsilon, \delta_\varepsilon) + C'(L)\ d_\varepsilon{}^{\frac{N-1}{2}} \end{aligned}$$

Finally from the continuity of $\sigma_1^L(\alpha, ., .)$:

$$\limsup_{\varepsilon \longrightarrow 0} F_\varepsilon(u_\varepsilon) \leq \sigma_1^L(\alpha, \gamma, \delta)$$

which reduces to the inequality of (ii) when $L \longrightarrow \infty$. □

4.1. Lowerbound for the Γ-limit of the energy

We are going to prove the assertion (i) of theorem 2.1.

Let (u_ε) be a sequence in $L^2(\Omega)$ such that:

$$u_\varepsilon \longrightarrow u \text{ in } L^2(\Omega)\ , \quad l = \liminf_{\varepsilon \longrightarrow 0} E^\varepsilon(u_\varepsilon) < +\infty \tag{4.13}$$

Define the Borel non-negative measure μ_ε on $\overline{\Omega}$ by setting:

$$\mu_\varepsilon(A) = F_\varepsilon(u_\varepsilon, A)$$

From (4.13), the sequence (μ_ε) is bounded and tight on $\overline{\Omega}$ ($\overline{\Omega}$ is compact). We can write, possibly only for a subsequence still denoted by ε:

$$\liminf_{\varepsilon \longrightarrow 0} E^\varepsilon(u_\varepsilon) = \lim_{\varepsilon \longrightarrow 0} E^\varepsilon(u_\varepsilon) = l$$

$$\mu_\varepsilon \longrightarrow \mu_0 \text{ for the narrow convergence on } \overline{\Omega}$$

As *supp* $\mu_\varepsilon \subset \overline{\Omega_\varepsilon}$, we have *supp* $\mu_0 \subset \overline{\Omega_0}$, hence:

$$l = \lim_{\varepsilon \longrightarrow 0} \mu_\varepsilon(\overline{\Omega}) = \mu_0(\overline{\Omega_0}) \tag{4.14}$$

In fact, by using the narrow convergence of μ_ε, for every Borel subset A of $\overline{\Omega}$ such that $\mu_0(\partial A) = 0$, one has:

$$\lim_{\varepsilon \longrightarrow 0} F_\varepsilon(u_\varepsilon, A) = \lim_{\varepsilon \longrightarrow 0} \mu_\varepsilon(A) = \mu_0(A) \tag{4.15}$$

Since $u_\varepsilon = 0$ on $\Omega\backslash\Omega_\varepsilon$ and:

$$E^\varepsilon(u_\varepsilon) \geq \varepsilon \int_{\Omega_0} |Du_\varepsilon|^2 \, dx + \frac{1}{\varepsilon}\int_{\Omega_0} W(u_\varepsilon)\, dx \ ,$$

we already know that the limit u lies in $BV(\Omega)$ and has the form: $u = \alpha_1 \, 1_{A\cap\Omega_0} + \alpha_2 \, 1_{\Omega_0\backslash A}$ where A is a subset of Ω_0 with finite perimeter. Moreover, by applying Modica's results [2], on the open set $\Omega^\eta = \Omega \cap \{x_N > \eta\}$, with $\eta > 0$, one gets:

$$\liminf_{\varepsilon \longrightarrow 0} \ F_\varepsilon(u_\varepsilon, \Omega^\eta) \geq \frac{c}{\alpha_2 - \alpha_1} \int_{\Omega^\eta} |Du| \, dx + + \int_{\Gamma' \cap \Omega^\eta} \widehat{\sigma}(u^+) \, dH^{N-1} \qquad (4.16)$$

where c and $\widehat{\sigma}$ are given by (2.1) and (2.3). Choosing a sequence η tending to 0 such that $\mu_0(\{x_N = \eta\}) = 0$ and using (4.15) and (4.16) we are led to:

$$\begin{aligned} l = \mu_0(\overline{\Omega_0}) &= \mu_0(\Gamma_0) + \lim_{\eta \longrightarrow 0} \mu_0(\Omega^\eta) \\ &\geq \mu_0(\Gamma_0) + \frac{c}{\alpha_2 - \alpha_1} \int_{\Omega_0} |Du| \, dx + \int_{\Gamma'} \widehat{\sigma}(u^+) \, dH^{N-1} \end{aligned}$$

So the lowerbound of theorem 2.1 is obtained provided we can prove:

$$\mu_0 \, 1_{\Gamma_0} \geq \widehat{\widehat{\sigma}}(u^+) \ H^{N-1}(\Gamma_0 \cap .)$$

Denoting $\theta_0 = H^{N-1}(\Gamma_0 \cap .)$, the last inequality reads as:

$$\frac{d\mu_0}{d\theta_0}(x) \geq \widehat{\widehat{\sigma}}(u^+(x)) \ , \quad \theta_0 \text{ a.e. } x \in \Gamma_0 \qquad (4.17)$$

Let Q_δ be the interval $]-\delta/2, +\delta/2[^N$. By the Besicovitch differentiation theorem, one has:

$$H^{N-1} \text{ a.e. } x \in \Gamma_0 \ , \quad \frac{d\mu_0}{d\theta_0} = \lim_{\delta \longrightarrow 0} \frac{\mu_0(Q_\delta)}{\delta^{N-1}} \qquad (4.18)$$

On the other hand, as the limit u of u_ε satisfies $u(x) \in \{\alpha_1, \ \alpha_2\}$ a.e. on $\Omega\backslash\Omega_0$, the traces on Γ_0 from $x_N > 0$ and from $x_N < 0$ satisfy :

$$u^+ \in \{\{\alpha_1, \ \alpha_2\} \quad H^{N-1} \text{ a.e. } x \in \Gamma_0 \ , \quad u^- = 0$$

Let us fix $x_0 \in \Gamma_0$ such that the equality in (4.18) holds and let us set:

$$\begin{aligned} w_{\varepsilon,\delta}(y) &= u_\varepsilon(x_0 + \delta y) \\ w_{0,\delta}(y) &= u(x_0 + \delta y) \end{aligned}$$

It is well known that for H^{N-1} a.e. $x_0 \in \Gamma_0$, one has:

$$w_{0,\delta} \longrightarrow u_0 \ \text{ in } L^1(Q) \ \text{ as } \delta \longrightarrow 0 \qquad (4.19)$$

where:

$$\begin{aligned} &u_0(y) = \alpha_i \ \text{ if } y_N > 0 \ \text{ and } u^+(x_0) = \alpha_i \qquad (4.20) \\ &u_0(y) = 0 \ \text{ if } y_N < 0 \end{aligned}$$

From now on we assume that x_0 has been chosen in such a way that (4.18) and (4.19) hold.

Let us go on with the blow-up argument at x_0 as done in another situation by Fonseca-Müller [8]. Take a sequence (δ_k) tending to 0 such that $\mu_0(\partial Q_{\delta_k}) = 0$. Owing to (4.15) one has:

$$\frac{d\mu_0}{d\theta_0}(x_0) = \lim_{k \longrightarrow +\infty} \lim_{\varepsilon \longrightarrow 0} \frac{F_{\varepsilon, d_\varepsilon, h_\varepsilon}(u_\varepsilon, x_0 + Q_{\delta_k})}{\delta_k^{N-1}} \tag{4.21}$$

Assuming that $\frac{d\mu_0}{d\theta_0}(x_0) < +\infty$ (otherwise the inequality (4.17) is trivial), we can choose for every k, some $\varepsilon(k) > 0$ such that:

$$\varepsilon_k = \frac{\varepsilon}{\delta_k} \leq \frac{1}{k} \tag{4.22}$$

$$\| w_{\varepsilon, \delta_k} - u_0 \|_{L^1(Q)} \leq \| w_{0, \delta_k} - u_0 \|_{L^1(Q)} + \frac{1}{k} \tag{4.23}$$

$$\frac{F_{\varepsilon, d_\varepsilon, h_\varepsilon}(u_\varepsilon, Q_{\delta_k})}{\delta_k^{N-1}} \leq \frac{d\mu_0}{d\theta_0}(x_0) + \frac{1}{k} \tag{4.24}$$

For this ε (depending on k), let us set:

$$v_k(y) = u_\varepsilon(x_0 + \delta_k y)$$

$$d_k = \frac{d_\varepsilon}{\delta_k}, \quad h_k = \frac{h_\varepsilon}{\delta_k}$$

By lemma 4.1 we may rewrite (4.22) as:

$$F_{\varepsilon_k, d_k, h_k}(v_k, Q) \leq \frac{d\mu_0}{d\theta_0}(x_0) + \frac{1}{k}$$

We notice that $d_k \longrightarrow 0$, $h_k \longrightarrow 0$, $\varepsilon_k \longrightarrow 0$, while $h_k/d_k \longrightarrow \delta$, $\varepsilon_k/d_k \longrightarrow \gamma$. Since $v_k \longrightarrow u_0$ in $L^1(Q)$ where u_0 has the particular form given by (4.20), we may apply lemma 4.3 (i) with $\Omega = Q$. Combined with (4.24) that yields (4.17):

$$\frac{d\mu_0}{d\theta_0}(x_0) \geq \liminf_{k \longrightarrow +\infty} F_{\alpha_k, d_k, h_k}(v_k, Q) \geq \widehat{\sigma}(u^+(x_0))$$

□

4.2 Upperbound for the Γ-limit

We prove the assertion (ii) of theorem 2.1 in case $u = \alpha_1 1_{\Omega_0 \cap A} + \alpha_2 1_{\Omega_0 \setminus A}$ where A is an open set of $\mathbb{R}^N$ with smooth boundary ∂A such that $H^{N-1}(\partial A \cap \partial \Omega_0) = 0$. The conclusion in the general case is then deduced by an approximation procedure for functions u such that $E^0(u) < +\infty$. This procedure is completely described in Modica [2] or Bouchitté [9] to which we refer for this part of the proof.

Set $A_1 = A \cap \Omega$ and $A_2 = \Omega \backslash A$. For every $\eta > 0$ let us define:

$$\Gamma_\eta^i = \{x \in \Gamma_0 \cap A_i \ ; \ \ d(x, \partial\Omega) > \eta, \ d(x, \partial A) > \eta\}$$
$$\Sigma_\eta^i = \Gamma_\eta^i \times [-\eta/2, \eta/2]$$

Noticing that

$$H^{N-1}(\Gamma_\varepsilon \backslash (\Sigma_\eta^1 \cup \Sigma_\eta^2)) \leq \frac{h_\varepsilon}{d_\varepsilon} \ H^{N-1}(\Gamma_0 \backslash (\Gamma_\eta^1 \cup \Gamma_\eta^2)),$$
$$\lim_{\eta \longrightarrow 0} H^{N-1}(\Gamma_0 \backslash (\Gamma_\eta^1 \cup \Gamma_\eta^2)) = H^{N-1}(\Gamma_0 \cap (\Gamma' \cup \partial A)) = 0,$$

we can write:

$$\limsup_{\varepsilon \longrightarrow 0} H^{N-1}(\Gamma_\varepsilon \backslash (\Sigma_\eta^1 \cup \Sigma_\eta^2)) = o(\eta) \tag{4.25}$$

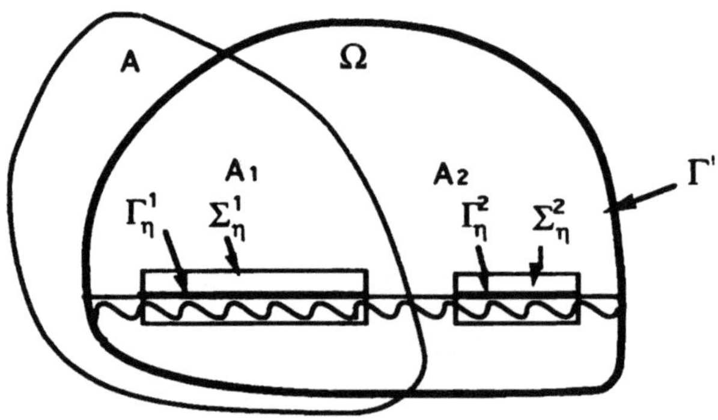

figure 4

Let us apply Modica's construction [2] to approach $\tilde{u} = \alpha_1 \ 1_{A_1} + \alpha_2 \ 1_{A_2}$ on the open subset $\Omega^{-\eta} = \Omega \cap \{x_N > -\eta\}$ taking into account the boundary energy on Γ'. We find a sequence $\tilde{u}_\varepsilon$ in $L^2(\Omega^{-\eta})$ such that:

$$\tilde{u}_\varepsilon \longrightarrow \tilde{u} \quad \text{in } L^2(\Omega^{-\eta}) \tag{4.26}$$

$$\tilde{u}_\varepsilon \quad \text{is bounded in} \quad L^\infty(\Omega^{-\eta}) \tag{4.27}$$

$$\tilde{u}_\varepsilon = \tilde{u} \quad \text{if} \quad d(x, \partial A) > \eta \quad \text{and} \quad d(x, \partial\Omega^{-\eta}) > \eta \tag{4.28}$$

$$F_\varepsilon(\tilde{u}_\varepsilon, \Omega^{-\eta} \cup \Gamma' \backslash \Gamma_\varepsilon) \longrightarrow l_\eta \quad \text{as} \quad \varepsilon \longrightarrow 0 \quad \text{where :} \tag{4.29}$$

$$l_\eta = \frac{c}{\alpha_2 - \alpha_1} \int_{\Omega^{-\eta}} |D\tilde{u}| + \int_{\Gamma'} \widehat{\sigma}(u) \ dH^{N-1} \tag{4.30}$$

(To obtain (4.27), we notice that, according to the growth condition on W (H2d), the infimum associated with the definition (2.3) of $\widehat{\sigma}(\alpha_i)$ is reached for some value s_i; then we can choose $\tilde{u}_\varepsilon$ so that $\tilde{u}_\varepsilon \in [\alpha_1, \alpha_2] \cup [s_1, s_2]$)

From (4.28), we get $\tilde{u}_\varepsilon = \alpha_i$ on Σ_η^i so that we can modify $\tilde{u}_\varepsilon$ inside Σ_η^i using the lemma 4.3 (ii). Let us consider a covering of $\Sigma_\eta^1 \cup \Sigma_\eta^2$ by cells of size η by setting:

$$Q_\eta^k = \eta(k + Q) \ , \quad k \in Z^{N-1}$$
$$I_\eta^i = \{k \in Z^{N-1} \ ; \quad Q_\eta^k \subset \Sigma_\eta^i\}$$

To simplify, we assume that $\Sigma_\eta^i = \sum_{k \in I_\eta^i} Q_\eta^k$

By an easy rescaling, lemma 4.3 (ii) leads to the existence for each $k \in I_\eta^i$ of a sequence (w_ε^k) such that:

$$w_\varepsilon^k \longrightarrow \alpha_i \text{ in } L^2(Q_\eta^k) \ , \quad w_\varepsilon^k = \alpha_i \text{ on } \partial Q_\eta^k \cap \{x_N > 0\} \tag{4.31}$$
$$\lim_{\varepsilon \longrightarrow 0} F_\varepsilon(w_\varepsilon^k, Q_\eta^k) = \widehat{\widehat{\sigma}}(\alpha_i)\ \eta^{N-1}$$

Define:

$$u_\varepsilon = \begin{cases} \tilde{u}_\varepsilon & \text{on} \quad \Omega_\varepsilon \backslash \Sigma_\eta^1 \cup \Sigma_\eta^2 \\ w_\varepsilon^k & \text{on} \quad Q_\eta^k \ , \quad k \in I_\eta^1 \cup I_\eta^2 \\ 0 & \text{on} \quad \Omega \backslash \Omega_\varepsilon \end{cases}$$

It is easy to check that $u_\varepsilon \longrightarrow u$ in $L^2(\Omega)$. Moreover:

$$F_\varepsilon(u_\varepsilon, \overline{\Omega_\varepsilon}) \leq F_\varepsilon(\tilde{u}_\varepsilon, \Omega^{-\eta} \cup \Gamma' \backslash \Gamma_\varepsilon) + F_\varepsilon(\tilde{u}_\varepsilon, \Gamma_\varepsilon \backslash (\Sigma_\eta^1 \cup \Sigma_\eta^2)) + \\ + \sum_{k \in I_\eta^1 \cup I_\eta^2} F_\varepsilon(w_\varepsilon^k, Q_\eta^k) \tag{4.32}$$

From (4.27), $\sigma(\tilde{u}_\varepsilon)$ is bounded. Thus from (4.29), (4.31) and (4.32):

$$\limsup_{\varepsilon \longrightarrow 0} F_\varepsilon(u_\varepsilon, \overline{\Omega_\varepsilon}) \leq l_\eta + o(\eta) + \int_{\Gamma_\eta^1 \cup \Gamma_\eta^2} \widehat{\widehat{\sigma}}(u^+)\ dH^{N-1} \tag{4.33}$$

We conclude by passing to the limit as $\eta \longrightarrow 0$ using (4.30) and the fact that $\int_{\Gamma_0} |D\tilde{u}| = 0$ due to $H^{N-1}(\partial A \cap \Gamma_0) = 0$.

Finally, in order to fit the constraint on the total mass ($\int_\Omega u_\varepsilon\ dx = \int_\Omega u\ dx$), we use Modica's method which consists in changing u_ε slightly inside one of the two phases (see [2]). □

Acknowledgements: The research of the first author is part of the project "EURHomogenization", contract SC1-CT91-0732 of the program SCIENCE of the Commission of the European Communities.

References

[1] Cahn J.W. and Hilliard J.E., Free energy of a non-uniform system. J. Chem. Phy. 31, 3, (1959), 688-699.

[2] Modica L., Gradient theory of phase transitions with boundary contact energy. Ann. Inst. H. Poincare Anal. Non Lineaire , 5, (1987), 453-486.

[3] Alberti G., Ambrosio L. and Buttazzo G., Singular perturbation problems with a compact support semilinear term, to appear.

[4] Seppecher P. Equilibrium of Cahn-Hilliard fluid on a wall: influence of the wetting properties of the fluid upon the stability of a thin liquid film, Eur. J. Mech., B/Fluids, 12, 1, (1993).

[5] Cox R. G., The spreading of a liquid on a rough solid surface, J. Fluid Mech., 131, (1983), 1-26.

[6] Giusti E., Minimal surfaces and functions of bounded variation, Birkhäuser, Basel (1984).

[7] Müller S., Homogenization of nonconvex integral functionnals and cellular elastic materials, Arch. Rational Mech. Anal., 99 (1987), 189-212.

[8] Fonseca I. and Müller S., Quasiconvex integrands and lowersemicontinuity. To appear in SIAM J. Math. Anal.

[9] Bouchitté G. Singular perturbation of variational problems arising from a two-phase transition model, Appl. Math. Optim., 21, (1990), 289-314.

[10] Modica L., The gradient theory of phase transitions and the minimal interface criterium, Arch. Rational Mech. Anal., 98 (1987), 123-142.

Mathematical models of phase boundaries in alloys: Phase field and Sharp interface

G. Caginalp and W. Xie

Abstract. We consider a phase-field model of a binary mixture or alloy which has a phase boundary. The model identifies all macroscopic parameters and the interface thickness, ϵ. In the limit as ϵ approaches zero, a new two-phase alloy solidification model (with a sharp interface) is obtained. For small concentrations, in the sense defined precisely in the text, we recover the classical sharp interface problems, the theory of which is reviewed. We obtain, in the simplest phase-field system, a new (nonlinear) interface relation for concentration c which is discontinuous across the interface and subject to

$$\left[\ln \frac{c}{1-c}\right]_-^+ = -2M$$

$$-\sigma(\alpha v + \kappa) = [s]_E\left[T - T_B - \frac{T_A - T_B}{2M} \ln \frac{1-c^+}{1-c^-}\right]$$

where the physical parameters are defined precisely in the text.

1. Introduction

The study of free boundaries in recent years can be grouped broadly into two categories: (a) sharp interface problems in which one or more variables (or their derivatives) are typically discontinuous across an interface; and (b) systems of parabolic equations in which the interface is specified by a level set of one of the variables. In the mid-1980's a close relationship was established between these two types of models ([C1, 2]). Similar issues have also been discussed for a broad spectrum of free boundary problems (see [Gl] for a survey).

An important example of (a) is the surface tension and kinetics model (or modified Stefan model) which describes solidification. The material, which occupies a spatial region, $\Omega \subset R^d$, may be in either of two phases, e.g. liquid and solid, separated by an interface, $\Gamma(t) \subset R^{d-1}$, is described by the equations

$$\rho C_v T_t = K_1 \Delta T \quad \text{in} \quad \Omega \setminus \Gamma(t) \tag{1.1}$$

$$\rho l v = -K_1 [\nabla T \cdot n]_-^+ \quad \text{on} \quad \Gamma(t) \tag{1.2}$$

$$T - T_M = -\frac{\sigma}{[s]_E}(\kappa + \alpha v) \quad \text{on} \quad \Gamma(t) \tag{1.3}$$

where T is (absolute) temperature, ρ the density, C_v the specific heat per mass, K_1 the thermal conductivity, l the latent heat per mass, σ the surface tension, α the relaxation scaling, s the entropy per unit volume and $[s]_-^+$ is the difference between the solid and liquid phases. The (normal) velocity to the interface $\Gamma(t)$ at (x, t) is denoted v while the sum of principle curvatures at (x, t) on $\Gamma(t)$ is denoted $\kappa(x, t)$ or κ. Thus the sharp interface problem can be stated as finding $T(x, t)$ and $\Gamma(t)$ in suitable spaces subject to initial and boundary conditions.

A special case of (1.1)-(1.3), known as the classical Stefan model, is obtained by setting $\sigma = 0$, so that $T - T_M = 0$ at the interface. We note that the difference between the two models ($\sigma \neq 0$ and $\sigma = 0$) is actually quite profound for two reasons. Setting $\sigma = 0$ has the effect of (1) eliminating an important length scale in the problem, which is crucial in terms of the stability of the interface, and (2) enabling the sign $T(x, t)$ to determine the phase of the material, i.e., $T > 0$ implies the material is in the liquid phase.

An alternative formulation of phase boundary problems along the lines of (b) is the phase field model in which a phase or "order" parameter, $\phi(x, t)$ is employed. In the distinguished limit of the surface tension and kinetics model, (1.1)-(1.3), one can write the parabolic system for (T, ϕ) as

$$C_v T_t + \frac{l}{2}\phi_t = \frac{K_1}{\rho}\Delta T \tag{1.4}$$

$$\alpha\epsilon^2\phi_t = \epsilon^2\Delta\phi + \frac{1}{2}(\phi - \phi^3) + \epsilon\frac{[s]_E}{3\sigma}(T - T_M) \tag{1.5}$$

where ϵ represents the thickness of the interfacial layer and all other variables are the same as previously defined. The interface in (1.4) and (1.5) is now defined implicitly as the level curve where ϕ vanishes.

The theoretical link between the phase field model (1.4) and (1.5) and sharp interface models such as (1.1)-(1.3) was established using formal matched asymptotics in [C1, 2] and rigorously proven under various conditions in [CC]. A number of theorems have also been proven for the special case in which the phase equation (1.5) is considered independently for fixed T (see [ESS] and references therein). While the theoretical results are concerned with the limit $\epsilon \to 0$, computations have shown that the phase field equations exhibit an interface which is close to the sharp interface problem even when ϵ is relatively large [CS].

In the present paper, we discuss a generalization of this model to the case of alloys. A related approach has been used in the isothermal case in [WBM1]. We will use a simplified set of equations which are derived in [CJ] and studied in mathematical detail in [CX]. Our aim is to write the simplest system of equations of phase field type which has the characteristic behavior of alloys and can be shown through asymptotic analysis to have the appropriate scaling and coefficients. Furthermore, the analysis then leads to a new sharp interface alloy problem which generalizes equations which are currently studied.

2. The Alloy Problem

The problem of phase boundaries involving mixture of two materials (e.g. alloys or impurities) is of great practical interest and presents challenging theoretical issues which we review briefly in this section. We consider a substance (e.g., a binary alloy) comprised of a mixture of material A and material B, with concentration $c(x,t) \in (0,1)$ denoting the fraction of material A in the mixture. Two aspects of this problem are particularly fascinating from a mathematical perspective. (a) There is a jump in the concentration itself, and not just in the gradient, as in temperature. (b) The diffusion of impurities is generally much smaller in the solid phase compared with the liquid, and leads to partly degenerate equations in the limit of zero diffusion in the solid.

An understanding of (a) is best accomplished by means of the (c,T) phase diagram (Figure 1). A clearly written reference is [S1]. We focus on the $c=0$ side of the diagram. A complete phase diagram for an "ideal" mixture or alloy [Lu] is shown in Figure 2. The intriguing aspect of this phase diagram is that liquid and solid are not separated by a single curve, but by two curves (called liquidus and solidus), which merge at the melting temperature T_B of the pure B material ($c=0$). To a reasonable approximation for small c, one may assume these two curves are linear. To be specific, we will discuss the phase diagram displayed in Figure 1(a). At a fixed temperature $T_E < T_B$ the liquidus line intersects with $T = T_E$ at a point C_L while the solidus intersects at C_S. Thus one expects an interface at equilibrium at temperature T_E to exhibit concentration C_S on the liquidus side and C_L on the solid. That is, the concentration is a piecewise constant function with a jump between C_L and C_S at the phase boundary.

While the (equilibrium) phase diagram explains (a), it is irrelevant to (b) since the latter is a dynamical phenomena, exhibited in the (concentration) diffusion equation below. The dynamical modelling of the alloy problem is typically accomplished by means of a (sharp interface) free boundary problem which often accounts for only the liquid phase ("one phase problem") thereby neglecting the diffusion of impurities in the solid which is so small. Also, the concentration of impurities is assumed to be sufficiently small that one does not need to regard the thermodynamic variables such as latent heat as a function of c.

THE TWO-PHASE (SHARP INTERFACE) PROBLEM: Thus, a reasonable two-phase description of the dilute binary alloy, may be written by expanding on the usual discussion of the one-phase problem (e.g., [S1]) as follows.

The (dilute) two-phase alloy model: We consider the coupled system for (T,c,Γ) satisfying (1.1), (1.2) and

$$c_t = K_2^{\pm}\Delta c \quad \text{in} \quad \Omega \setminus \Gamma(t) \tag{2.1}$$

$$-v[c]_-^+ = K_2[\nabla c \cdot n]_-^+ \quad \text{on} \quad \Gamma(t) \tag{2.2}$$

$$T - T_B = -\frac{\sigma}{[s]_E}(\kappa + \alpha v) + \frac{1}{2}(m_l c_+ + m_s c_-) \quad \text{on} \quad \Gamma(t) \tag{2.3}$$

$$m_l c_+ = m_s c_- \quad \text{on} \quad \Gamma(t) \tag{2.4}$$

where T_B is the melting temperature of the pure B material (i.e., T_M in (1.3)) and c_+ represents the limit of c from the liquid side (and c_- the solid side) while $[\cdot]_-^+$ is the difference between the limits from the liquid and solid sides. Also, $K_2^\pm$ are the diffusivity of the solute in the liquid and solid phases, respectively, and all other parameters are as defined in Section 1.

We note that the last term in (2.3) has been written to reflect the symmetry of the situation since clearly (in equilibrium) one has $m_l c_+ = m_s c_-$ ($c_+ = c_L$, $c_- = c_S$ in equilibrium). In fact, equation (2.3), which is a generalization of the interface relation (1.3), can be understood very simply in terms of the basic phase diagram (Figure 1). If we consider a stationary planar interface, then $\kappa = v = 0$ in (2.3) so that the value of T is given by

$$T - T_B = \frac{1}{2}(m_l c_+ + m_s c_-) = m_l c_L = m_s c_S. \tag{2.5}$$

This means that a temperature T is compatible as the coexistence of liquid and solid phases if the liquid has concentration c_L and the solid c_S, as shown by the horizontal line in Figure (1a) or (1b). Thus, (2.3) represents a modification of (1.3) in which the equlibrium alloy effects are combined linearly with the Gibbs-Thomson (namely the $\sigma\kappa$ term) and the kinetic undercooling ($\alpha\sigma v$ term) effects.

THE ONE-PHASE (SHARP INTERFACE) PROBLEM: The single-phase version of (2.1)-(2.3) is obtained by suppressing the role of concentration in the solid while maintaining some of the key relationships in the phase diagram. In particular, the relation (2.3) is replaced by the single phase version

$$T - T_B = -\frac{\sigma}{[s]_E}(\kappa + \alpha v) + m_l c \quad \text{on} \quad \Gamma(t) \tag{2.3$'$}$$

where we can unambiguously write c (rather than c_+ or c_-) since we are only concerned with concentration in the liquid. Furthermore, by making the approximation that the jump in c in the dynamics, $[c]_-^+$, is equal to that of the statics, i.e., $[c]_-^+ = c_L - c_S$ and using the relation

$$j \equiv \frac{c_S}{c_L} \quad \text{or} \quad c_L - c_S = (1-j)c_L, \tag{2.5a,b}$$

one can write the analog of (2.2) as

$$-(1-j)vc = K_2^+ \nabla c \cdot n \quad \text{on} \quad \Gamma(t) \tag{2.2$'$}$$

where c again denotes concentration in the liquid (i.e., c_+ on the interface). The right hand side of (2.2′) has been further simplified by assuming that the limit of ∇c on the interface from the solid side can be neglected. The rationale for this would be that the physical effect of diffusion and conservation of mass are adequately represented. In other words, equation (2.2) expresses the idea that the excess concentration, which is deposited (or absorbed) at the interface as a result of different concentrations in the two phases, must diffuse into the material in order to conserve mass. The modification to (2.3′) implies the hope that allowing this excess to diffuse into the liquid alone will not change the physical picture drastically. Thus the issues involved in a "one phase alloy" problem are similar to those in the "one-phase Stefan" problems, and the difficulties demonstrated in [O] are also similar. Nevertheless, our objective is not to endorse the "one-phase" approach but merely to demonstrate its relationship to the other models, and mention that its use leads to some exact solutions and stability calculations (see, for example [MS], [S1]). The problem is usually written as a system which is coupled to a two-phase temperature system but can easily be modified so that a system with truly one phase results.

The (dilute) one-phase alloy model: Consider a system for (T, c, Γ) satisfying (1.1), (1.2), (2.2′), (2.3′) and

$$c_t = K_2^+ \Delta c \quad \text{in liquid.} \tag{2.1'}$$

This model accounts for heat conduction in liquid and solid but impurity diffusion only in liquid. For the *completely one phase problem*, consider (1.1) also to be valid in the liquid only, while (1.2) is modified to

$$-\rho l v = K_1 \nabla T \cdot n \quad \text{on} \quad \Gamma(t). \tag{1.2'}$$

Thus the completely one-phase model involves a time dependent domain, with part of the boundary consisting of $\Gamma(t)$.

In addition to the similarities between the one-phase alloy and one-phase Stefan models, there are significant differences. In terms of thermal diffusion, the differences between the two phases is usually not vast, so that the neglecting of one of the two phases is a mathematical convenience, and any degeneracies which may aries are mathematical artifacts. For the alloy problem, however, there usually is a large difference in the diffusivity of solute, i.e., $K_2^- \ll K_2^+$, and K_2^- is generally close to zero. Thus, at least part of the degeneracy, embodied in (2.1) as K_2^- vanishes, is generic to the physical problem, and presents a fundamental issue which is independent of the particular mathematical approach.

It is worth noting that both the one-phase and two-phase models defined above are based on a linearized phase diagram such as Figure 1 and do not incorporate the physics of Figure 2.

In the subsequent section, we consider a phase field approach to this problem in which K_2 has a transition layer between the values K_2^- in the solid to K_2^+ in the liquid, and $K_2^- \neq 0$. The model embodies the entire phase diagram shown in Figure 2.

3. A Thermal-Alloy Phase Field Model

We consider a model based on the phase field ideas of Section 1 which incorporates the physics of binary alloys in addition to thermal properties. Our aim here is to study the simplest set of equations which (a) exhibit the proper behavior as the interface thickness, ϵ, approaches zero, and (b) identify all of the material parameters.

We use a formulation consisting of a free energy density, F, and a nonconserved equation for ϕ,

$$\tau \phi_t = -\frac{\delta F}{\delta \phi} \tag{3.1}$$

and a conserved equation

$$c_t = \nabla \cdot K_2(\phi)c(1-c)\nabla\Big(\frac{\delta F}{\delta c}\Big) \tag{3.2}$$

coupled with an energy balance equation, analogous to (1.4). The energy balance equation and (3.1) lead to the phase field equations for a pure material, (1.4), (1.5). Here τ is a relaxation time (which will be identified in terms of the macroscopics later) and K_2 is the solute diffusivity. The mobility term $c(1-c)$ has also been used in [WBM1] who study a formulation based on (3.1) and (3.2) alone for isothermal problems. However, our scaling and asymptotics differ from theirs. A fourth order concentration equation has also been coupled with a phase equation in a related paper [WBM2].

Using a simplified free energy [CJ] and an additional physical approximation one obtains the (simplest possible) system

$$\alpha\epsilon^2\phi_t = \epsilon^2\Delta\phi + \frac{1}{2}(\phi-\phi^3) + \frac{\epsilon}{3\sigma}[s]_E\Big\{T - T_A c - T_B(1-c)\Big\} \tag{3.3}$$

$$C_v T_t + \frac{l}{2}\phi_t = \nabla \cdot K_1 \nabla T \tag{3.4}$$

$$c_t = \nabla \cdot K_2(\phi)c(1-c)\nabla\Big[M\phi + \ln\frac{c}{1-c}\Big] \tag{3.5}$$

where α, σ, C_v, l and K_1 are the physical parameters defined in Section 1 for the pure B material. The interface thickness is once again given by ϵ. Also $K_2(\phi)$ is the solute diffusion coefficient, so that $K_2(\pm 1) = K_2^{\pm}$ and $M \equiv \frac{1}{2}\ln\frac{m_l}{m_s}$. The additional simplification involves replacing $N = -\frac{1}{2}[s]_E(T_B - T_A)$ with M and incorporating the RT coefficient of $\ln\frac{c}{1-c}$ into K_2, thereby omitting crossterms involving products of ∇c and $T^{-1}\nabla T$. Note that for small c, these crossterms would be second order in c, so the difference is significant only for c of order 1.

In this section, an asymptotic analysis for $\epsilon \ll 1$ will be carried out for layered solutions of the above system (3.3)-(3.5). A complete determination of the solution will involve initial and boundary conditions; however, they will be left unspecified here, as our primary attention is rather on the laws of motion of the interface. A supplementary discussion in some special cases will be presented in the last section. (We will specify the boundary and initial conditions at that time). Our analysis here treats motion of interface as a moving internal layer of width $O(\epsilon)$. The interfacial curve Γ is defined as the set of points in domain Ω at which $\phi = 0$. Setting $x = (x_1, x_2)$, we define $r(x,t,\epsilon)$ in a neighbourhood of Γ to be the sign distance from x to Γ, with the convention that $+$ is the direction of positive ϕ and $-$ is the direction of negative ϕ. Note that for any small $\epsilon > 0$, the system (3.3)-(3.5) is parabolic and has regular solutions [CX]. It is reasonable to suppose that Γ is a smooth function in a neighbourhood of Γ. The interface Γ itself may be represented by the equation

$$r(x,t,\epsilon) = 0 \quad \text{and} \quad |\nabla r| = 1 \quad \text{and} \quad \Delta r = \kappa \tag{3.7}$$

on interface Γ, where κ is the curvature of Γ.

We define a function $s(x,t,\epsilon)$ so that (r,s) is a local orthogonal coordinate system in the neighbourhood of Γ, such that on Γ, s measures arclength from some point depending smoothly on t. We set up outer expansions for the functions c, T and ϕ, i.e.,

$$c(x,t,\epsilon) = c^0(x,t) + \epsilon c^1(x,t) + \cdots \tag{3.8a}$$
$$T(x,t,\epsilon) = T^0(x,t) + \epsilon T^1(x,t) + \cdots \tag{3.8b}$$
$$\phi(x,t,\epsilon) = \phi^0(x,t) + \epsilon \phi^1(x,t) + \cdots . \tag{3.8c}$$

The terms on the right of these expansions may be discontinuous at $r = 0$.

The inner expansion proceed by using the stretching transformation $z \equiv r/\epsilon$ and considering c, T and ϕ as functions of the variables z, s and t and $c(x,t,\epsilon) = \hat{c}(z,s,t,\epsilon)$, etc.. Thus

$$\hat{c}(z,s,t,\epsilon) = \hat{c}^0(z,s,t) + \epsilon \hat{c}^1(z,s,t) + \cdots \tag{3.9a}$$
$$\hat{T}(z,s,t,\epsilon) = \hat{T}^0(z,s,t) + \epsilon \hat{T}^1(z,s,t) + \cdots \tag{3.9b}$$
$$\hat{\phi}(z,s,t,\epsilon) = \hat{\phi}^0(z,s,t) + \epsilon \hat{\phi}^1(z,s,t) + \cdots . \tag{3.9c}$$

By the definition of r, we have

$$\hat{\phi}(0,s,t,\epsilon) \equiv 0. \tag{3.10}$$

In the sequel, we will use the notation $f|_{\Gamma\pm}$ to denote the limiting values of f as Γ is approached from the side where $r > 0$ or $r < 0$ respectively. The notation $f_r|_{\Gamma\pm}$ is defined similarly. The variables ϕ and c will exhibit transition layers as shown in Figure 3, while $\nabla T \cdot n$ exhibits a similar transition layer behavior.

The main results in this section are the following (see Figure 4).

Proposition 3.1. *In limit as $\epsilon \to 0$, there exists a formal asymptotic solution, (ϕ, T, c), of the phase field alloy model (3.3)-(3.5) which is governed by the following sharp interface model*

$$\begin{cases} C_v T_t = K_1 \Delta T \quad \textit{in} \quad \Omega \setminus \Gamma(t) \\ c_t = K_2^{\pm} \Delta c \quad \textit{in} \quad \Omega \setminus \Gamma(t) \\ [T]_-^+ = 0 \quad \textit{on} \quad \Gamma(t) \\ [K_1 \nabla T \cdot n]_-^+ = -lv \quad \textit{on} \quad \Gamma(t) \\ [\ln \dfrac{c}{1-c}]_-^+ = -2M \quad \textit{on} \quad \Gamma(t) \\ [K_2 \nabla c \cdot n]_-^+ = -[c]_-^+ v \quad \textit{on} \quad \Gamma(t) \\ \alpha v + \kappa = \dfrac{[s]_E}{-\sigma} \Big[T - T_B - \dfrac{T_A - T_B}{2M} \ln \dfrac{1-c_+}{1-c_-} \Big] \quad \textit{on} \quad \Gamma(t). \end{cases} \tag{3.11}$$

Thus, (3.11) is a new sharp interface alloy problem which describes the entire (c, T) phase diagram, unlike the alloy models described in Section 2.

Proposition 3.2. *In limit as $\epsilon \to 0$ and for small concentration c, there exists a formal asymptotic solution, (ϕ, T, c), (with linearization for concentration c) of the phase field* *alloy model (3.3)-(3.5) which is governed by the two-phase alloy problem*

$$\begin{cases} C_v T_t = K_1 \Delta T \quad in \quad \Omega \setminus \Gamma(t) \\ c_t = K_2^{\pm} \Delta c \quad in \quad \Omega \setminus \Gamma(t) \\ [T]_-^+ = 0 \quad on \quad \Gamma(t) \\ [K_1 \nabla T \cdot n]_-^+ = -lv \quad on \quad \Gamma(t) \\ \dfrac{c^+}{c^-} = e^{-2M} \quad on \quad \Gamma(t) \\ [K_2 \nabla c \cdot n]_-^+ = -[c]_-^+ v \quad on \quad \Gamma(t) \\ \alpha v + \kappa = \dfrac{[s]_E}{-\sigma} \Big[T - T_B - \dfrac{T_A - T_B}{2M}(c_- - c_+) \Big] \quad on \quad \Gamma(t). \end{cases} \tag{3.12}$$

Verification of Proposition 3.1. To obtain the first two equations, we use the outer expansion (3.8).

<u>The outer solution.</u>

Setting (3.8) into (3.3)-(3.5) and equating coefficients of corresponding powers of ϵ, we obtain

$O(1)$ system:

$$\phi^0 - (\phi^0)^3 = 0 \tag{3.13}$$

$$C_v T_t^0 = K_1 \Delta T^0 \tag{3.14}$$

$$c_t^0 = \nabla \cdot K_2^{\pm} c^0 (1 - c^0) \nabla \Big(\ln \frac{c^0}{1 - c^0} \Big) \tag{3.15}$$

where we take $\phi_-^0 = -1$ and $\phi_+^0 = 1$.

$O(\epsilon)$ system:

$$\Big(1 - 3(\phi^0)^2\Big)\phi^1 + \frac{2[s]_E}{3\sigma} \Big\{ T^0 - T_A c^0 - T_B (1 - c^0) \Big\} = 0 \tag{3.16}$$

$$C_v T_t^1 + \frac{l}{2} \phi_t^1 = K_1 \Delta T^1 \tag{3.17}$$

$$\begin{aligned} c_t^1 &= \nabla \cdot K_2^{\pm} c^0 (1 - c^0) \nabla \Big[M \phi^1 + \frac{c^1}{c^0 (1 - c^0)} \Big] \\ &\quad + \nabla \cdot K_2^{\pm} c^1 (1 - 2c^0) \nabla \Big(\ln \frac{c^0}{1 - c^0} \Big) \\ &\quad + \nabla \cdot K_2'(\phi^0) \phi^1 c^0 (1 - c^0) \nabla \Big(\ln \frac{c^0}{1 - c^0} \Big). \end{aligned} \tag{3.18}$$

Thus the first two equations follow from the $O(1)$ system of the outer solution. To obtain the interface conditions, we need to use the stretched inner expansion (3.9).

<u>The inner solution.</u>

Using the (r, s) coordinate system one may write the Laplacian as

$$\Delta u = u_{rr} + \Delta r u_r + |\nabla s|^2 u_{ss} + \Delta s u_s \tag{3.19}$$

and the time derivative u_t as $u_t + r_t u_r + s_t u_s$. In terms of (r,s,t), the system (3.3)-(3.5) can be written as follows.

$$\alpha\epsilon^2\Big[\phi_t+\phi_r r_t+\phi_s s_t\Big]=\epsilon^2\Big[\phi_{rr}+|\nabla s|^2\phi_{ss}+\Delta r\phi_r+\Delta s\phi_s\Big]$$
$$+\frac{1}{2}(\phi-\phi^3)+\epsilon\frac{[s]_E}{3\sigma}\Big\{T-T_Ac-T_B(1-c)\Big\} \tag{3.20}$$
$$C_v\Big[T_t+r_tT_r+s_tT_s\Big]+\frac{l}{2}\Big[\phi_t+r_t\phi_r+s_t\phi_s\Big]$$
$$=K_1\Big[T_{rr}+|\nabla s|^2T_{ss}+\Delta rT_r+\Delta sT_s\Big] \tag{3.21}$$
$$\Big[c_t+r_tc_r+s_tc_s\Big]=K_2(\phi)c(1-c)\Big\{F_{1rr}+|\nabla s|^2F_{1ss}+\Delta rF_{1r}+\Delta sF_{1s}\Big\}$$
$$+\Big\{K_2(\phi)c(1-c)\Big\}_rF_{1r}+|\nabla s|^2\Big\{K_2(\phi)c(1-c)\Big\}_sF_{1s}, \tag{3.22}$$

where

$$F_1\equiv F_1(\phi,c)\equiv M\phi+\ln\frac{c}{1-c}.$$

Scaling the r variable by use of $z=r/\epsilon$, one may write (3.20)-(3.22) as follows.

$$\hat\phi_{zz}+\frac{1}{2}(\hat\phi-\hat\phi^3)+\epsilon\Big[-\alpha r_t\hat\phi_z+\Delta r\hat\phi_z+\frac{[s]_E}{3\sigma}\Big(\hat T-T_A\hat c-T_B(1-\hat c)\Big)\Big]$$
$$+\epsilon^2\Big[|\nabla s|^2\hat\phi_{ss}+\Delta s\hat\phi_s-\alpha\hat\phi_t-\alpha s_t\hat\phi_s\Big]=0 \tag{3.23}$$
$$K_1\hat T_{zz}+\epsilon\Big[K_1\Delta r\hat T_z-\frac{l}{2}r_t\hat\phi_z-C_vr_t\hat T_z\Big]$$
$$+\epsilon^2\Big[K_1(|\nabla s|^2\hat T_{ss}+\Delta s\hat T_s)-C_v(\hat T_t+s_t\hat T_s)-\frac{l}{2}(\hat\phi_t+s_t\hat\phi_s)\Big]=0 \tag{3.24}$$
$$\Big\{K_2(\hat\phi)\hat c(1-\hat c)\Big(F_1(\hat\phi,\hat c)\Big)_z\Big\}_z+\epsilon\Big[\Delta rK_2(\hat\phi)\hat c(1-\hat c)\Big(F_1(\hat\phi,\hat c)\Big)_z-r_t\hat c_z\Big]$$
$$+\epsilon^2\Big[K_2(\hat\phi)\hat c(1-\hat c)\Big(|\nabla s|^2F_{1ss}+\Delta sF_{1s}\Big)$$
$$+|\nabla s|^2\Big(K_2(\hat\phi)\hat c(1-\hat c)\Big)_sF_{1s}-\hat c_t-s_t\hat c_s\Big]=0. \tag{3.25}$$

Substituting (3.9) into (3.23)-(3.25), we obtain

$O(1)$ system:

$$\hat\phi^0_{zz}+\frac{1}{2}\Big(\hat\phi^0-(\hat\phi^0)^3\Big)=0, \tag{3.26}$$
$$\hat T^0_{zz}=0, \tag{3.27}$$
$$\Big\{K_2(\hat\phi^0)\hat c^0(1-\hat c^0)\Big(M\hat\phi^0+\ln\frac{\hat c^0}{1-\hat c^0}\Big)_z\Big\}_z=0. \tag{3.28}$$

We seek to construct bounded solutions of the $O(1)$ equations (3.27) and (3.28) above with $0 < \hat{c}^0 < 1$. The solutions to (3.26)-(3.28) which satisfy this requirement are

$$\hat{T}^0(z,s,t) = c_1(s,t) \tag{3.29}$$

$$M\hat{\phi}^0 + \ln\frac{\hat{c}^0}{1-\hat{c}^0} = c_2(s,t) \tag{3.30}$$

where $c_1(s,t)$ and $c_2(s,t)$ are constants with respect to the variable z, so that c_2 may also depend on $\hat{T}^0$. Next, one has the

$O(\epsilon)$ system:

$$\hat{\phi}^1_{zz} + \frac{1}{2}\Big(1 - 3(\hat{\phi}^0)^2\Big)\hat{\phi}^1 = -\alpha r^0_t \hat{\phi}^0_z + \Delta r^0 \hat{\phi}^0_z + \frac{[s]_E}{3\sigma}\Big\{\hat{T}^0 - T_A\hat{c}^0 - T_B(1-\hat{c}^0)\Big\}, \tag{3.31}$$

$$K_1\hat{T}^1_{zz} - \frac{l}{2}r^0_t\hat{\phi}^0_z = 0, \tag{3.32}$$

$$\Big\{K_2(\hat{\phi}^0)\hat{c}^0(1-\hat{c}^0)\Big(M\hat{\phi}^1 + \frac{\hat{c}^1}{\hat{c}^0(1-\hat{c}^0)}\Big)_z\Big\}_z = r^0_t\hat{c}^0_z, \tag{3.33}$$

where the equalities (3.29) and (3.30) have been used.

Note that equation (3.26) and conditions $\hat{\phi}^0(-\infty) = -1$, $\hat{\phi}^0(0) = 0$, $\hat{\phi}^0(\infty) = 1$ imply that

$$\hat{\phi}^0(z) = \tanh\frac{z}{2}. \tag{3.35}$$

Integrating the equations (3.32) and (3.33), one obtains

$$K_1\hat{T}^1_z + \frac{l}{2}r^0_t\hat{\phi}^0 = c_3(s,t) \tag{3.36}$$

$$K_2(\hat{\phi}^0)\hat{c}^0(1-\hat{c}^0)\Big(M\hat{\phi}^1 + \frac{\hat{c}^1}{\hat{c}^0(1-\hat{c}^0)}\Big)_z = r^0_t\hat{c}^0 + c_4(s,t). \tag{3.37}$$

After using the matching conditions (see, e.g., [CF]), we obtain from equations (3.30), (3.36) and (3.37) that

$$K_1 T^0_r|_{\Gamma\pm} \pm \frac{l}{2}r^0_t|_{\Gamma\pm} = c_3(s,t) \tag{3.38}$$

$$\pm M + \ln\frac{c^0}{1-c^0}|_{\Gamma\pm} = c_2(s,t) \tag{3.39}$$

$$K_2^{\pm}c^0_r|_{\Gamma\pm} = r^0_t c^0|_{\Gamma\pm} + c_4(s,t). \tag{3.40}$$

Noticing also that the normal velocity v (dropping the superscript) is given by $-r_t$, we may write $-r^0_t = v^0$ in the above equations (3.38)-(3.40). Subtracting the equation with the minus sign from that with the plus sign from (3.38)-(3.40) respectively, we obtain the interface conditions to lowest order. These are

$$K_1[T^0_r]^+_- = -lv^0 \tag{3.41}$$

$$\left[\ln \frac{c^0}{1-c^0}\right]_-^+ = -2M \tag{3.42}$$

$$\left[K_2 c_r^0\right]_-^+ = -v^0 [c^0]_-^+. \tag{3.43}$$

These give the interface conditions for first order approximation of the solution (except the last interface condition in (3.11)).

To obtain the last interface condition of (3.11), we note that $\hat{\phi}_z^0$ satisfies

$$\hat{\phi}_{zz}^1 + \frac{1}{2}\left[1 - 3(\hat{\phi}^0)^2\right]\hat{\phi}^1 = 0,$$

i.e., $\hat{\phi}_z^0$ is an eigenfunction of the linear operator $L\psi \equiv \psi_{zz} + \frac{1}{2}\left[1 - 3(\hat{\phi}^0)^2\right]\psi = 0$ corresponding to the eigenvalue 0 and it is simple, so that the solvability condition gives

$$-(\alpha v^0 + \Delta r^0)\int_{R^1} (\hat{\phi}_z^0)^2 dz = \int_{R^1} \frac{[s]_E}{3\sigma}\left[\hat{T}^0 - T_A\hat{c}^0 - T_B(1-\hat{c}^0)\right]\hat{\phi}_z^0 dz. \tag{3.44}$$

By using equation (3.28), we can calculate the integral of the right hand side in (3.44), that is

$$\begin{aligned}&\int_{R^1} \frac{[s]_E}{3\sigma}\left[\hat{T}^0 - T_A\hat{c}^0 - T_B(1-\hat{c}^0)\right]\hat{\phi}_z^0 dz\\ &= \frac{[s]_E}{3\sigma}\left[2\hat{T}^0 - 2T_B - \frac{1}{M}(T_A - T_B)\ln\frac{1-\hat{c}^0(\infty)}{1-\hat{c}^0(-\infty)}\right].\end{aligned} \tag{3.45}$$

From (3.44), (3.45) and the matching conditions, we may obtain, by use of $\Delta r^0 = \kappa^0$, the interface relation which generalizes Gibbs-Thomson,

$$-(\alpha v^0 + \kappa^0) = \frac{[s]_E}{2\sigma}\left[2T^0 - 2T_B - \frac{1}{M}(T_A - T_B)\ln\frac{1-c_+^0}{1-c_-^0}\right] \tag{3.46}$$

on $\Gamma(t)$, where we have used $\int_{R^1}(\hat{\phi}_z^0)^2 dz = 2/3$. This completes verification of Proposition 3.1.

Verification of Proposition 3.2. Note that for small values of c_+^0, c_-^0, one has

$$\begin{aligned}\left[\ln\frac{c^0}{1-c^0}\right]_-^+ &= \ln\frac{c_+^0(1-c_-^0)}{c_-^0(1-c_+^0)}\\ &\simeq \ln\frac{c_+^0}{c_-^0}\end{aligned}$$

and

$$\ln\frac{1-c_+^0}{1-c_-^0} \simeq c_-^0 - c_+^0.$$

Then the Proposition 3.2 follows immediately from the proof of Proposition 3.1.

Finally, a more detailed phase field alloy model, with cross terms in the temperature and concentration equations, can be stated as

$$\begin{cases} \alpha\epsilon^2\phi_t = \epsilon^2\Delta\phi + \frac{1}{2}(\phi - \phi^3) + \frac{\epsilon}{3\sigma}[s]_E\Big\{T - T_A c - T_B(1-c)\Big\} \\ C_v T_t + \frac{1}{2}\Big(l(c)\phi_t - (Q + [l]_{B,A}\phi)c_t\Big) = \nabla \cdot K_1 \nabla T \\ c_t = \nabla \cdot K_2(\phi)c(1-c)\nabla\Big[N\phi + RT\ln\frac{c}{1-c}\Big] \end{cases} \tag{3.47}$$

where

$$\begin{cases} l(c) = \Big(T_A c + T_B(1-c)\Big)[s]_E \\ [l]_{B,A} = (T_B - T_A)[s]_E \\ N = \frac{1}{2}[s]_E(T_A - T_B) \\ Q = 2(T_B - T_A)\Big(C_v + \frac{s^l + s^s}{2}\Big) \end{cases} \tag{3.48}$$

and R is Boltzmann's constant and the other parameters are the same as in (3.1)-(3.5), and s^s and s^l are the entropy densities in solid and liquid respectively.

By the same method as employed in verification of Proposition 3.1, we have the following asymptotic result.

Proposition 3.3. *In limit as $\epsilon \to 0$, there exists a formal asymptotic solution, (ϕ, T, c), of phase field alloy model (3.47) which is governed by the following two-phase sharp interface alloy model*

$$\begin{cases} C_v T_t - \frac{1}{2}\Big(Q \pm [l]_{B,A}\Big)c_t = K_1\Delta T \quad in \quad \Omega \setminus \Gamma(t) \\ c_t = \nabla K_2^{\pm} c(1-c)\nabla\Big(RT\ln\frac{c}{1-c}\Big) \quad in \quad \Omega \setminus \Gamma(t) \end{cases} \tag{3.49}$$

and on the interface $\Gamma(t)$

$$\begin{cases} [T]_-^+ = 0 \\ [K_1\nabla T \cdot n]_-^+ = \frac{1}{2}\Big(Q[c]_-^+ + [l]_{B,A}(c^+ + c^-) - 2T_B[s]_E\Big)v \\ RT[\ln\frac{c}{1-c}]_-^+ = -2N \\ [K_2 RT\nabla c \cdot n]_-^+ + [RK_2 c(1-c)\Big(\ln\frac{c}{1-c}\Big)\nabla T \cdot n]_-^+ = [c]_-^+ v \\ \alpha v + \kappa = \frac{[s]_E}{-\sigma}\Big[T - T_B - \frac{R(T_A - T_B)}{2N}T\ln\frac{1-c^+}{1-c^-}\Big]. \end{cases} \tag{3.50}$$

Note that equations (3.47) differ from the simplest version (3.3)-(3.5) in that all equilibrium constants including N, are obtain from free energy considerations, and cross terms in c and T are present. Note that the T dependence is simpler in the interface conditions and the two equations agree in the small c or $1 - c$ limit.

4. Equilibrium Phase Diagram Retrieval

The equations (3.3)-(3.5) are obtained directly from the free energies and so must contain all of the information necessary to reconstruct the (c, T) phase diagram. By specializing the calculations to the equilibrium case, we retrieve the phase diagram for the simplest phase field alloy equation (3.3)-(3.5).

We note that, from Proposition 3.1, in equilibrium, one has the interface relation

$$2M(T - T_B) = (T_A - T_B)\ln\frac{1-c^+}{1-c^-}. \tag{4.1}$$

Combining the above equation, (4.1), with the equation (3.42) in Section 3, we can obtain c^+ and c^- in terms of T by solving the above two equations, that is

$$c^- = \frac{1-\hat{M}}{\hat{M}e^{-2M}(1-e^{2M})} \tag{4.2}$$

$$c^+ = \frac{1-\hat{M}}{1-e^{2M}} \tag{4.3}$$

where

$$\hat{M} = e^{\frac{2M}{T_A - T_B}(T-T_B)}. \tag{4.4}$$

The expressions on the right hand sides of (4.2) and (4.3) are two nonlinear functions of T, denoted by $c^{\pm}(T)$, with

$$c^{\pm}(T_B) = 0, \quad c^{\pm}(T_A) = 1 \quad \text{and} \quad c^-(T) < c^+(T)$$

for T between T_A and T_B (verify in each case as $T_A < T_B$, $T_A > T_B$ and $M < 0$, $M > 0$, respectively). Also, the derivatives of $c^{\pm}(T)$ with respect to T are

$$(c^+)'(T) = \frac{-2M}{(1-e^{2M})(T_A - T_B)}e^{\frac{2M}{T_A-T_B}(T-T_B)}$$

$$(c^-)'(T) = \frac{-2Me^{2M}}{(1-e^{2M})(T_A - T_B)}e^{\frac{-2M}{T_A-T_B}(T-T_B)}$$

which are both negative if $T_A < T_B$ and $M \neq 0$ and both positive if $T_A > T_B$ and $M \neq 0$. Thus, equations (4.2) and (4.3) determine the equilibrium phase diagram completely as shown in Figure 2.

Note also that $c^{\pm} \to 0$ as $T \to T_B$, we can linearize the functions on the right hand sides of (4.2) and (4.3) by using the Taylor expansion at point $T = T_B$. This gives

$$c^- \simeq \frac{1}{m_s}(T - T_B) \tag{4.5}$$

$$c^+ \simeq \frac{1}{m_l}(T - T_B) \tag{4.6}$$

as $T \to T_B$, where

$$m_s = \frac{(1-e^{2M})(T_A - T_B)}{-2Me^{2M}} \tag{4.7}$$

$$m_l = \frac{(1-e^{2M})(T_A - T_B)}{-2M}. \tag{4.8}$$

This recovers the typical (linear) equilibrium phase diagram for small c used in [S1, 2], [O] and [X].

Similarly, by the 'symmetry' of the equation for c, we can linearize the functions on the right hand sides of (4.2) and (4.3) at point $T = T_A$ to recover the linear equilibrium phase diagram for small $1 - c$.

5. A PDE System with Constant Temperature

In this section, we state a rigorous result for a system of ϕ and c (neglecting the temperature equation). It is convenient mathematically to replace, without loss of generality, the physical parameters by constants. We then consider the following system.

$$\phi_t = \Delta\phi + \frac{2}{\epsilon^2}\Big\{\phi(1-\phi^2) + \epsilon(a+bc)\Big\} \tag{5.1}$$

$$c_t = \nabla \cdot \Big\{c(1-c)\nabla\Big(M\phi + \ln\frac{c}{1-c}\Big)\Big\} \tag{5.2}$$

in $Q_T \equiv \Omega \times (0,T)$, subject to the suitable boundary and initial conditions. In particular, if we let Ω be a one-dimensional domain, e.g., $\Omega = (-1,1)$, then we impose the following boundary and initial condition

$$\phi(x,0) = \tanh\frac{x}{\epsilon}, \quad x \in (-1,1) \tag{5.3}$$

$$\phi(\pm 1,t) = \tanh\frac{\pm 1}{\epsilon}, \quad t \in (0,T) \tag{5.4}$$

$$c(x,0) = c_0(x), \quad x \in (-1,1) \tag{5.5}$$

$$c_x(\pm 1,t)) = 0, \quad t \in (0,T) \tag{5.6}$$

where a, b, and M are constants, and the function $c_0(x) \in C^1[0,1]$ with $0 < c_0(x) < 1$ and $c_{0x}(\pm 1) = 0$ and $\epsilon > 0$ is a small parameter.

The main results in this section are the following two theorems. The proof can be found in [CX].

Theorem 5.1. *Let $c_0(x) \in C^1[0,1]$ with $0 < c_0(x) < 1$ and $c_{0x}(\pm 1) = 0$. Then for any $\epsilon > 0$, there exists a unique solution $(\phi^\epsilon(x,t), c^\epsilon(x,t))$ to the problem (5.1)-(5.6). The solution is smooth in Q_T.*

Theorem 5.2. *In limit $\epsilon \to 0^+$, there exists a pair of functions $(S(t), c(x,t))$ defined on $(0,T_0)$ with $-1 < S(t) < 1$ as $t \in (0,T_0)$, such that*

$$c_t = c_{xx} \quad in \quad Q_T \setminus \Big\{(S(t),t) \mid 0 < t < T_0\Big\}$$

with boundary and interface conditions

$$c(x,0) = c_0(x), \quad x \in (-1,1)$$

$$
\begin{aligned}
&c_x(\pm 1,t) = 0, \quad t \in (0,T_0)\\
&[c_x]_-^+ = -[c]_-^+ S'(t), \quad t \in (0,T_0)\\
&\frac{2}{3}S'(t) = -2a - \frac{b}{M}\ln\frac{1-c^+}{1-c^-}, \quad t \in \mathcal{E}\\
&\left[\ln\frac{c}{1-c}\right]_-^+ = -2M, \quad t \in \mathcal{E}
\end{aligned}
$$

where

$$
\mathcal{E} = \left\{ t \in (0,T_0) \ \mid \ c^\pm(S(t),t) \neq 0,1 \right\}
$$

$$
T_0 = \sup\left\{ \ t \ \mid \ -1 < S(\tau) < 1, \tau \le t \ \right\}
$$

and

$$
\phi^\epsilon(x,t) \to \begin{cases} -1 & \text{if } \ x < S(t) \\ +1 & \text{if } \ x > S(t) \end{cases}
$$

as $\epsilon \to 0^+$.

Remark: Theorem 5.2 is the rigorous justification of Proposition 3.1 in the one-dimensional case under the initial and boundary conditions considered above.

Acknowledgements: It is a pleasure to thank Prof. J. Jones for helpful discussions. Interactions with B. Boettinger, J. Cahn, S. Coriell, G. McFadden, A. Wheeler are also gratefully acknowledged. This work has been supported by NSF Grant DMS-9002242 and NIST Grant 60NANBOD1026.

References

[C1] G. Caginalp, *Mathematical models of phase boundaries,* Material Instabilities in Continuum Problems and Related Mathematical Problems (J. Ball, ed.), Symposium 1985-1986, Heriot-Watt, Oxford, (1988), 35-52.

[C2] G. Caginalp, *Stefan and Hele-shaw type models as asymptotic limits of phase field equations,* Physics Rev. A, **39** (1989), 887-896.

[CC] G. Caginalp & X. Chen, *Phase field equations in the singular limit of sharp interface problems,* The IMA Volumes in Mathematics and its Applications, **43** (1991), 1-28.

[CF] G. Caginalp & P. Fife, *Dynamics of layered interface arising from phase boundaries,* SIAM J. Appl. Math., **48** (1988), 506-518.

[CJ] G. Caginalp & J. Jones, *A derivation of phase field models of thermal alloys,* Univesity of Pittsburgh Preprint.

[CS] G. Caginalp & E. Socolovsky, *Phase field computations of single-needle crystals, crystal growth and motion by mean curvature,* submitted to SIAM J. Scientific and Statistical Computing.

[CX] G. Caginalp & W. Xie, *A mathematical analysis of the phase field alloy model,* University of Pittsburgh Preprint.

[ESS] L.C. Evans, M. Soner & P. Souganidis, *Phase transitions and generalized motion by mean curvature,* Comm. Pure Appl. Math., **45** (1992), 1097-1123.

[Fe] J. Fehribach, *Pertubation methods for solid diffusion in a Stefan Problem*, SIAM J. Math. Anal., **19** (1988), 86-99.

[Gl] J. Glimm, *The continuous structure of discontinuities*, Proc. of Nice conference (Jan. 1988).

[Lu] C. Lupis, *Chemical Thermodynamics of Materials*, North-Holland, New York (1983).

[MS] W. Mullins & R. Sekerka, *Stability of a planar interface during solidfication of a binary alloy*, J. Appl. Phys., **35** (1964), 445-451.

[O] J. Ockendon, *Linear and nonlinear stability of a class of moving boundary problems*, Free Boundary Problems, Pavia 1979, 443-478.

[S1] R. Sekerka, *Melt growth* in Int. School of Crystallography 7th course: Interfacial aspects of phase transformation, Erice-Trapani, Siccly (Ed. B. Mutaftschiev), D. Reidel Publ (1982), 489-508.

[S2] R. Sekerka, *On the modeling of solid-liquid interface dynamics*, Proc. Darkin Conf. U.S. Steel Res. Lab Monroeville, PA (1976), 301-321.

[WBM1] A. Wheeler, W. Boettinger & G. McFadden, *A phase-field model for isothermal phase trasitions in binary alloys*, Phys. Rev. A, **45** (1992), 7424-7439.

[WBM2] ——, *A phase field model of solute trapping during solidification*, NIST Preprint.

[X] W. Xie, *Phase transition in binary systems and the binary alloy problem*, Math. Meth. in Appl. Sci., **15** (1992), 205-221.

Captions to the Figures

Figure 1: Idealized phase diagrams in the neighborhood of $(c, T) = (0, T_B)$ which display linear liquidus and solidus lines which separate the single phase regions from the coexistence region. This idealization is inherent in typical sharp interface alloy models (one-phase or two-phase) and represents a local approximation to the phase diagram displayed in Figure 2. Figure 1(a) assumes negative slopes for the liquidus and solidus lines while Figure 1(b) assume they are positive. The sign of the slopes determines which phase has higher solute concentration. In Figure 1(a), solute from the liquid must be rejected as the material freezes. The phase field alloy model we discuss is not restricted to constant slopes, but asymptotically approaches constant slopes in the neighborhood of small c.

Figure 2: The phase diagram for the range of c values displaying the melting temperatures T_A and T_B for the pure materials A and B. The liquidus is given by the upper curve while the solidus is given by the lower curve. The extent to which the liquidus and solidus are separated from the line connecting $(0, T_B)$ and $(1, T_A)$ is determined by the physical parameters. The phase diagram can be obtained from the free energy through the phase field alloy model (in equilibrium) so that the liquidus is $c^+(T)$ and the solidus is $c^-(T)$.

Figure 3: Profiles of ϕ and c across the interface. In phase field models, a phase change corresponds to a transition layer in the order parameter, ϕ, while temperature T has a transition layer in its gradient, the concentration c has a transition layer which is similar to that of ϕ. However, while ϕ always has a transition from (approximately) -1 to 1, the transition in c (between the two positive values c_S and c_L) depends on the temperature, in accordance with the phase diagram of Figure 2.

Figure 4: Relationship between different alloy models. The most general phase field model (upper left box) leads to a number of different limiting sets of equations, some of which are classical, e.g., the traditional one-phase (dilute binary alloy), and some are introduced in this paper, e.g., the two-phase alloy model without cross terms, in which the conditions involving c^+ and c^- (not the derivatives of c) are the new interface relations replacing the Gibbs-Thomson condition. With various simplifications and idealization, we obtain the "simplest" phase field alloy model which retains the key characteristics of an alloy. These idealizations consists of fixing, for example, the surface tension σ, for the mixture, which is accurate for pairs of materials which similar surface tension between phases, or for small concentrations of one material in another. The solid lines and arrows between the boxes indicate a mathematical relationship in the sense of a theorem for simpler geometries and matched asymptotic for more complicated geometries. The dashed lines indicate physical idealizations or approximations.

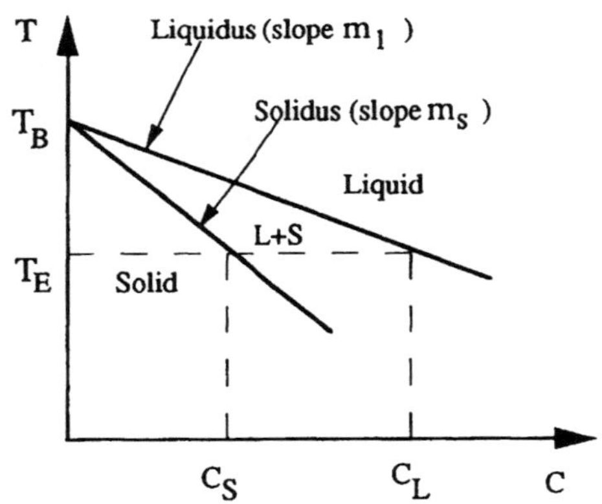

Figure 1(a)

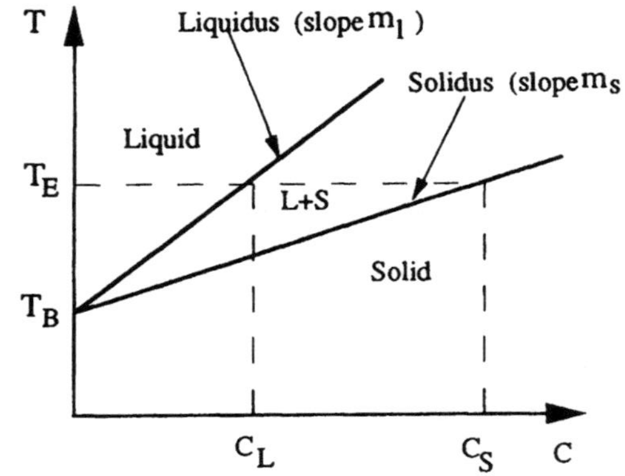

Figure 1(b)

FIGURE 1: Idealized phase diagrames illustrating the equilibrium

(See fully written captions).

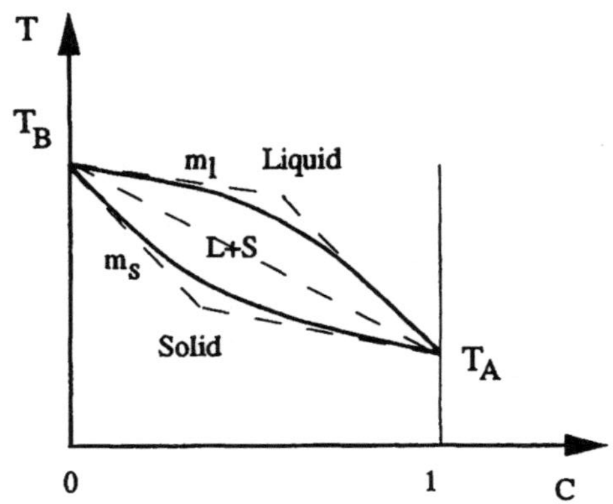

Figure 2

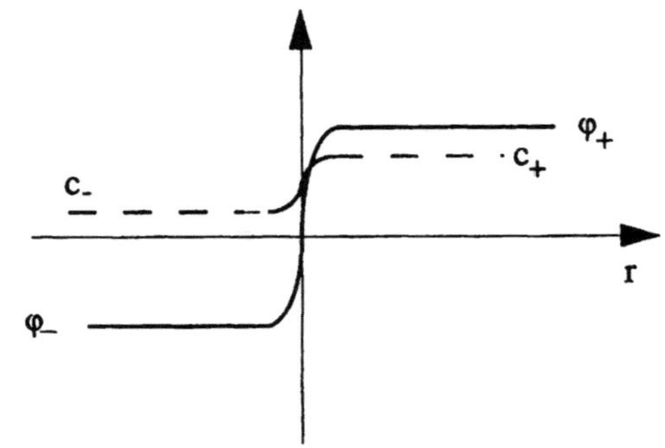

Figure 3

FIGURE 2: Phase diagram illustrating the equilibrium.

FIGURE 3: Profiles of c and φ as ε small.

FIGURE 4: PHASE FIELD ALLOY MODEL AND ITS DISTINGUISHED LIMITS

Phase field alloy model

$$\tau_1 \phi_t = -\frac{\delta F}{\delta \phi}$$

$$T_t - \nabla \cdot K_1 \nabla T = H(\phi, c, \phi_t, c_t)$$

$$\tau_2 c_t = \nabla \cdot K_2 c(1-c) \nabla \left(\frac{\delta F}{\delta c} \right)$$

↓ Appropriate F, H and physical idealization

Simpler phase field alloy equations

$$\alpha \epsilon^2 \phi_t = \epsilon^2 \Delta \phi + \frac{1}{2}(\phi - \phi^3) + \frac{\epsilon}{3\sigma}[s]_E \left\{ T - T_A c - T_B(1-c) \right\}$$

$$C_v T_t + \frac{1}{2}\left(l(c)\phi_t - (Q + [l]_{B,A}\phi) c_t \right) = \nabla \cdot K_1 \nabla T$$

$$c_t = \nabla \cdot K_2(\phi) c(1-c) \nabla \left[N\phi + RT \ln \frac{c}{1-c} \right]$$

→ $\varepsilon \to 0$: Two-phase alloy model with cross term

↓ Physical approximation

Simplest phase field alloy equations

$$\alpha \epsilon^2 \phi_t = \epsilon^2 \Delta \phi + \frac{1}{2}(\phi - \phi^3) + \frac{\epsilon}{3\sigma}[s]_E \left\{ T - T_A c - T_B(1-c) \right\}$$

$$C_v T_t + \frac{l}{2}\phi_t = \nabla \cdot K_1 \nabla T$$

$$c_t = \nabla \cdot K_2(\phi) c(1-c) \nabla \left[M\phi + \ln \frac{c}{1-c} \right]$$

→ $\varepsilon \to 0$: Two-phase alloy model without cross term

↓ $c \to 0$

Pure phase field equations

$$\alpha \epsilon^2 \phi_t = \epsilon^2 \Delta \phi + \frac{1}{2}(\phi - \phi^3) + \frac{\epsilon}{3\sigma}[s]_E (T - T_M)$$

$$C_v T_t + \frac{l}{2}\phi_t = \nabla \cdot K_1 \nabla T$$

↓ $\varepsilon \to 0$

Pure surface tension and kinetics model

$$C_v T_t = K_1 \Delta T \quad \text{in} \quad \Omega \setminus \Gamma(t)$$

$$[K_1 T_n]_-^+ = -lv \quad \text{on} \quad \Gamma(t)$$

$$-\sigma(\alpha v + \kappa) = [s]_E (T - T_M) \quad \text{on} \quad \Gamma(t)$$

Two-phase alloy model with cross term

$$C_v T_t - \frac{1}{2}\left(Q \pm [l]_{B,A} \right) c_t = K_1 \Delta T \quad \text{in} \quad \Omega \setminus \Gamma(t)$$

$$c_t = \nabla \cdot K_2^{\pm} c(1-c) \nabla \left(RT \ln \frac{c}{1-c} \right) \quad \text{in} \quad \Omega \setminus \Gamma(t)$$

$$[K_1 T_n]_-^+ = \frac{1}{2}\left(Q[c]_-^+ + [l]_{B,A}(c^+ + c^-) - 2T_B [s]_E \right) v \text{ on } \Gamma(t)$$

$$RT[\ln \frac{c}{1-c}]_-^+ = -2N \quad \text{on} \quad \Gamma(t)$$

$$[K_2 T c_n]_-^+ + [K_2 c(1-c)\left(\ln \frac{c}{1-c} \right) T_n]_-^+ = \frac{1}{R}[c]_-^+ v \text{ on } \Gamma(t)$$

$$\alpha v + \kappa = \frac{[s]_E}{-\sigma}\left[T - T_B - \frac{R(T_A - T_B)}{2N} T \ln \frac{1-c^+}{1-c^-} \right] \text{ on } \Gamma(t)$$

Two-phase alloy model without cross term

$$C_v T_t = K_1 \Delta T \quad \text{in} \quad \Omega \setminus \Gamma(t)$$

$$c_t = K_2^{\pm} \Delta c \quad \text{in} \quad \Omega \setminus \Gamma(t)$$

$$[K_1 T_n]_-^+ = -lv \quad \text{on} \quad \Gamma(t)$$

$$[\ln \frac{c}{1-c}]_-^+ = -2M \quad \text{on} \quad \Gamma(t)$$

$$[K_2 c_n]_-^+ = -[c]_-^+ v \quad \text{on} \quad \Gamma(t)$$

$$\alpha v + \kappa = \frac{[s]_E}{-\sigma}\left[T - T_B - \frac{T_A - T_B}{2M} \ln \frac{1-c^+}{1-c^-} \right] \quad \text{on} \quad \Gamma(t)$$

↓ Small c approximation (dilute impurities)

Dilute binary alloy model

$$C_v T_t = K_1 \Delta T \quad \text{in} \quad \Omega \setminus \Gamma(t)$$

$$c_t = K_2^{\pm} \Delta c \quad \text{in} \quad \Omega \setminus \Gamma(t)$$

$$[K_1 T_n]_-^+ = -lv \quad \text{on} \quad \Gamma(t)$$

$$c^+ = c^- e^{-2M} \quad \text{on} \quad \Gamma(t)$$

$$[K_2 c_n]_-^+ = -[c]_-^+ v \quad \text{on} \quad \Gamma(t)$$

$$\alpha v + \kappa = \frac{[s]_E}{-\sigma}\left[T - T_B - \frac{T_A - T_B}{2M}(c^- - c^+) \right] \quad \text{on} \quad \Gamma(t)$$

↓ Neglect solid phase

Traditional one-phase (dilute) binary alloy model

$$C_v T_t = K_1 \Delta T \quad \text{in} \quad \Omega(t)$$

$$c_t = K_2 \Delta c \quad \text{in} \quad \Omega(t)$$

$$K_1 T_n = -lv \quad \text{on} \quad \Gamma(t)$$

$$K_2 c_n = -(1-j)cv \quad \text{on} \quad \Gamma(t)$$

$$T - T_B = \frac{-\sigma}{[s]_E}(\alpha v + \kappa) + m_l c \quad \text{on} \quad \Gamma(t)$$

(Dilute binary alloy model) ⇢ $c \to 0$ ⇢ Pure surface tension and kinetics model

$$l(c) = \left(T_A c + T_B(1-c) \right)[s]_E$$

$$[l]_{B,A} = (T_B - T_A)[s]_E$$

$$Q = 2(T_B - T_A)\left(C_v + \frac{s^s + s^l}{2} \right)$$

$$j = \frac{m_l}{m_s}, \quad M = \frac{1}{2}\ln \frac{m_l}{m_s}$$

$$N = \frac{1}{2}[s]_E (T_A - T_B)$$

$T_n \equiv \nabla T \cdot n$, etc., $[T_n]_-^+ \equiv$ jump in T_n, etc.

New ideas in Calculus of Variations and Geometric Measure Theory

Ennio De Giorgi

Abstract. The aim of this conference is to state some results, problems and conjectures which seem to me particularly interesting. I have tried to formulate these problems in a self contained way, in order to make them understandable even by non specialists in Geometric Measure Theory. To this aim, let us begin by introducing the α-dimensional Hausdorff measure.

1991 Mathematics Subject Classification: 35A15, 35K99, 49M10.

Definition 1. Let M be a metric space, and let $E \subset M$. For any $\alpha > 0$ we define

$$\mathcal{H}^\alpha_M(E) = \frac{[\Gamma(1/2)]^\alpha}{2^\alpha \Gamma(1+\alpha/2)} \sup_{\delta>0} \inf \left\{ \sum_{i=1}^\infty \operatorname{diam}^\alpha(E_i) : E \subset \bigcup_{i\in I} E_i,\ \operatorname{diam}(E_i) < \delta \right\},$$

where $\Gamma(r) = \int_0^\infty e^{-t} t^{r-1}\, dt$ is the Euler function. We also define $\mathcal{H}^0_M(E)$ to be the cardinality of E if E is a finite set, $+\infty$ otherwise. In case $E = \mathbf{R}^n$ endowed with the euclidean metric, we set $\mathcal{H}^\alpha_{\mathbf{R}^n}(E) = \mathcal{H}^\alpha(E)$.

We can now state an existence result.

Theorem 1. If $f \in L^1(\mathbf{R}^n) \cap L^\infty_{\mathrm{loc}}(\mathbf{R}^n)$ then there exists

$$\min_E \left\{ \int_E f(x)dx + \mathcal{H}^{n-1}(\partial E); E = \overline{E} \subset \mathbf{R}^n,\ \partial E = \partial \mathring{E} \right\}.$$

Remark 1. Theorem 1 says nothing about uniqueness of minimizers, but some results concerning generic uniqueness of minimizers have been pointed out to me by B.White (see [28], [29]).

The proof of Theorem 1 is based on several results of the theory of sets of finite perimeter (see for instance [23], [24]). Recently Modica and Mortola discovered how the variational problem of Theorem 1 can be approximated by simpler variational problems. This approximation can be studied in the framework of Γ-convergence. One of the consequences of [25] is the following result.

Theorem 2. Let be $\varphi \in C_0^\infty(\mathbf{R}^n \times \mathbf{R})$, $-\infty < a < b < +\infty$; then

$$\lim_{\lambda\to+\infty} \left\{ \min_u \int_{\mathbf{R}^n} \left[\varphi(x, u(x)) + \frac{|\nabla u|^2}{\lambda} + \lambda(u-a)^2(u-b)^2\right] dx \right\} =$$

$$= \min_E \left\{ \gamma(a,b)\mathcal{H}^{n-1}(\partial E) + \int_E \varphi(x,b)dx + \int_{\mathbf{R}^n \setminus E} \varphi(x,a)dx \right\},$$

where $\gamma(a,b) = 2\int_a^b (t-a)(b-t)\,dt$.

Remark 2. More generally, one could ask (see for instance [1], [2]) for what polynomials P the following limit exists:

$$\lim_{\lambda \to +\infty} \left\{ \min_u \int_{\mathbf{R}^n} \left[\varphi\big(x, u(x)\big) + \frac{|\nabla u|^2}{\lambda} + \lambda P(u(x)) \right] dx \right\}.$$

To the approximating functionals of Modica-Mortola we can associate the evolution equations

$$\frac{\partial u}{\partial t} = \frac{\Delta_x u}{\lambda} - \lambda(u-a)(u-b)(2u-a-b) - \frac{\partial \varphi}{\partial u}(x,u).$$

In view of the Modica-Mortola theorem, it is natural to think that these equations can be used to approximate flows by mean curvature or, more generally, flows along the first variation of functionals like those of Theorem 2. A very general conjecture in this direction is the following.

Conjecture 1. Let be $n, k \in \mathbf{N}$, $n \geq 1$, $k \geq 1$, and let $f : \mathbf{R}^n \to \mathbf{R}^k$, $\varphi : \mathbf{R}^k \to \mathbf{R}$ be smooth periodic functions. For every $c \in \mathbf{R}^k$, let $u_c(x,t,\lambda)$ be the classical solution (periodic in the variables $x_1, \ldots, x_n$) of the equation (or system):

$$\frac{\partial u}{\partial t} = \Delta_x u - \lambda^2 \nabla \varphi(u)$$

with the initial conditions

$$u(x,0,\lambda) = f(x) + c.$$

Then:

$$\text{for a.e. } (x,t,c) \quad \exists \lim_{\lambda \to +\infty} u_c(x,t,\lambda)$$

and

$$\text{for a.e. } (x,c) \quad \exists \lim_{\substack{\lambda \to +\infty \\ t \to +\infty}} u_c(x,t,\lambda)$$

Remark 3. Similar conjectures have been stated in [8] and [9], replacing the periodicity assumption on f or φ by suitable assumptions on their growth at infinity. In the case $k=1$, $\varphi(y) = \left(y^2 - 1\right)^2$ in the previous conjecture is that one already discussed, and it is tied to mean curvature motion. If f is periodic, the behaviour as $\lambda \to +\infty$ of $u_c(x,t,\lambda)$ could be very interesting in the theory of stationary periodic sets.

The case $k > 1$, $\varphi(y) = \left(|y|^2 - 1\right)^2$ seems to be related to the evolution of harmonic maps; the case $k > 1$ with a polynomial φ having more than 2 minimum points seems to be related to the evolution of minimal partitions. Finally, polynomials like

$$\varphi(y) = \left[|y - \xi_1|^2 - \rho_1^2\right]^2 \left[|y - \xi_2|^2 - \rho_2^2\right]^2$$

should lead to the study of interesting problems where the difficulties relative to both partition problems and harmonic maps are present.

Let us now present more specific conjectures concerning a particular case of conjecture 1. Let $n = 3$, $k = 1$ and let $u(x,t,\lambda)$ be the periodic solution of the initial value problem

$$(1)_\alpha \qquad \begin{cases} \dfrac{\partial u}{\partial t} = \Delta_x u - \lambda^2 \sin u \\ \\ u(x,0,\lambda) = \sin x_1 + \sin x_2 + \sin(\alpha x_3) \end{cases}$$

Conjecture 2. There exists $\beta > 0$ such that if $0 < \alpha < \beta$ then

$$\lim_{\substack{\lambda \to +\infty \\ t \to +\infty}} u(x,t,\lambda) = \pi \frac{\sin(\alpha x_3)}{|\sin(\alpha x_3)|}.$$

Conjecture 3. There exists $\gamma > 0$ such that if $\alpha > \gamma$ and

$$A_\alpha^+ = \left\{ x \in \mathbf{R}^3 : \lim_{\substack{\lambda \to +\infty \\ t \to +\infty}} u(x,t,\lambda) = \pi \right\},$$

then $\sin x_1 = \sin x_2 = 1$ implies that (x_1, x_2, x_3) belongs to the interior of A_α^+.

Conjecture 4. There exists $\alpha_0 > 0$ such that for $\beta = \alpha_0$ it holds conjecture 1 and for $\gamma = \alpha_0$ it holds conjecture 2.

Remark 4. In the case $\gamma < 1$, conjecture 3 should include the example pointed out to me by B.White during the conference; it concerns the case $\alpha = 1$, and the $\lim\limits_{\substack{\lambda \to +\infty \\ t \to +\infty}} u(x,t,\lambda)$ exists for a.e. x and is a function constant in two open sets separated by a minimal Weierstrass surface.

If β, γ exist as in conjectures 1, 2, then other conjectures could be formulated, concerning the solution $u_c(x,t,\lambda)$ of the following initial value problem in $\mathbf{R}^3 \times [0,+\infty[$:

$$(2)_\alpha \qquad \begin{cases} \dfrac{\partial u}{\partial t} = \Delta_x u - \lambda^2 \sin u \\ \\ u(x,0,\lambda) = \sin x_1 + \sin x_2 + \sin(\alpha x_3) + c \end{cases}$$

Conjecture 5. (Local stability) If $0 < \alpha < \beta$ or $\alpha > \gamma$ and u, u_c are the solutions of $(1)_\alpha$, $(2)_\alpha$ respectively, then for a.e. (x,t) it holds

$$\lim_{\substack{\lambda \to +\infty \\ c \to 0}} u_c(x,t,\lambda) = \lim_{\lambda \to +\infty} u(x,t,\lambda).$$

Conjecture 6. (Asymptotic stability) If $0 < \alpha < \beta$ and u, u_c are the solutions of $(1)_\alpha$, $(2)_\alpha$ respectively, then for every x such that $\sin(\alpha x_3) \neq 0$ it holds

$$\lim_{\substack{\lambda \to +\infty \\ t \to +\infty \\ c \to 0}} u_c(x,t,\lambda) = \lim_{\substack{\lambda \to +\infty \\ t \to +\infty}} u(x,t,\lambda) = \pi \frac{\sin(\alpha x_3)}{|\sin(\alpha x_3)|}.$$

Conjecture 7. If $\alpha > \gamma$ and u_c is the solution of problem $(2)_\alpha$, then:

$$\lim_{c\to 0^+} \liminf_{\substack{\lambda\to+\infty \\ t\to+\infty}} u_c(x,t,\lambda) = \pi$$

and

$$\lim_{c\to 0^-} \limsup_{\substack{\lambda\to+\infty \\ t\to+\infty}} u_c(x,t,\lambda) = -\pi$$

for every $x \in \mathbf{R}^3$.

Remark 5. Conjectures 1, 3, 6, 7 indicate a great difference in the asymptotic behaviour of u of $(1)_\alpha$ and $(2)_\alpha$ between the cases $0 < \alpha < \beta$ and $\gamma < \alpha$. If Conjecture 4 holds, the case $\alpha = \alpha_0$ should be particularly interesting.

Conjecture 8. Let be α_0 the constant given by conjecture 4, and u_c be the solution of problem $(2)_{\alpha_0}$; then

$$\mathcal{H}^4\left(\left\{(x,t) \in \mathbf{R}^3\times]0,+\infty[:\ \liminf_{\substack{\lambda\to+\infty \\ c\to 0}} u_c(x,t,\lambda) < \limsup_{\substack{\lambda\to+\infty \\ c\to 0}} u_c(x,t,\lambda)\right\}\right) = +\infty.$$

Remark 6. After proving (or disproving) the above conjectures, one could look for a finer analysis, *e.g.* trying to calculate or characterize in some way the various limits. Moreover, since such limits, as functions of x, or (x,t), (x,α), (x,t,α), (x,t,α,c) are expected to be discontinuous along some analytic manifold with singularities, it would be very interesting to check such expectation and classify as precisely as possible the singularities. In such a way problems $(1)_\alpha$ and $(2)_\alpha$ could yield a good example of the questions that it is reasonable to raise in the general setting of Conjecture 1.

Beside the above treated evolution problems, which give rise to "fast evolution", other problems have been studied (see *e.g.* [22]) that can be obtained by allowing the time scale to vary with the parameter λ and give rise to "slow evolution", *i.e.* equations of the form:

$$\frac{\partial\psi_\lambda}{\partial t} = \frac{1}{\lambda}\Delta_x\psi_\lambda - \lambda\nabla P(\psi_\lambda)$$

so that $\lim_{\lambda\to+\infty}\psi_\lambda(x,t)$ is equivalent to $\lim_{\lambda\to+\infty}u_\lambda(x,\lambda^{-1}t)$, where u_λ solves

$$\frac{\partial u_\lambda}{\partial t} = \Delta_x u_\lambda - \lambda^2\nabla P(u_\lambda);$$

in this connection, we present the following

Conjecture 9. For $\lambda > 0$ and $q \notin \{0,-1,-2\}$ the $\lim_{\lambda\to+\infty} u_\lambda(x,\lambda^q t)$ exists and is independent of t; in particular:
if $q < -2$ then $\lim_{\lambda\to+\infty} u(x,\lambda^q t) = \lim_{\substack{\lambda\to+\infty \\ t\to 0^+}} u_\lambda(x,\lambda^{-2}t)$
if $-2 < q < -1$ then $\lim_{\lambda\to+\infty} u_\lambda(x,\lambda^q t) = \lim_{\substack{\lambda\to+\infty \\ t\to+\infty}} u_\lambda(x,\lambda^{-2}t) = \lim_{\substack{\lambda\to+\infty \\ t\to 0^+}} u_\lambda(x,\lambda^{-1}t)$

if $-1 < q < 0$ then $\lim\limits_{\lambda\to+\infty} u_\lambda(x,\lambda^q t) = \lim\limits_{\substack{\lambda\to+\infty\\ t\to+\infty}} u_\lambda(x,\lambda^{-1}t) = \lim\limits_{\substack{\lambda\to+\infty\\ t\to 0^+}} u_\lambda(x,t)$

if $q > 0$ then $\lim\limits_{\lambda\to+\infty} u_\lambda(x,\lambda^q t) = \lim\limits_{\substack{\lambda\to+\infty\\ t\to+\infty}} u_\lambda(x,t)$.

Remark 7. If any stationary point of P is either an absolute maximum or an absolute minimum, it is very likely that the limit of conjecture 9 dos not depend on t even when $q = -1$.

The above quoted result of L.Modica and S.Mortola (Theorem 2) has had many generalisations, which lead to corresponding generalisations of the (fast and slow) evolution equations considered above. To such generalisations should correspond very general notions of "variation" (including as particular cases the classical notion of mean curvature) and of "movement along the variation" (which should include the classical notion of mean curvature motion). We present now some definitions, which are a first proposal to be discussed in view of such generalisations.

Definition 2. Let be $\Omega \subset \mathbf{R}^n$ open, and, for $\lambda \in \mathbf{R}$, $\psi_\lambda : \Omega \to \Omega$. We write $\psi \in DefC_0^\infty(\Omega)$ (ψ is a $C_0^\infty(\Omega)$ deformation of Ω) if for any $x \in \Omega$ $\psi_0(x) = x$ and, set $\Phi(x,\lambda) = x - \psi_\lambda(x)$, it holds $\Phi \in \left[C_0^\infty(\Omega \times \mathbf{R})\right]^n$.

Definition 3. (Sub-variation by deformation) Let be $\Omega \subset \mathbf{R}^n$ open, F a functional with values in $\mathbf{R} \cup \{-\infty, +\infty\}$, u a function. If the restriction of u to Ω, let it be u_Ω, lies in the domain of F, and $F(u_\Omega) \in \mathbf{R}$, we call $Var^-(F,u,\Omega)$ the set of the pairs (α,ν) such that ν is a positive Borel measure on Ω, $\alpha \in \left[L^1_{loc}(\Omega,\nu)\right]^n$, and, for any $\psi \in DefC_0^\infty(\Omega)$, set

$$F^-(u\circ\psi_\lambda) = \begin{cases} F(u\circ\psi_\lambda) & \text{if } u\circ\psi_\lambda \in \mathrm{dom}F;\\ +\infty & \text{otherwise}\end{cases}$$

the following holds:

$$\liminf_{\lambda\to 0}\frac{1}{|\lambda|}\Big\{F^-(u\circ\psi_\lambda) - F(u\circ\psi_0) - \int_\Omega \alpha(x)\big[\psi_\lambda(x) - x\big]d\nu(x)\Big\} \geq 0.$$

If $u_\Omega \notin \mathrm{dom}F$ or $F(u_\Omega) \notin \mathbf{R}$, we set $Var^-(F,u,\Omega) = \emptyset$. If for some $(\alpha,\nu) \in Var^-(F,u,\Omega)$ it holds $\int |\alpha| d\nu = 0$ we write $0 \in Var^-(F,u,\Omega)$.

Remark 8. If E is a minimizer set given by Theorem 1, and for every $w \in L^\infty(\mathbf{R}^n)$ we call S_w the discontinuity set of w and set

$$F(w) = \int_{\mathbf{R}^n} f(x)w(x)dx + \mathcal{H}^{n-1}(S_w),$$

then $0 \in Var^-(F,\chi_E,\mathbf{R}^n)$.

Definition 4. In the hypotheses of Definition 2, we set

$$Var^+(F,u,\Omega) = \Big\{(\alpha,\nu):\ (-\alpha,\nu) \in Var^-(-F,u,\Omega)\Big\},$$

and $Var(F,u,\Omega) = Var^+(F,u,\Omega) \cap Var^-(F,u,\Omega)$.

Remark 9. In the hypotheses of Remark 8, if $f \in C^0(\mathbf{R}^n)$, then $0 \in Var(F, \chi_E, \mathbf{R}^n)$.

Remark 10. The definition of $Var(F,u,\Omega)$ may be localized, setting

$$(\alpha,\nu) \in Var_{loc}(F,u,\Omega)$$

if and only if for any open set $A \subset\subset \Omega$ it holds $(\alpha_A, \nu_{\mathcal{B}(A)}) \in Var(F(\cdot,A),u,A)$, where $\nu_{\mathcal{B}(A)}$ is the restriction of ν to the Borel subsets of A.

Definition 5. (Movement of measures) Let $\Omega \subset \mathbf{R}^n \times \mathbf{R}$ be open; for every $t \in \mathbf{R}$ set $\Omega_t = \{x \in \mathbf{R}^n : (x,t) \in \Omega\}$ and let μ_t, ν_t be Borel measures on Ω. Suppose that for every $\varphi \in C_0^\infty(\Omega)$ and for every $t \in \mathbf{R}$:

$$\int_{\Omega_t} |\varphi(x,t)| d\mu_t(x) < +\infty,$$

$$t \mapsto \int_{\Omega_t} \varphi(x,t) d\mu_t(x) \text{ is absolutely continuous.}$$

For $i = 0,\dots,n$, let $v_i : \Omega \times \mathbf{R} \to \mathbf{R}$ be functions such that $v_{it} \in L^1_{loc}(\Omega_t, \nu_t)$ for a.e. t, where $v_{it}(x) = v_i(x,t)$. Then we write $\big((v_0,\dots,v_n),\nu\big) \in \mathcal{MM}(\Omega,\mu)$ if and only if the following holds:

$$\frac{d}{dt}\int_{\Omega_t} \varphi(x,t) d\mu_t(x) =$$

$$= \int_{\Omega_t} \frac{\partial\varphi}{\partial t}(x,t) d\mu_t(x) + \int_{\Omega_t} \varphi(x,t) v_0(x,t) d\nu_t(x) + \int_{\Omega_t} \sum_{i=1}^n \frac{\partial\varphi}{\partial x_i}(x,t) v_{it}(x) d\nu_t(x).$$

Definition 6. (Movement by variation) In the hypotheses of definitions 3, 4, we write $u \in Mvar(F,u,\Omega)$ if and only if there exist $(\alpha_1,\dots,\alpha_n,\mu)$ such that
for a.e. t: $(\alpha_t,\nu_t) \in Var_{loc}\big(F_t, u(\cdot,t), \Omega_t\big)$;
for any t, $A \subset\subset \Omega_t$: $\mu_t(A) = F_t(u(\cdot,t),A)$;
$\big((-|\alpha|^2,\alpha_1,\dots,\alpha_n),\nu\big) \in \mathcal{MM}(\mu,\Omega)$.

References

[1] Ambrosio, L., Metric space valued functions of bounded variation, Ann. Sc. Norm. Sup. 17 (1990), 439-478.

[2] Baldo, S., Minimal interface criterion for phase transitions in mixtures of Cahn-Hilliard fluids, Ann. Inst. H. Poincarè 7 (1990), 67-90.

[3] Barles, G., H.M. Soner, P.E. Souganidis, Front propagation and phase field theory, SIAM J. Cont. and Opt., to appear.

[4] Brakke, K.A., The motion of a surface by its mean curvature, Princeton U.P., Princeton, N.J., 1978.

[5] Bronsard, L., R.V. Kohn, On the slowness of phase boundary motion in one space dimension, Comm. Pure Appl. Math. 43 (1990), 983-997.

[6] Bronsard, L., R.V. Kohn, Motion by mean curvature as the singular limit of Ginzburg-Landau Dynamics, J. Diff. Eqs. 90 (1991),211-237.

[7] Chen, Y.G., Y. Giga, S.Goto, Uniqueness and existence of viscosity solutions of generalized mean curvature flow equations, J. Diff. Geom. 33 (1991), 749-786.

[8] De Giorgi, E., Some conjectures on flow by mean curvature, in: Methods of real analysis and partial differential equations, M.L. Benevento, T. Bruno, C. Sbordone eds., Liguori, Napoli 1990.

[9] De Giorgi, E., Conjectures on limits of some quasilinear parabolic equations and flow by mean curvature, in: Partial differential equations and related subjects, M. Miranda ed., Pitman Res. Notes in Math. 269 (1992), 85-95.

[10] De Mottoni, P., M. Schatzman, Evolution géometrique d'interfaces, C. R. Ac. Sci. Paris 309 (1989), 453-458.

[11] Ecker, K., G. Huisken, Mean curvature evolution of entire graphs, Ann. of Math. 130 (1989), 453-471.

[12] Evans, L.C., H.M.Soner, P.E. Souganidis, Phase transition and generalized motion by mean curvature, Comm. Pure Appl. Math., to appear.

[13] Evans, L.C., J. Spruck, Motion of level sets by mean curvature I, J. Diff. Geom. 33 (1991), 635-681.

[14] Evans, L.C., J. Spruck, Motion of level sets by mean curvature II, Trans. Am. Math. Soc. 330 (1992), 321-332.

[15] Evans, L.C., J. Spruck, Motion of level sets by mean curvature III, J. Geom. Anal., to appear.

[16] Evans, L.C., J. Spruck, Motion of level sets by mean curvature IV, preprint.

[17] Ilmanen, T., Generalized flow of sets by mean curvature on a manifold, Indiana Univ. Math. J., to appear.

[18] Ilmanen, T., The level-set flow on a manifold, Proc of the 1990 Summer Inst. in Diff. Geom., R.E. Greene and S.T. Yau eds., Am. Math. Soc., 1992.

[19] Ilmanen, T., Elliptic regularization and partial regularity for motion by mean curvature, preprint, 1991.

[20] Ilmanen, T., Convergence of the Allen-Cahn Equation to Brakke's motion by mean curvature, preprint, 1991.

[21] Ilmanen, T., Convergence to Brakke's motion by mean curvature, preprint, 1992.

[22] Keller, J.B., J. Rubinstein, P. Sternberg, Fast reaction, slow diffusion and curve shortening, SIAM J. Appl. Math. 49 (1989), 116-133.

[23] Massari, U., Esistenza e regolarità delle ipersuperfici di curvatura media assegnata in $\mathbf{R}^n$, Arch. Rational Mech. and Anal. 55 (1974), 357-382.

[24] Massari, U., M. Miranda, Minimal surfaces of codimension 1, North Holland, Amsterdam 1984.

[25] Modica, L., Mortola, S., Un esempio di Γ^--convergenza, Boll. U.M.I. 14-B (1977), 285-299.

[26] Soner, H.M., Motion of a set by the curvature of its boundary, J. Diff. Eq., to appear.

[27] Soner, H.M., P.E. Souganidis, Uniqueness and singularities of cylindrically symmetric surfaces moving by mean curvature, preprint.

[28] White, B., Spaces of m-dimensional surfaces that are stationary for a parametric elliptic integrand, Indiana Univ. Math. J. 36 (1987), 567-602.

[29] White, B., Spaces of minimal submanifolds for varying riemannian metrics, Indiana Univ. Math. J. 40, (1991), 161-200.

The minimum time problem and its applications to front propagation

M. Falcone

Abstract. We present an application of a numerical scheme for the minimum time problem to the approximation of a front propagating in the normal direction with velocity $c(x) > 0$. The front at time t corresponds to a level curve of the viscosity solution of the Hamilton–Jacobi–Bellman equation associated to the minimum time problem. We prove an a priori estimate which guarantees the convergence of the scheme to the viscosity solution and we derive an estimate of the Hausdorff distance between the level sets of the exact and the approximate solution.

Keywords. Level sets, viscosity solutions, front propagation, numerical schemes.

1991 Mathematics Subject Classification. Primary 65M25, 49L25; Secondary 35Q80.

1. Introduction

We deal with the problem of detecting numerically the evolution of a front which moves in the normal direction with a prescribed speed $c(x) > 0$ starting from a given initial position Γ. We will use to this end the relation established in Falcone–Giorgi–Loreti [FGL] between this problem and a particular version of the classical minimum time problem. Roughly speaking, in this interpretation the front corresponds to the boundary of particular level sets for a rescaled minimum time function v, these are the reachable sets at time t in the control terminology. This connection allows to apply the algorithm for the minimum time problem studied by Bardi-Falcone [BF1], [BF2], [BF3] and Lanucara [L]. We refer in particular to [BF1] for the proof that the time-discrete scheme converges to the viscosity solution of the first order Hamilton–Jacobi–Bellman equation associated to the minimum time problem, i.e. converges to v. We present here a fully-discrete version of the algorithm and we prove that, for the settings corresponding to the front propagation model, it gives a first order approximation of v and of its level sets (see the results in Sections 3 and 4). An interesting feature of this approach is that we solve a stationary problem (instead of an evolutive problem) saving CPU time. As a consequence of the level set approach we can follow the changes in the topology of the front (for example we can give an approximation of several fronts merging together). The convergence does not rely on regularity assumptions on the fronts at times $t > 0$ and the algorithm can be used to determine the evolution of fronts starting from very general sets. In practice, any compact set can be accepted even if the a priori estimates require some restrictions.

We should mention that the theory of viscosity solution for Hamilton–Jacobi equations and the level set approach have been applied to the study of the evolution of fronts by several authors starting from Barles [Ba], who treated the case of a constant speed of propagation. For the evolution of fronts moving by mean curvature or even more general flows we refer to Barles-Soner-Souganidis[BSS], Evans-Spruk [ESp1], [ESp2], Chen-Giga-Goto [CGG] and to the references therein.

As far as the numerical approximation is concerned several strategies have been adopted, all dealing with evolutive problems. Osher and Sethian (see [OS], [S1], [S2], [S3]) have used high-order methods derived from ENO-schemes to study numerically first order Hamilton-Jacobi equations. Finite elements techniques have been used to give approximations to the solutions of the degenerate parabolic equation related to the mean curvature flow and to establish estimates for the approximation of the fronts before the onset of singularities. We refer in particular to the works by Bellettini–Paolini–Verdi [BPV1], [BPV2], Nochetto–Paolini–Verdi [NPV] and Dziuk [D] for this approach. At present our scheme does not apply to the approximation of the mean curvature flow. In fact, to treat that problem it would be necessary to know a priori the curvature at each point of the domain of computation and, of course, this is possible only in very special cases. However, starting from the curvature of the initial shape, one can try to compute an approximate curvature of the front at each step of the algorithm and use this information in the next step. We will study this approach in a future work.

The paper is organized as follows.
In Section 2 we give a brief presentation of the minimum time problem and we explain its relation with the front propagation model. Section 3 is devoted to the fully discrete scheme and to the proof of convergence via an a priori estimate. In Section 4 we establish the estimate between the front and its approximation. Finally, in Section 5 we present some numerical tests.

2. The minimum time model for front propagation

We are given the following system of controlled ordinary differential equations

$$\begin{cases} \dot{y}(t) = b(y(t), \alpha(t)) \\ y(0) = x \end{cases} \tag{2.1}$$

where $y \in \mathbb{R}^N$,

$$\alpha(\cdot) \in \mathcal{A} \equiv L^\infty([0, +\infty[; A) \tag{2.2}$$

and $A \subset \mathbb{R}^M$ is compact. A compact set $\mathcal{T} \subset \mathbb{R}^N$, *the target*, is also given and the minimum time function is defined as

$$T(x) \equiv \inf_{\alpha(\cdot)\in\mathcal{A}} \{t \in \mathbb{R}_+ : y_x(t,\alpha) \in \mathcal{T}\}. \tag{2.3}$$

The minimum time problem consists in determining T and, possibly, an optimal control $\alpha^*(\cdot) \in \mathcal{A}$ where the minimum is attained. Notice that T will not be defined everywhere. In fact, its domain will be the so called *reachable set*

$$\mathcal{R} \equiv \{x \in \mathbb{R}^N : \exists \overline{\alpha} \in \mathcal{A} \quad \text{and } \overline{t} \in \mathbb{R}_+ \text{ such that } y_x(\overline{t},\overline{\alpha}) \in \mathcal{T}\}\ . \tag{2.4}$$

As one can see by its definition, $\mathcal{R}$ depends strictly on the dynamics (1.1) and on the *admissible control set* A so that one cannot consider $\mathcal{R}$ as given even if (2.1), (2.2) and $\mathcal{T}$ are the data of the problem.

We will assume that there exists a positive constant L_b such that:

$$(A1) \qquad \begin{cases} \text{i)}\ \ |b(x,a) - b(y,a)| \le L_b|x-y|, \quad \forall a \in A \\ \text{ii)}\ \forall x \in \partial\mathcal{T},\ \exists \overline{a} \in A \ \text{ such that } b(x,\overline{a})\cdot\eta(x) < 0\ . \end{cases}$$

where η denotes the exterior normal. We refer to Bardi [B] and Bardi-Falcone [BF1] for the results related to the solution of this problem via the dynamic programming approach and for a time-discrete scheme converging to T. We just mention that the direct application of the dynamic programming principle leads to a free boundary problem for a first order Hamilton–Jacobi equation, since that equation is valid only on $\mathcal{R}$. This difficulty can be solved introducing the Kruzkov's [K] change of variable

$$v(x) \equiv 1 - e^{-T(x)} \tag{2.5}$$

and extending v over $\mathbb{R}^N$ setting $v \equiv 1$ on $\mathbb{R}^N \setminus \mathcal{R}$. Assumptions (A1 i,ii) guarantee that v is the unique viscosity solution of

$$(HJ) \qquad \begin{cases} v(x) = \min\limits_{a\in A}[\,b(x,a)\cdot\nabla v(x)] + 1 & \text{on } \mathbb{R}^N \setminus \mathcal{T} \\ v(x) = 0 & \text{on } \mathcal{T}. \end{cases}$$

and that it is locally Lipschitz continuous. By v one can easily get $\mathcal{R}$ since, by (2.4) and (2.5), we have $\mathcal{R} = \{x : v(x) < 1\}$. Notice that (2.5) implies $0 \le v \le 1$.

The minimum time problem has been used in [FGL] to study a front propagation model which applies to the evolution of the interfaces in problems of combustion ([Ba], [S1]), crystal growth (see [TCH] and references therein), chemical reactions ([SS]) and many other problems where the fronts evolves isotropically in the normal direction. We will use here the terminology of the combustion model.

Assume that the space is filled with a flammable gas and that the burnt region at the initial time (say $t = 0$) is a compact set Ω with smooth boundary Γ. At time $t = 0$ the particles are ignited along $\Gamma \equiv \partial\Omega$ and we assume that the flame propagation is normal to the flame front with speed $c(x)$. The problem is to determine the burnt region at time $t > 0$, Ω_t. Γ_t will denote the *flame front* at time t, i.e. $\Gamma_t \equiv \partial\Omega_t$. We will assume that

$$c : \mathbb{R}^N \to \mathbb{R} \text{ is Lispchitz continuous and } c(x) > 0 \text{ on } \mathbb{R}^N. \tag{2.6}$$

In [FGL] we proved that the burnt region at time t corresponds to a level set of v. More precisely,

$$\Omega_t = \{x \in \mathbb{R}^N : v(x) \leq 1 - e^{-t}\} \tag{2.7}$$

where v is the viscosity solution of (HJ) corresponding to the settings

$$A \equiv B(0,1)\ , \qquad b(x,a) = -c(x)\,a \quad \text{and } \mathcal{T} \equiv \Omega\ . \tag{2.8}$$

The basic reason which allows to describe the evolution of the front in the normal direction by the above minimum time problem is that A is a ball centered at the origin. At a point $\overline{x}$ the minimum in (HJ) will be attained for $a^*(\overline{x}) = \nabla v(\overline{x})/|\nabla v(\overline{x})|$ which is the direction normal to the surface $v(x) = v(\overline{x})$. This means that, when (2.8) holds true, (HJ) can be written as

$$\begin{cases} v(x) + c(x)|\nabla v(x)| = 1 & \text{in } \mathbb{R}^N \setminus \Omega \\ v(x) = 0 & \text{in } \Omega. \end{cases} \tag{2.9}$$

We just mention that the above relation between the minimum time problem and the (isotropic) propagation of fronts remains valid also in more general situations (see [BF1], [FGL]). In fact, we can treat the case of a piecewise smooth target, eventually add an advection term, deal with an Ω with empty interior and weaken the assumptions on c treating also the case $c(x) \geq 0$. Moreover the above theory allows to prove easily that, for any t, Ω_t is a compact connected subset of $\mathbb{R}^N$ provided Ω has the same properties.

The study of the anisotropic propagation leads to a differential game model which has been investigated in [So] (see also [ES]). This formulation has a numerical interest since one can adopt the approximation scheme developed in [BFS]. A future work will be devoted to the numerical analysis of the anisotropic normal propagation using these techniques.

3. The fully discrete scheme

In [BF1] we have introduced the following time–discrete scheme for (HJ),

$$\begin{cases} v_h(x) = \min\limits_{a\in A}[\beta\, v_h(x + hb(x,a))] + 1 - \beta & \text{in } \mathbb{R}^N \setminus \mathcal{T} \\ v_h(x) = 0 & \text{in } \mathcal{T} \end{cases} \tag{HJ_h}$$

where $\beta \equiv e^{-h}$ and h is a positive time step. We proved that v_h converges locally uniformly to v in $\mathbb{R}^N$, provided (A1) is satisfied and $\mathcal{T}$ has a piecewise regular boundary. Moreover, on any compact subset Q of $\mathbb{R}^N$ such that there exists an open–loop optimal controls having bounded total variation we obtain the estimate

$$|v(x) - v_h(x)| \leq Ch\ , \qquad \text{for any } x \in Q \text{ and } h \leq \overline{h}, \tag{3.1}$$

i.e. (HJ$_h$) defines a first order approximation scheme (see [BF2] for details). It is worthwhile to notice that in the case of the flame front propagation model corresponding to (2.1), (2.2) and (2.8) we have $\mathcal{R} = \mathbb{R}^N$ and (3.1) holds for any compact subset of $\mathbb{R}^N$ containing Ω due to a result by Hajek [H]. In particular, when $c(x) \equiv c$ the optimal controls for (2.1) are constant in time (their value depends only on the initial position once Ω is given) and the optimal trajectories are straight lines connecting the initial position to Ω.

We will construct a space discretization of (HJ_h) in order to obtain an algorithm for the computation of v. Let us fix in $\mathbb{R}^N$ a rectangle Q containing $\mathcal{T} = \Omega$ in its interior. We build a regular triangulation of Q and denote by X the set of its nodes x_i, $i \in I \equiv \{1, \dots, N\}$, and by $\mathcal{S}$ the set of the simplices S_j, $j \in J \equiv \{1, \dots, L\}$. We set

$$I_0 \equiv \{i \in I : \quad x_i \in \mathcal{T}\} \tag{3.2}$$

$$I_{\text{in}} \equiv \{i \in I \setminus I_0 : \exists\, a \in A \quad \text{such that} \quad x_i + hb(x_i, a) \in Q\} \tag{3.3}$$

$$I_{\text{out}} \equiv \{i \in I \setminus I_0 : x_i + hb(x_i, a) \notin Q, \ \forall\, a \in A\} \tag{3.4}$$

and define our fully discrete approximation scheme

$$(HJ_h^k) \qquad \begin{cases} w(x_i) = \min\limits_{a \in A}[\beta\, w(x_i + hb(x_i, a))] + 1 - \beta & \text{for } i \in I_{\text{in}} \\ w(x_i) = 0 & \text{for } i \in I_0 \\ w(x_i) = 1 & \text{for } i \in I_{\text{out}}. \end{cases}$$

where we set $w(x_i + hb(x_i, a)) = 1$ if $x_i + hb(x_i, a) \notin Q$. We denote by W^k the set

$$W^k \equiv \{w \in C(Q) : \nabla w(x) = \text{constant}\ , \quad \forall x \in S_j\}.$$

Theorem 3.1. *(HJ_h^k) has a unique solution $w \in W^k$ such that $w : Q \to [0, 1]$.*

Proof. Define

$$(Fu)(x_i) \equiv \min_{a \in A}[\beta u(x_i + hb(x_i, a)] + 1 - \beta\ . \tag{3.5}$$

For any $u \in W^k$, if $x_i + hb(x_i, a) \in Q$ we have

$$u(x_i + hb(x_i, a)) = \sum_{j \in I} \lambda_{ij} u(x_j) \tag{3.6}$$

where the coefficients λ_{ij} are the baricentric coordinates of $x_i + hb(x_i, a)$ with respect to the simplex $S \in \mathcal{S}$ which contains it, that is

$$x_i + hb(x_i, a) = \sum_{j \in I} \lambda_{ij} x_j\ , \quad \lambda_i \geq 0\ , \quad \sum_{j \in I} \lambda_{ij} = 1\ .$$

Since $u(x) \geq v(x)$ in Q implies

$$(Fu)(x_i) \geq (Fv)(x_i) \qquad \forall i \in I\ , \tag{3.7}$$

for any $0 \leq u \leq 1$ we have

$$1 - \beta \leq (Fu)(x_i) \leq 1 \qquad \forall i \in I\ . \tag{3.8}$$

By (3.6) and (3.8) the problem is reduced to a finite dimensional non linear fixed point problem for $F : [0,1]^N \to [0,1]^N$. It is easy to check that the fully discrete operator F is a contraction map for any positive h (the contraction coefficient being β), so that the existence and uniqueness of the fixed point U^* are guaranteed. The solution w is obtained by linear interpolation from $U^* = (w(x_1), \ldots, w(x_N))$. □

Remark 3.1 In practice, we will start the fixed point procedure from

$$U_0 = \begin{cases} 0 & \text{for } i \in I_0, \\ 1 & \text{for } i \in I \setminus I_0. \end{cases} \tag{3.9}$$

By the monotonicity property of F this guarantees that the fixed point sequence converges monotonically decreasing to the solution so that we can apply an acceleration technique, originally developed for other optimal control problems (see [CDF]), to speed up the convergence when $\beta \approx 1$.

Remark 3.2 The set I_{out} will be empty when we apply (HJ_h^k) to the study of front propagation, i.e. to the minimum time problem corresponding to the settings (2.8). As a consequence the algorithm does not impose any "boundary condition" to v on ∂Q. The only active boundary condition will be the one on Ω and the approximation of v will be accurate up to the boundary of the domain of computation (see the convergence results at the end of this section, Remark 3.3 and the numerical experiments in Section 5).

We turn now to prove that the fully discrete algorithm converges of v. For simplicity, we will give the proof for $c(x) = c > 0$ (see Remark 3.3 for the general case). Let us define

$$\mathcal{R}_0^h \equiv \Omega \tag{3.10}$$

$$\mathcal{R}_n^h \equiv \left\{ x \in \mathbb{R}^N \setminus \bigcup_{j=0}^{n-1} \mathcal{R}_j : \exists a \in A \quad \text{such that} \quad x - hca \in \mathcal{R}_{n-1}^h \right\}. \tag{3.11}$$

The above sets are the level sets of v_h, remark that v_h jumps on the boundary of each of them (see Lemma 3.3). We will use the simplified notation $\mathcal{R}_n$ whenever we do not need to take into account the dependence on h of $\mathcal{R}_n^h$. We need the following "consistency" assumption on the triangulation.

Definition 3.2. *We say that a triangulation is "consistent" if, for any $n \in \mathbb{N}$ and any $x_i \in \mathcal{R}_n \cap X$ there exist a control $\overline{a}_i \in A$ and a node $z_i \in \mathcal{R}_{n-1} \cap X$ such that $x_i - hc\overline{a}_i = z_i$.*

The above assumption requires that one can connect nodes in the level set $\mathcal{R}_n$ to nodes in $\mathcal{R}_{n-1}$. Notice that the vectorfield always spans a neighbourhood of the origin of radius ch so that it will be possible to reach $\mathcal{R}_{n-1}$ from any node in $\mathcal{R}_n$. However to be "consistent" a triangulation should have enough points in each level set; this can be obtained eventually refining the triangulation.

Let v, v_h and w denote respectively the solution of (HJ) with the settings (2.8) and the solutions of the corresponding problems (HJ_h) and (HJ_h^k). $V(S)$ will denote the set of the vertices of a simplex S and $d(\cdot,\Omega)$ will be the Euclidean distance from Ω.

Lemma 3.3. *Let (2.8) hold, Q and Ω be compact and $\Omega \subset Q$.*

$$a)\quad \|v\|_{L^\infty(Q)} < 1, \quad \forall Q \subset \mathbb{R}^N \text{ compact};$$

$$b)\quad v_h(x) \le v_h(y)\,, \quad \forall x \in \bigcup_{j=0}^{n} \mathcal{R}_j\,, \quad \forall y \in \mathbb{R}^N \setminus \bigcup_{j=0}^{n} \mathcal{R}_j;$$

$$c)\quad v_h(x) = 1 - e^{-nh} \quad \forall x \in \mathcal{R}_n\,, \quad \forall n \in \mathbb{N}.$$

Moreover, if $k \le ch$ and the triangulation is "consistent":

$$d)\quad 1 - e^{-(n-1)h} \le w(x) \le 1 - e^{-(n+1)h} \qquad \forall x \in \mathcal{R}_n \cap X\,, \quad n \ge 1;$$

$$e)\quad |v_h(x) - w(x)| \le h, \quad \forall x \in \mathcal{R}_n,\ n \in \mathbb{N}.$$

Proof. a) Since $b(x,a) \equiv -ca$ and $a \in B(0,1)$ we have $T(x) = d(x,\Omega)/c$ which implies

$$\|T\|_{L^\infty(Q)} \le \frac{1}{c} \max_{x \in Q} d(x,\Omega) < +\infty$$

for any compact set Q. Then the statement follows by (2.5).
b) By induction. For $n = 0$ the statement is true since

$$0 = v_h(x) \le v_h(y) \qquad \forall x \in \Omega\,, \quad \forall y \in \mathbb{R}^N \setminus \Omega\,.$$

Let $x \in \mathcal{R}_{n+1}$, the minimum in (HJ$_h$) will be attained for $\overline{a} \in A$ such that $x - hc\overline{a} \in \mathcal{R}_n$. If $y \in \mathbb{R}^N \setminus \bigcup_{j=0}^{n+1} \mathcal{R}_j$, then $y - hca \in \mathbb{R}^N \setminus \bigcup_{j=0}^{n} \mathcal{R}_j$, for any $a \in A$, so that

$$v_h(x) = \beta v_h(x - hc\overline{a}) + 1 - \beta \le \beta v_h(y - hca) + 1 - \beta\,, \quad \forall a \in A$$

and we get the result by (HJ$_h$).
c) By definition $\mathcal{R}_n \cap \mathcal{R}_m = \emptyset$ for $n \ne m$ and $\bigcup_{j=0}^{\infty} \mathcal{R}_j = \mathbb{R}^N$. Since v_h is the solution of (HJ$_h$), the statement is true for $n = 0$ and using (b) it can be easily proved by induction for any $n \in \mathbb{N}$.
d) Since the triangulation is "consistent", for any $x_i \in \mathcal{R}_n \cap X$ we can find $n+1$ vertices $x_i^{(q)}, q \in \{0,\cdots,n\}$ such that $x_i^{(0)} = x_i$ and $x_i^{(q)} \in \mathcal{R}_{n-q}$. Then,

$$\begin{aligned} w(x_i) &\le \beta w(x_i^{(1)}) + 1 - \beta \le \beta^2 w(x_i^{(2)}) + 1 - \beta^2 \le \dots \\ &\le \beta^n w(x_i^{(n)}) + 1 - \beta^n = 1 - \beta^n. \end{aligned} \tag{3.12}$$

Moreover, it is immediately seen that $w(x_i) = 1-\beta$ for any $x_i \in \mathcal{R}_1 \cap X$ and $w(x) \ge 0$ for any $x \in \mathcal{R}_1$ (notice that the equality in the last relation is valid for any $x \in S$ such that $V(S) \subset \mathcal{R}_0$). This implies, $w(x_i) \ge 1-\beta$ for any $x_i \in \mathcal{R}_2 \cap X$ and $w(x) \ge 1-\beta$

for any $x \in \mathcal{R}_2$. Continuing by recursion we obtain

$$w(x) \geq 1 - \beta^{n-1} \qquad \text{for any } x \in \mathcal{R}_n, \forall n \in \mathbb{N}. \tag{3.13}$$

Since $k \leq ch$ and the distance between $\mathcal{R}_n$ and $\mathcal{R}_{n\pm 2}$ is greater than ch, if $x \in \mathcal{R}_n$ and S is the simplex containing x, $V(S) \subset \mathcal{R}_{n-1} \cap \mathcal{R}_n \cap \mathcal{R}_{n+1}$. The statement follows from the fact that $w(x)$ is obtained by convex combination of the values at nodes $x_i \in V(S)$.
e) The proof is a direct consequence of (c) and (d). □

By Remark 3.2, the proof of the following theorem is straightforward recalling that w is Lipschitz continuous.

Theorem 3.4. *Let (2.8) hold, Q and Ω be compact and $\Omega \subset Q$. Assume that for any $n \in \mathbb{N}$, $S \in \mathcal{S}$ and $x \in S \cap \mathcal{R}_n$ there exist $y \in S$ and two positive constants C_1 and C_2 such that*

$$|v_h(x) - w(y)| \leq C_1 h + C_2 k. \tag{3.14}$$

Then, for any compact subset Q of $\mathbb{R}^N$, there exist two positive constants C_3, C_4 such that

$$\|v - w\|_{L^\infty(Q)} \leq C_3 h + C_4 k \equiv \rho(h, k). \tag{3.15}$$

Coupling Theorem 3.4 with Lemma 3.3 we get the uniform convergence to v of the fully discrete scheme.

Corollary 3.5. *Let (2.8) hold, Q be a compact set of $\mathbb{R}^N$ containing Ω in its interior, Ω be a compact set with non empty interior and piecewise regular boundary, $0 < k \leq ch$ and the triangulation be "consistent". Then, w converges uniformly to v as h and k tend to 0 and the order of convergence is 1.*

Remark 3.3 The general case of a positive non constant speed $c(x) > 0$ can be solved following the same arguments with the new definition of the level sets

$$\mathcal{R}_0 \equiv \Omega \tag{3.16}$$

$$\mathcal{R}_n \equiv \{x \in \mathbb{R}^N \setminus \bigcup_{j=1}^{n-1} \mathcal{R}_j : x - hc(x)a \in \mathcal{R}_{n-1}\} \tag{3.17}$$

and requiring that the grid is sufficiently fine so that $x \in S \cap \mathcal{R}_n$ implies $V(S) \subset \mathcal{R}_{n-1} \cup \mathcal{R}_n \cup \mathcal{R}_{n+1}$. This assumption holds true for $k \leq \bar{c}h$ where $\bar{c} \equiv \min\{c(x) : x \in Q \setminus \text{int}(\Omega)\}$.

We can also weaken the hypotheses on the triangulation. In fact, one can always reach $\mathcal{R}_{n-1}$ from any node in $\mathcal{R}_n$ so that it suffices to assume that a simplex S for which $S \cap \mathcal{R}_n \neq \emptyset$ has at least one node in $\mathcal{R}_n$. Repeating the argument used in the proof of Lemma 3.3(d) we should add an error Ck at each step (notice that their number is finite). This calculation still gives the estimate (3.14) so that (3.15) holds true.

4. Level sets approximation

By the above results on the convergence of the algorithm and taking into account the properties of our vectorfield we are now able to prove an estimate of the distance between the level sets of the exact and the approximate solution. Notice that the uniform convergence of a sequence of function $\{f_n\}$ to f does not guarantee in general that the level sets of f_n stay close to those of f.

Let us recall, for reader's convenience, that the *Hausdorff distance* between two compact sets X and Y is defined by

$$d_H(X,Y) \equiv \max\left\{\max_{x\in X} d(x,Y), \max_{y\in Y} d(X,y)\right\} \tag{4.1.}$$

We will denote by Ω_t and $\hat{\Omega}_t$ the following level sets of v and w,

$$\Omega_t \equiv \{x \in \mathbb{R}^N : v(x) \leq 1 - e^{-t}\} \text{ and } \hat{\Omega}_t \equiv \{x \in \mathbb{R}^N : w(x) \leq 1 - e^{-t}\} \tag{4.2}$$

and by $\rho(h,k)$ the right hand side of (3.15). In this section we will write w_Δ instead of w whenever we want to make explicit the dependence of w on the time and space discretization and we will use the short notation $\lim_{\Delta\to 0} w_\Delta$ to identify the limit of w for h and k tending to 0.

Theorem 4.1.
a) There exists $\overline{\rho} > 0$ such that for any $z \in \Omega_t$ and for any h and k such that $\rho(h,k) \leq \overline{\rho}$ there is at least one point $\hat{z} \in \hat{\Omega}_t$: $d(z,\hat{z}) \leq \rho(h,k)$.
b) Let the triangulation be "consistent" and $k \leq ch$. Then there exists $\overline{\rho} > 0$ such that for any $\hat{z} \in \hat{\Omega}_t$ and for any h and k such that $\rho(h,k) \leq \overline{\rho}$ there is at least one point $z \in \Omega_t$: $d(z,\hat{z}) \leq \rho(h,k)$.

Proof. a) Assume that for any positive constant $\overline{\rho}$ and h, k such that $\rho(h,k) \leq \overline{\rho}$ there exists a point $z \in \Omega_t$ such that $B(z,\rho) \cap \hat{\Omega}_t = \emptyset$. Let $z \in \text{int}(\Omega_t)$. Then we can construct a sequence $\{z_n\}$ such that $z_n \in B(z, \frac{1}{n})$ and $w_\Delta(z_n) > 1 - e^{-t}$. Since w_Δ converges uniformly to v and v is Lipschitz continuous we have

$$\lim_{\substack{\Delta\to 0\\ n\to\infty}} \|v(z) - w_\Delta(z_n)\| = 0$$

which leads to a contradiction due to the fact that, by construction,

$$\lim_{\substack{\Delta\to 0\\ n\to\infty}} w_\Delta(z_n) \geq 1 - e^{-t} > v(z).$$

Then the statement is true for the interior points of Ω_t. Let $z \in \partial\Omega_t$. We can take a sequence $\{z_n\}$ of points belonging to $\text{int}(\Omega_t)$ and converging to z. By the above argument, we can find a corresponding sequence $\{\hat{z}_n\}$ of points belonging to $\hat{\Omega}_t$ such that $d(z_n, \hat{z}_n)) \leq \rho$. Since $\{\hat{z}_n\}$ is bounded and $\hat{\Omega}_t$ is closed we can pass to the limit on a subsequence obtaining a point $\hat{z} \in \hat{\Omega}_t$ such that $d(z,\hat{z}) \leq \rho$.

b) Let $\hat{z} \in \hat{\Omega}_t$ and n be such that $\hat{z} \in \mathcal{R}_n$. We take a point $x \in B(\hat{z}, 2ch) \cap \mathcal{R}_{n-2}$. For any $x \in \mathcal{R}_{n-2}$ we know that, by definition, $(n-3)ch \leq d(x,\Omega) \leq (n-2)ch$ which implies

$$1 - e^{-(n-3)h} \leq v(x) \leq 1 - e^{-(n-2)h}.$$

By Lemma 3.3(d) we can conclude

$$v(x) \leq 1 - e^{-(n-2)h} \leq w(x) \leq 1 - e^{-t}.$$

Since $d(x, \hat{z}) \leq 2ch$, for h such that $2ch \leq \overline{\rho}$ we can take $z = x$ and the proof is complete. □

By the above theorem, we can immediately obtain the following estimate on the level sets

Corollary 4.2.. *Let the triangulation be "consistent" and $k \leq ch$. Then, for h and k small enough*

$$\|v - w\|_\infty \leq C_1 h + C_2 k \quad \text{implies} \quad d_H(\Omega_t, \hat{\Omega}_t) \leq C_1 h + C_2 k.$$

5. Numerical results

As we said in the introduction one advantage of our approach is that we solve a stationary problem. This means that numerically we work on a fixed triangulation and we do not "follow" the front during its evolution, so that the saving in terms of memory allocations and/or of CPU time can be relevant also with respect to the algorithms dealing with the evolutive problem (5.1). Another interesting feature of this method is that the informations on the fronts are available altogether at the end, i.e. when we have the approximate solution w and we can cut it by the hyperplanes $z = 1 - e^{-t}$. The information on a level set can be obtained, of course, only if that level set is contained entirely in the domain Q where the solution has been computed. However, we point out that the boundary condition we impose on w is such that it does not affect the solution, i.e. the approximation of the fronts is accurate up to the boundary ∂Q as it can be observed looking at the following tests (see also Remark 3.2). This implies that if more informations are needed one should look for the solution on a rectangle Q_1 larger than Q but it is not necessary to recompute the solution on Q. In fact, one can restart the algorithm imposing the values already obtained on the boundary ∂Q and compute the solution only on $Q_1 \setminus Q$. The following results have been obtained working on a simple structured triangulation and for $c(x) = c$. Moreover, to simplify the solution of the minimization problem which appears in (HJ$_h^k$), we made a discretization of the unit ball A with 37 points (one at the origin and one each 10 degrees on the boundary) and we just computed the minimum by comparison. In practice, there is no need to compare all the values at each node if one takes into account the geometry of the problem. The following examples show that the approximation of the front shape is good even for rather big mesh sizes and for a poor discretization of the control set A.

Figure 1 shows w and the corresponding level sets for $\Omega = B(0,\varepsilon)$ ($k = 0.06$ and $h = 0.01$). Figure 2 shows the evolution of a rhomb, i.e. an Ω with non empty interior and piecewise regular boundary. The calculations for the rhomb have been made taking $k = 0.4$ and $h = 0.01$ so that the assumptions guaranteeing the estimate on the level sets are strongly violated: however the qualitative behaviour is correct. In Figure 3 are represented the approximate minimal time paths for the rhomb. The algorithm produces an approximate optimal control at any point of the grid without extra computations. The approximation of the optimal paths requires some extra work. By the equation and the knowledge of the approximate solution w we can compute the approximate optimal control associated to any point of our domain simply looking for the controls which give the minimum in the right-hand side of (HJ_h^k). The path is obtained by a standard one-step method for ordinary differential equations once we know the control a_n^* to be applied at the point y_n of the discrete trajectory.
Figure 4 and Figure 5 show the evolution corresponding to different Ω with an empty interior, they refer respectively to a segment ($k = 0.06$ and $h = 0.01$) and to an L-shaped curve ($k = 0.25$ and $h = 0.01$).

Another interesting feature of the level set approach is the possibility to follow the topological changes of the front. Figure 6 shows the merging of four circular fronts moving with the same speed ($k = 0.06$ and $h = 0.01$).
We remark that the choice of the initial condition is crucial when using a numerical approximation scheme based on the evolutive equation such as in [S2], [S3], since the evolutive problem is

$$(5.1) \qquad \begin{cases} v_t(x,t) + v(x,t) + c(x)\,|\nabla_x v(x,t)| = 0 & \text{in } \mathbb{R}^N \setminus \Omega \\ v(x,0) = \Phi(x) & \text{in } \Omega\ . \end{cases}$$

A typical choice is to take as Φ a modified distance from Ω which is negative in Ω, zero on $\partial\Omega$ and positive outside Ω. In this way, the front at time t will correspond to the "zero level set" of $v(x,t)$ and the burnt region to the set $\{x : v(x,t) \le 0\}$. When Ω has a complicated shape or it is disconnected (as in the last example) this construction can be an hard work. On the contrary, the minimum time algorithm can easily treat these situations since it only needs Ω.

Acknowledgments
I want to thank P. Lanucara who has greatly contributed to the development of the code for the solution of the minimum time problem.

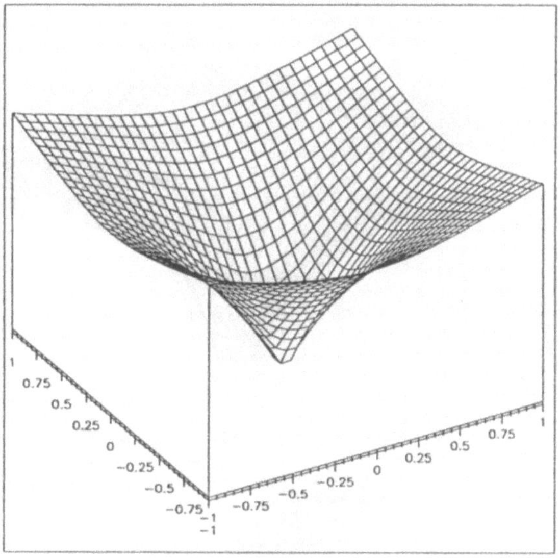

a)

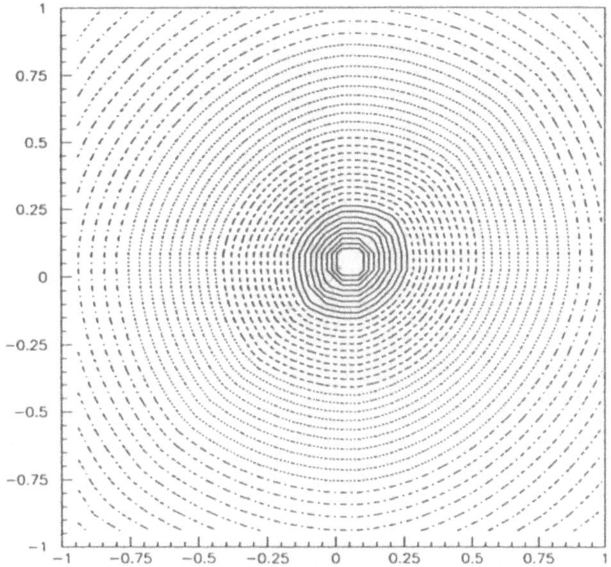

b)

FIGURE 1. The evolution of a ball.
a) the approximate value function
b) its level curves

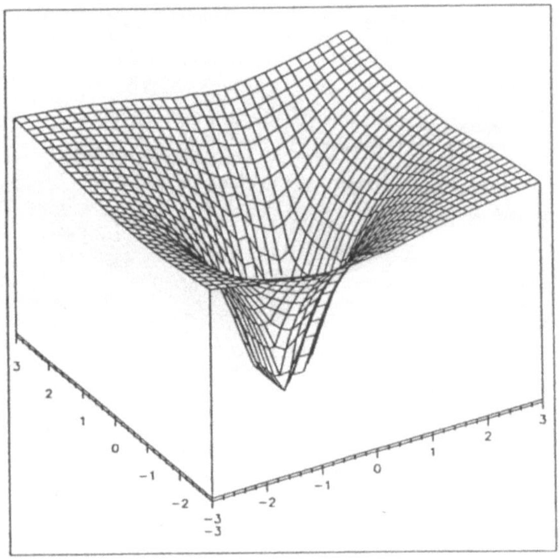

a)

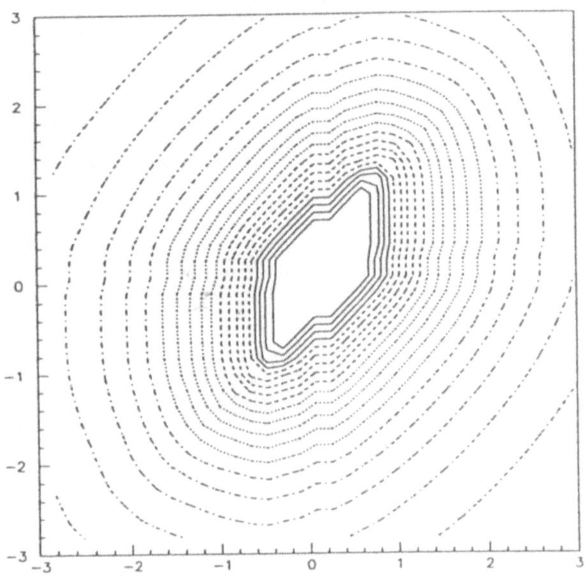

b)

FIGURE 2. The evolution of a rhomb

a) the approximate value function

b) its level curves

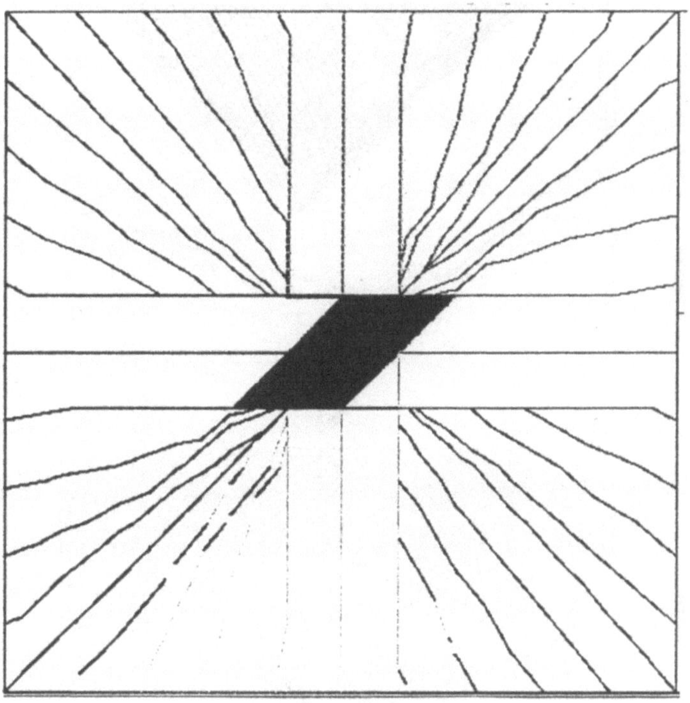

FIGURE 3. Minimal time paths for the rhomb

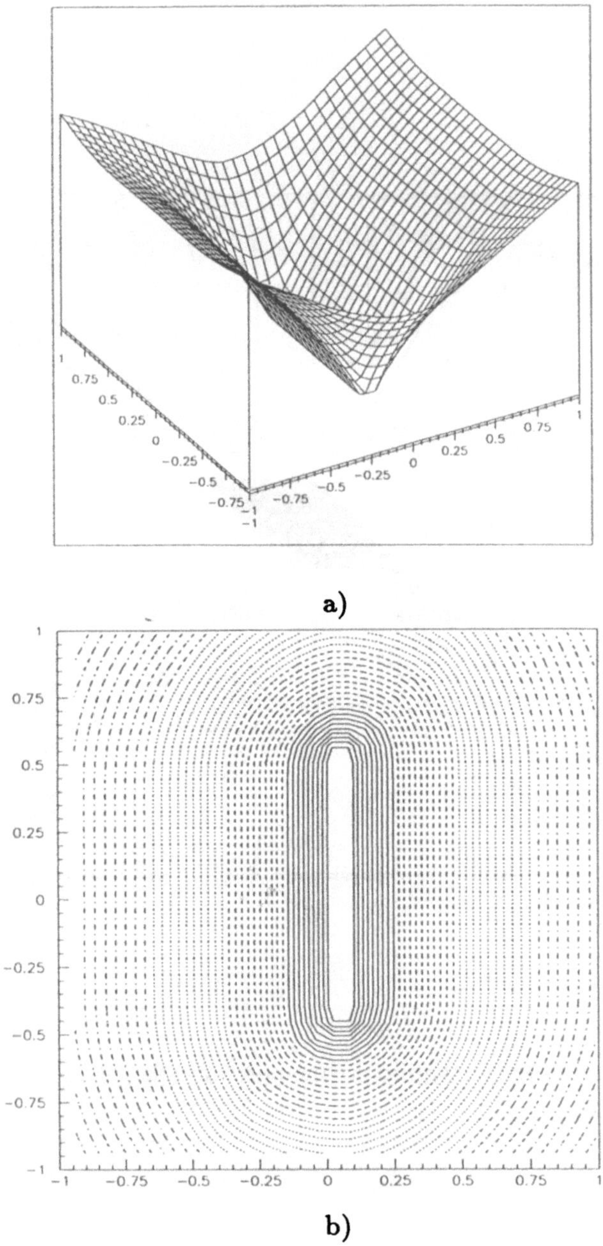

FIGURE 4. Evolution of a vertical segment

a) the approximate value function

b) its level curves

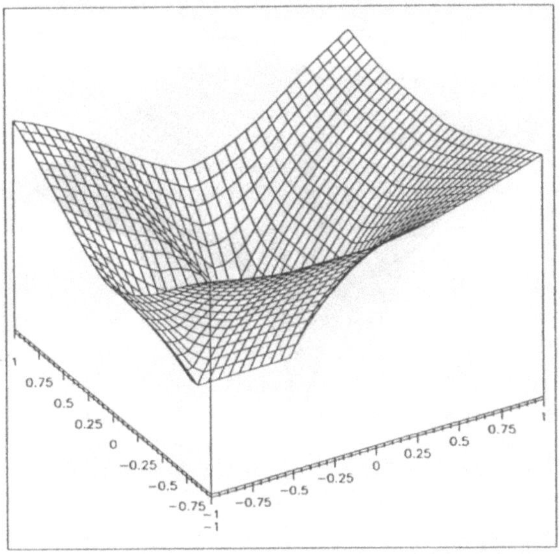

a)

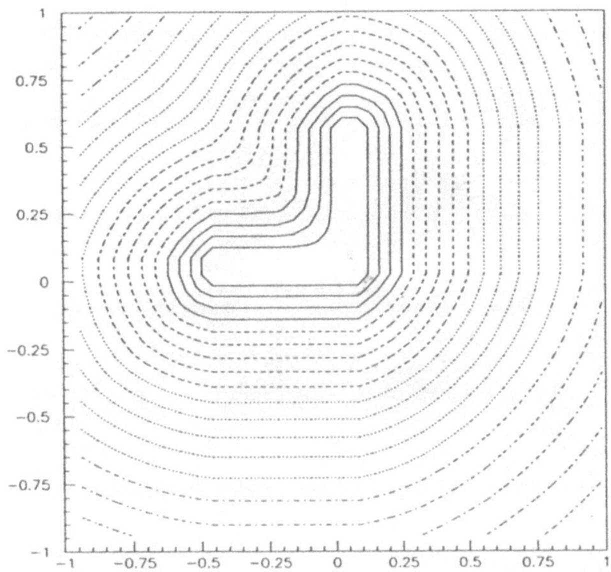

b)

FIGURE 5. Evolution of an L−shaped curve

a) the approximate value function

b) its level curves

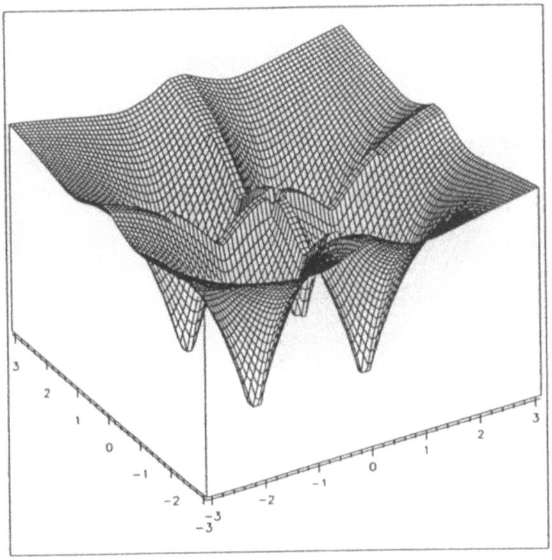

a)

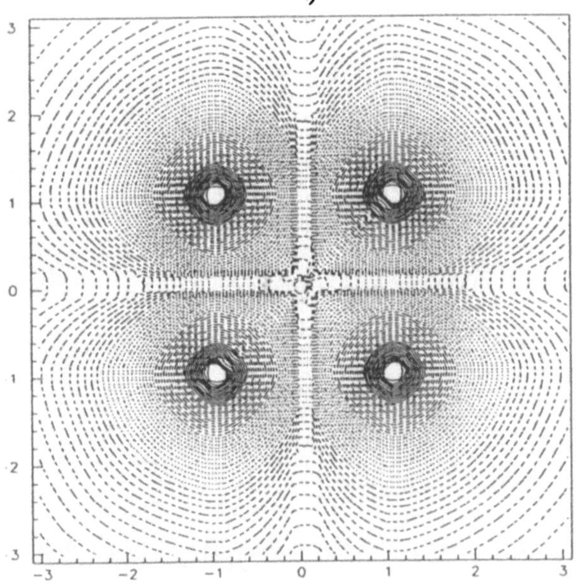

b)

FIGURE 6. Merging of four circular fronts

a) the approximate value function

b) its level curves

References

[B] Bardi M., *A boundary value problem for the minimum time function*, SIAM. J. Control Optim **27**, 1989, 776–785.

[Ba] Barles G., *Remarks on a flame propagation model*, INRIA Report n. 464, 1985.

[BF1] Bardi M., Falcone M., *An approximation scheme for the minimum time function*, SIAM J. Control Optim., **28** (1990), 950-965.

[BF2] Bardi M., Falcone M., *Discrete approximation of the minimal time function for systems with regular optimal trajectories*, in A. Bensoussan, J.L. Lions (Eds.), Analysis and Optimization of Systems, Lecture Notes in Control and Information Sciences, n. 144, Springer–Verlag, 1990, 103–112.

[BF3] Bardi M., Falcone M., *Numerical solution of the minimum time problem*, in preparation.

[BFS] Bardi M., Falcone M., Soravia P., *Fully discrete schemes for the value function of pursuit-evasion games*, to appear on Annals of Dynamic Games

[BSS] Barles G., Soner H.M., Souganidis P., *Fronts propagations and phase field theory*, preprint, 1991.

[BPV1] Bellettini G., Paolini M., Verdi C., *Front-tracking and variational methods to approximate interfaces with prescribed mean curvature*, in P. Neittaanmäki, Proceedings of the Conference on "Numerical methods for free boundary problems", Internat. Series of Numer. Math., n. 99, Birkäuser, Basel, 1991.

[BPV2] Bellettini G., Paolini M., Verdi C., *Numerical minimization of geometrical type problems related to calculus of variations*, Calcolo **27** (1990), 251-278

[CGG] Chen Y.G., Giga Y., Goto S.,*Uniqueness and existence of viscosity solutions of generalized mean curvature flow equations*, J. Differential Geometry, **33**, (1991), 749-786.

[D] Dziuk G., *An algorithm for evolutionary surfaces*, Num. Math. **58** (1991), 603–611.

[ES] Evans L.C., Souganidis P.E., *Differential games and representation formulas for solutions of Hamilton–Jacobi–Isaacs equations*, Ind. Univ. Math. J., **33** (1984), 773–797.

[ESp1] Evans L.C., Spruk J.,*Motion of level sets by mean curvature I*, J. Differential Geometry, **33**, (1991), 635-681.

[ESp2] Evans L.C., Spruk J.,*Motion of level sets by mean curvature II*, Trans. AMS, **330**, (1992), 321-332.

[FGL] Falcone M., Giorgi T. and Loreti P., *Level sets of viscosity solutions and applications*, Quaderno I.A.C., n.22,1990 revised version: May 1992.

[H] Hajek O., *Terminal manifolds and switching locus*, Math. System Theory, **6** (1973), 289–301.

[HL] Hermes H., La Salle J.P., *Functional analysis and time optimal control*, Academic Press, New York, 1969.

[K] Kruzkov S.N., *Generalized solutions of the Hamilton–Jacobi equations of the eikonal type*, Math. USSR-Sb. **27** (1975), 406-445.

[L] Lanucara P., *Approssimazione numerica del problema di tempo minimo*, Tesi di laurea, Università di Roma "La Sapienza", 1990.

[Li] Lions P.L., *Generalized solutions of Hamilton–Jacobi equation*, Pitman, London, 1982.

[NPV] Nochetto R.H., Paolini M., Verdi C., *Optimal interface error estimates for the mean curvature flow*, IAN-CNR, preprint, 1992.

[OS] Osher S., Sethian J.A., *Fronts propagating with curvature dependent speed: Algorithms on Hamilton–Jacobi Formulations*, J. Comp. Phys., **79** (1988), 12–49.

[S1] Sethian J.A., *An analysis of flame propagation*, Ph.D. Thesis, Lawrence Berkeley Laboratory, University of California.

[S2] Sethian J.A., *Curvature and the evolution of fronts*, Commun. Math. Phys., **101** (1985), 487-499.

[S3] Sethian J.A., *Numerical algorithms for propagating interfaces: Hamilton-Jacobi equations and conservation laws*, J. Differential Geometry, **31** (1990), 131-161.

[So] Soravia P., *Generalized motion of a front along its normal direction: a differential games approach*, to appear on Nonlinear Analysis T.M.A.

[SS] Sethian J.A., Strain J., *Crystal growth and dendritic solidification*, preprint, 1990.

[TCH] J.E. Taylor, J.W. Cahn and C.A. Handwerker, *Geometric models of crystal growth*, Acta Metallurgica Mater. **40** (1992), 1443-1474.

[T] J.E. Taylor, *Mean curvature and weighted mean curvature*, Acta Metallurgica Mater. **40** (1992), 1475-1485.

Author address
Dipartimento di Matematica, Università di Roma "La Sapienza", P. Aldo Moro 2, 00185-Roma. e-mail: falcone@sci.uniroma1.it

Planar motion of an anisotropic interface

Morton E. Gurtin

Abstract. The simplest theory for the motion of a phase interface in the plane, motion by curvature, is based on the relation $V = K$ with V the normal velocity and K the curvature. This paper discusses generalizations of that theory more appropriate to crystalline materials; such generalizations lead to the presence of corners, facets, and wrinklings.

1991 Mathematics Subject Classification: 80A22.

1. Background. The anisotropic evolution equation

There are situations of interest for which the motion of a phase interface is essentially independent of the behavior of the corresponding bulk phases. One of the first to model such motions was Mullins [35], who introduced the *curve-shortening equation*

$$V = K \tag{1}$$

to study the *planar* motion of grain boundaries. Here V is the normal velocity, with positive normal-direction outward from the boundary $\partial\Omega$ of the grain $\Omega = \Omega(t)$, and K is the curvature, with $K < 0$ when $\partial\Omega$ is a circle. The curve-shortening equation is a parabolic PDE with a large literature (cf. Brakke [8], Sethian [42], Abresch and Langer [1], Gage and Hamilton [23], Grayson [27], Osher and Sethian [38], Evans and Spruck [19, 20, 21], Chen, Giga, and Goto [13], Giga and Sato [24], Almgren, Taylor, and Wang [3], Taylor, Cahn, and Handwerker [44]). As shown by Gage and Hamilton [23] and Grayson [27], a boundary curve with evolution governed by (1), and of arbitrary initial shape, shrinks to a point in finite time, with asymptotic shape a circle.

The curve-shortening equation also arises as an approximation to the Landau-Ginzburg equation, a result established formally by Allen and Cahn [2] and Rubinstein, Sternberg, and Keller [41], and then rigorously by Barles, Soner, and Souganidis [9], Evans, Soner, and Souganidis [18], Chen [12], and De Mottoni and Schatzman [16].

Mullins's theory was generalized by Gurtin [28] and Angenent and Gurtin [5] to include anisotropy and the possibility of a difference in bulk energies between phases. The resulting equation is

$$b(\theta)V = g(\theta)K - U, \tag{2}$$

where θ, the normal-angle, is the counterclockwise angle from a fixed axis to the outward normal of $\partial\Omega$. Here

$$g(\theta) = f(\theta) + f''(\theta) \tag{3}$$

with $f(\theta) > 0$ the interfacial energy; U is the relative energy of the material in Ω; and $b(\theta) > 0$, the kinetic modulus, is a material function. The presence of the angle θ reflects anisotropy, and the particular form in which f appears in (3) is a consequence of thermodynamics. In fact, a consequence of (2) and (3) is the thermodynamic inequality

$$(d/dt)\Big\{\int_{\partial\Omega(t)} f(\theta)ds + U\text{area}(\Omega(t))\Big\} = -\int_{\partial\Omega(t)} b(\theta)V^2 ds, \tag{4}$$

showing that $b(\theta)V^2$ represents the rate of energy dissipation, per unit length.

In writing (2) we have neglected the dependence of $b(\theta)$ on V, a dependence not ruled out by thermodynamics and considered important by material scientists (Owen, Schoen, and Srinivasan [40], Hillert [32]). The behavior of (2) with $b(\theta)$ replaced by $b(\theta, V) > 0$ remains an open problem.

2. Backward parabolicity. Corners, facets, and wrinklings

When

$$g(\theta) > 0, \tag{5}$$

(2) is parabolic, and the underlying behavior, which is not much different than that for $V = K$, is well understood (Angenent [4], Chen, Giga, and Goto [13], Soner [43], Barles, Soner, and Souganidis [9]). What makes (2) nonstandard is the possibility of interfacial energies that satisfy

$$g(\theta) < 0 \tag{6}$$

for certain ranges of the angle θ, for in these ranges the evolution equations are *backward parabolic*. Such energies are not mathematical curiosities: material scientists give strong arguments in support of interfacial energies that satisfy $g(\theta) < 0$ for some values of θ (Gjostein [26], Cahn and Hoffman [14]).

Let

$$\mathbf{N}(\theta) = (cos\theta, sin\theta), \quad \mathbf{T}(\theta) = (sin\theta, -cos\theta), \tag{7}$$

so that $\mathbf{T}$ and $\mathbf{N}$ represent a unit tangent and normal to the interface when θ is its normal-angle. In analyzing energies with backward-parabolic angle-intervals an important concept is the **Frank diagram** $\mathcal{F}$, which is the polar diagram of $f(\theta)^{-1}$ (the curve defined in polar coordinates (r,θ) by $r = f(\theta)^{-1}$); $\mathcal{F}$ is locally strictly convex where $g(\theta) > 0$ and locally strictly concave where $g(\theta) < 0$. The importance of the Frank diagram becomes evident when one considers the homogeneous extension $\tilde{f}$ of f to $\mathbf{R}^2\backslash\{0\} : \tilde{f}(\alpha\mathbf{N}(\theta)) = \alpha f(\theta)$ for all angles θ and all $\alpha > 0$. Then $\mathcal{F}$ is the one-level set of $\tilde{f}$, so that convexity properties of $\tilde{f}$ are related to those of $\mathcal{F}$. In particular, $\tilde{f}$ is convex if and only if the region enclosed by $\mathcal{F}$ is convex. Further,

$$g(\theta) = \mathbf{T}(\theta) \cdot [\nabla\nabla\tilde{f}(\mathbf{N}(\theta))]\mathbf{T}(\theta), \tag{8}$$

which, to some extent, explains the form of (3).

A method of overcoming (6), introduced by Angenent and Gurtin [5], is to allow the interface to contain corners (jumps in angle) that exclude the backward-parabolic ranges of θ. In the presence of a corner the evolution equation (2) does not by itself characterize the motion of the interface; there is an additional condition requiring that the *capillary force*

$$\mathbf{C}\theta) = f(\theta)\mathbf{T}(\theta) + f'(\theta)\mathbf{N}(\theta) \quad \text{is} \quad \text{continuous.} \tag{9}$$

Thus for a corner corresponding to an angle jump from θ_1 to θ_2 we must have

$$\mathbf{C}(\theta_1) = \mathbf{C}(\theta_2), \tag{10}$$

which has an important consequence: the tangent line to $\mathcal{F}$ at θ_1 is also a tangent line to $\mathcal{F}$ at θ_2; that is, θ_1 and θ_2 must be angles of bitangency for the Frank diagram [5].

Let $C(\mathcal{F})$ denote the convexification of $\mathcal{F}$ (the boundary of the convex hull of $\mathcal{F}$), and let $F(\theta)$, the *convexified energy*, denote the energy whose Frank diagram is $C(\mathcal{F})$, so that $f(\theta) \geq F(\theta)$. Then angles θ with $f(\theta) = F(\theta)$ satisfy $g(\theta) \geq 0$; we refer to such angles as *globally stable*; we refer to angles θ with $f(\theta) > F(\theta)$ as *globally unstable*; and we refer to each maximal interval (θ_1, θ_2) of angles θ with $f(\theta) > F(\theta)$ as a *globally unstable angle-interval.* Then each globally unstable angle-interval (θ_1, θ_2) has θ_1 and θ_2 as angles of bitangency for $\mathcal{F}$ and hence as admissible angles for a corner.

Wrinklings consisting of facets with normal angle θ_1 alternating with facets of normal angle θ_2 are solutions of (2) and (9) provided the θ_1 and θ_2 facets evolve according to $V = -U/b(\theta_1)$ and $V = -U/b(\theta_2)$, respectively. Such wrinklings evolve as a rigid body with velocity ω defined by

$$\omega \cdot \mathbf{N}(\theta_1) = -b(\theta_1)^{-1}U, \quad \omega \cdot \mathbf{N}(\theta_2) = -b(\theta_2)^{-1}U \tag{11}$$

[5]. One also has the possibility of solutions involving curved sections with globally stable normal angles separated by wrinklings.

Local well-posedness of evolutions consistent with (2) and (9) - ensuing from initial curves consisting of globally stable sections separated by appropriate corners - is established by Angenent and Gurtin [6].

A limitation of the procedure described above is that it is inapplicable for an initial curve $\partial\Omega(0)$ with globally unstable normal angles. The next section confronts this difficulty.

3. Relaxed equation for initial data with globally unstable normal angles

Consider an initial curve $\partial\Omega(0)$ that has normal angles in a globally unstable interval (θ_1, θ_2). We formally consider Γ as being infinitesimally wrinkled, with each infinitesimal facet having either θ_1 or θ_2 as normal angle, an idea due to Cahn and Taylor (private communication with the author in 1990). The expansion

$$\mathbf{N}(\theta) = \mu_1(\theta)\mathbf{N}(\theta_1) + \mu_2(\theta)\mathbf{N}(\theta_2), \quad \theta \in (\theta_1, \theta_2), \tag{12}$$

then defines, for each i, the density $\mu_i(\theta)$ of θ_i-facets at any point of Γ with normal angle θ, with $\mu_i(\theta)$ measured per unit length of Γ.

The use of infinitesimal wrinklings is formally equivalent to replacing the interfacial energy $f(\theta)$ by the convexified energy $F(\theta)$, since

$$F(\theta) = \mu_1(\theta)f(\theta_1) + \mu_2(\theta)f(\theta_2), \quad \theta \in (\theta_1, \theta_2). \tag{13}$$

A further reason for the use of such wrinklings is that the initially wrinkled curve is more stable than the original curve:

$$\int_{\partial\Omega(0)} F(\theta)ds \leq \int_{\partial\Omega(0)} f(\theta)ds. \tag{14}$$

If we allow $\partial\Omega(t)$ to infinitesimally wrinkle in the same manner, we are led to the conclusion that the effective interfacial energy for the evolution should be $F(\theta)$, so that the *effective energy modulus* is given by

$$G(\theta) = F(\theta) + F''(\theta). \tag{15}$$

The next question we must answer is: What is an appropriate kinetic modulus for the infinitesimally wrinkled curve? As noted in the paragraph containing (11), if $\Gamma(t)$ is a finite wrinkling whose facets have θ_1 and θ_2 as normal angles, then $\Gamma(t)$ evolves as a rigid body with constant velocity ω (although $\Gamma(t)$ may shrink or grow tangentially). Since ω depends on the particular wrinkling only through θ_1 and θ_2, it seems reasonable to suppose that *infinitesimal wrinklings* with θ_1 and θ_2 as normal angles also evolve with rigid velocity ω, a supposition equivalent to replacing the kinetic modulus $b(\theta)$ between θ_1 and θ_2 by an effective modulus $B(\theta)$ that agrees with $b(\theta)$ at θ_1 and θ_2 and has polar diagram between θ_1 and θ_2 a straight line:

$$B(\theta)^{-1} = \mu_1(\theta)b(\theta_1)^{-1} + \mu_2(\theta)b(\theta_2)^{-1} \tag{16}$$

We are therefore led to an *effective kinetic modulus* $B(\theta)$ for all $\theta : B(\theta) > 0$ is continuous; $B(\theta) = b(\theta)$ for all globally stable angles θ; the polar diagram of $B(\theta)$ is a straight line over angle intervals with $f(\theta) > F(\theta)$.

This procedure, introduced by Gurtin [29], yields a *relaxed evolution equation*

$$B(\theta)V = G(\theta)K - U \tag{17}$$

with B and G the effective moduli corresponding to f and g. It is important to note that:

(a) The relaxed equation coincides with the original equation (2) at globally stable angles θ.

(b) Because of the construction of $G(\theta)$, no matter how smooth $f(\theta)$ is, $G(\theta)$ *will generally be discontinuous* whenever the angle θ changes from globally stable to globally unstable; this property of $G(\theta)$ renders the relaxed evolution equation nonstandard.

(c) $G(\theta) = 0$ whenever θ is globally unstable, so that (17) degenerates to hyperbolic at globally unstable angles.

The relaxed evolution equation (17) is studied by Gurtin, Soner, and Souganidis [30]; their main results are a theorem of existence and local uniqueness and a global comparison theorem for level-set solutions, a theorem established independently and Ohnuma and Sato [37].

Because of the lack of continuity of G as well as the degeneracy of (17) when $G = 0$, [30] discuss this equation within the weak framework of viscosity solutions. This approach to geometric equations, initiated by Evans and Spruck [19] and Chen, Giga, and Goto [13], and given an intrinsic form by Soner [43] and Barles, Soner, and Souganidis [9], is based on the use of level sets to characterize evolving curves, an idea due to Ohta, Jasnow, and Kawasaki [39], Sethian [42], Osher and Sethian [38], and Barles [7]. The difficulties concerning (17) result from the *discontinuous* nature of G; the degeneracy of the equation at angles θ with $G(\theta) = 0$ causes no great difficulty; were G continuous, most of the results would follow from those in [13].

[30] establishes the following results of physical interest:

(1)Viscosity solutions of (17) not only satisfy (2) away from corners, but, what is most interesting, such solutions automatically satisfy the force balance (11) across corners.

(2) If (θ_1, θ_2) is a globally unstable angle-interval,then, as shown by [5], a wedge whose sides have normal angles θ_1 and θ_2 and evolve according to $b(\theta_1)V = -U$ and $b(\theta_2)V = -U$, respectively, is a solution of the basic equations (2) and (9). [30] shows that the foregoing choice of the effective moduli G and B is the only possible choice if all such wedges are to be viscosity solutions of (17).

(3) For $U < 0$ and for the initially enclosed region Ω_0 large enough, $t^{-1}\Omega(t)$ converges to a dilation of the Wulff region for $1/B(\theta)$, a result conjectured in [5] and proved by Soner [43] for $G > 0$ and B with a convex polar diagram, and extended in [6] to general $B > 0$.

4. Fourth-order theories. Regularized evolution equation

Another method of analyzing behavior within the unstable range of angles is to regularize the evolution equation (2). Such a regularization, proposed in [5], was developed by Di Carlo, Podio-Guidugli, and Gurtin [17] and yields the equation

$$b(\theta)V = g(\theta)K - U - \epsilon(K_{ss} + \frac{1}{2}K^3), \tag{18}$$

where s denotes arc length and $\epsilon > 0$ is a small material parameter. The chief ingredient in the derivation of (18) is a *constitutive* dependence of interfacial energy on curvature, an idea that traces back to Gibbs [25] and Herring [31].

The equation (18) is very much like the Cahn-Hilliard equation (cf., [11]), with the angle-intervals on which $g(\theta) < 0$ playing the role of spinodals. When linearized about a flat interface at angle $\theta = \pi/2$, (18) has the form

$$u_t = \alpha u_{xx} - \delta u_{xxxx} \tag{19}$$

with $\delta > 0$, but with $\alpha < 0$ when $\theta = \pi/2$ lies in a spinodal. The linear equation (19) is exactly the equation used by Cahn [10] in his treatment of spinodal decomposition.

A somewhat similar model (Davi and Gurtin [15]), based on ideas of Mullins [36], allows for heat conduction within the interface with heat flow driven by changes in curvature; granted isotropy, it leads to the fourth-order parabolic equation

$$\beta V = -\alpha K_{ss}, \tag{20}$$

with $\alpha, \beta > 0$ material constants. This equation has been used by Mullins [36] with great success to model the formation of grooves at intersecting grains on exposed surfaces.

There are almost no analytical or numerical results for the evolution equations (18) and (20). In particular, it would be of great interest to use (18) to study the behavior of the interface within an angle interval that defines a corner of the unregularized equation (2).

ACKNOWLEDGMENT. This work was supported by the Army Research Office and the National Science Foundation.

References

[1] Abresch, U., J. Langer, The normalized curve shortening flow and homothetic solutions. J. Diff. Geom. 23 (1986), 175-196.

[2] Allen, S. M., J.W. Cahn, A macroscopic theory for antiphase boundary motion and its application to antiphase domain coarsening. Act. Metall. 27 (1979), 1085-1098.

[3] Almgren, F., J. E. Taylor, L. Wang, Curvature driven flows: a variational approach, SIAM J. Control Opt.. (issue dedicated to W. Fleming), Forthcoming.

[4] Angenent, S., Parabolic equations for curves on surfaces. I. Curves with p-integrable curvature, Ann. Math. 132 (1990), 451-483; II. Intersections, blowup, and generalized solutions, Ann. Math. 133 (1991), 171-217.

[5] Angenent, S., M. E. Gurtin, Multiphase thermomechanics with interfacial structure. 2. Evolution of an isothermal interface, Arch. Rational. Mech. Anal. 108 (1989), 323-391.

[6] Angenent, S., M. E. Gurtin, Anisotropic motion of a phase interface. Well-posedness of the initial-value problem and qualitative properties of the interface, J. reine angew. Math. Forthcoming.

[7] Barles, G., Remark on a flame propagation model, Rapport INRIA ♯ 464 (1985).

[8] Brakke, K. A., The Motion of a Surface by its Mean Curvature, Princeton University Press 1978.

[9] Barles, G., H. M. Soner, P. E. Souganidis, Front propagation and phase field theory, SIAM J. Cont. Optim. (issue dedicated to W. Fleming), Forthcoming.

[10] Cahn, J. W., Spinodal decomposition, Trans. Met. Soc. AIME 242 (1968), 166-180.

[11] Cahn, J. W., J. E. Hilliard, Spinodal decomposition: a reprise, Act. Metall. 19 (1971), 151-161.

[12] Chen, X. Generation and propagation of the interface for reaction diffusion equations, J. Diff. Eqts. 96 (1992), 116-141.

[13] Chen, Y.-G., Y. Giga, S. Goto, Uniqueness and existence of viscosity solutions of generalized mean curvature flow equations, J. Diff. Geom. 33 (1991), 749-786.

[14] Cahn, J. W., D. W. Hoffman, A vector thermodynamics for anisotropic surfaces 2: curved and faceted surfaces, Act. Metall. 22 (1974), 1205-1214.

[15] Davi, F., M. E. Gurtin, On the motion of a phase interface by surface diffusion, ZAMP 41 (1990), 782-811.

[16] DeMottoni, P., M. Schatzman, Development of surfaces in RN, Proc. Roy. Soc. Edinburgh, 116A (1990), 207-220.

[17] DiCarlo, A., Gurtin M. E, P. Podio-Guidugli, A regularized equation for anisotropic motion-by-curvature, SIAM J. Appl. Math. 52 (1992), 1111-1119.

[18] Evans, L. C., Soner, H. M., P. E. Souganidis, Phase transitions and generalized motion by mean curvature, Comm Pure Appl. Math. 45 (1992), 1097-1123.

[19] Evans, L. C., J. Spruck, Motion of level sets by mean curvature I, J. Diff. Geom. 33 (1991), 635-681.

[20] Evans, L. C., J. Spruck, Motion of level sets by mean curvature II, Trans. AMS, Forthcoming.

[21] Evans, L. C., J. Spruck, Motion of level sets by mean curvature III, J. Geom. Anal. Forthcoming.

[22] Fleming, W., H. M. Soner, Controlled Markov Processes and Viscosity Solutions, Springer-Verlag, Berlin 1993.

[23] Gage, M., R. S. Hamilton, The heat equation shrinking convex plane curves, J. Diff. Geom. 23 (1986), 69-95.

[24] Giga, Y., M. H. Sato, Generalized interface condition with the Neumann boundary condition, Proc. Japan Acad., Ser. A Math. 67 (1991), 263-266.

[25] Gibbs, J. W., On the equilibrium of heterogeneous substances, Trans. Conn. Acad. 3 (1876), 108-248, [(1878), 343-524]. Reprinted in: The Scientific Papers of J. Willard Gibbs, vol 1, Dover, New York 1961.

[26] Gjostein, N. A., Adsorption and surface energy 2: thermal faceting from minimization of surface energy, Act. Metall. 11 (1963), 969-977.

[27] Grayson, M. A., The heat equation shrinks embedded plane curves to round points, J. Diff. Geom. 26 (1987), 285-314.

[28] Gurtin, M. E., Multiphase thermomechanics with interfacial structure. 1. Heat conduction and the capillary balance law, Arch. Rational Mech. Anal. 104 (1988), 185-221.

[29] Gurtin, M. E., Thermomechanics of Evolving Phase Boundaries in the Plane, Oxford Press, Forthcoming.

[30] Gurtin, M. E., H. M. Soner, P. E. Souganidis, Anisotropic motion of an interface relaxed by the formation of infinitesimal wrinkles, J. Diff. Eqts. Submitted.

[31] Herring, C., Surface tension as a motivation for sintering, The Physics of Powder Metallurgy (ed. W. E. Kingston) McGraw-Hill, New York.

[32] Hillert, M., The role of interfaces in phase transformations. The Mechanism of Phase Transformations in Crystalline Solids, Institute for Metals, London 1969.

[33] Jensen, R. The maximum principle for viscosity solutions of second-order fully nonlinear partial differential equations, Arch. Rational Mech. Anal. 101 (1988), 1-27.

[34] Lasry, J. M., P.-L. Lions, A remark on regularization in Hilbert spaces, Israel J. Math. 55 (1986), 257-266.

[35] Mullins, W. W., Two-dimensional motion of idealized grain boundaries, J. Appl. Phys. 27 (1956), 900-904.

[36] Mullins, W. W., Theory of thermal grooving, J. Appl. Phys. 28 (1957), 333-339.

[37] Ohnuma, M., M. Sato, Singular degenerate parabolic equations with applications to geometric equations, Forthcoming.

[38] Osher, S., J. A. Sethian, Fronts propagating with curvature dependent speed, J. Comp. Phys. 79 (1988), 12-49.

[39] Ohta, T., D. Jasnow, K. Kawasaki, Universal scaling in the motion of random interfaces, Phys. Rev. Let. 49 (1982), 1223-1226.

[40] Owen, W. S., F. J. Schoen, G. R. Srinivasan, The growth of a plate of martensite, Phase Transformations, American Society for Metals, Metals Park, Ohio 1968.

[41] Rubinstein, J., P. Sternberg, J. B. Keller, Fast reaction, slow diffusion, and curve shortening, SIAM J. Appl. Math. 49, 116-133.

[42] Sethian, J. A., Curvature and evolution of fronts, Comm. Math. Phys. 101 (1985), 487-499.

[43] Soner, H. M., Motion of a set by the mean curvature of its boundary, J. Diff. Eqts., Forthcoming.

[44] Taylor, J., J. W. Cahn, C. A. Handwerker, Geometric models of crystal growth 1, Acta Met. 40 (1992), 1443-1474.

Singular limit for an Allen-Cahn equation with a nonlocal term

Danielle Hilhorst [*], *Elisabeth Logak* [†], *Yasumasa Nishiura* [‡]

1 Introduction

In this note, we study the limiting behavior, as $\epsilon \to 0$, of the solution u^ϵ of the problem (P_1^ϵ),

$$(P_1^\epsilon)\begin{cases} u_t = \epsilon\Delta u + \dfrac{1}{\epsilon}(1-u^2)(2u - \gamma\dfrac{1}{|\Omega|}\displaystyle\int_\Omega u) & \text{in } \Omega \times \mathbb{R}_+ \\ u_r = \dfrac{1}{\epsilon}(1-u^2) & \text{on } \partial\Omega \times \mathbb{R}_+ \\ u(x,0) = \tanh(\dfrac{|x|-r_0}{\epsilon}) & x \in \Omega. \end{cases}$$

We will also consider a related problem (P_2^ϵ) with a slightly different scaling, namely,

$$(P_2^\epsilon)\begin{cases} u_t = \Delta u + \dfrac{1}{\epsilon^2}(1-u^2)(2u - \epsilon\gamma\dfrac{1}{|\Omega|}\displaystyle\int_\Omega u) & \text{in } \Omega \times \mathbb{R}_+ \\ u_r = \dfrac{1}{\epsilon}(1-u^2) & \text{on } \partial\Omega \times \mathbb{R}_+ \\ u(x,0) = \tanh(\dfrac{|x|-r_0}{\epsilon}) & x \in \Omega \end{cases}$$

where $\Omega = \{x \in \mathbb{R}^n, r_1 < |x| < r_2\}$ is an annulus ($0 < r_1 < r_0 < r_2 < +\infty$) and γ is a positive constant ; we assume that $\gamma \in (0,2)$ in the case of Problem (P_1^ϵ).

As shown in section 6, the problems (P_1^ϵ) and (P_2^ϵ) can be obtained as the limits of systems of reaction-diffusion equations modelling for instance pattern formation in population dynamics or in chemical reactions or the micro-phase separation of block copolymer. The main feature is the nonlocal term represented by the integral of u, which is responsible for forming nontrivial stationary patterns. Without the nonlocal term, there are many results (e.g. by Bronsard and Kohn [2], X.Chen [4], de Mottoni and Schatzman [9], Evans, Soner and Souganidis [6]) proving the relation between

[*]Laboratoire d'Analyse Numérique, CNRS et Université Paris-Sud, 91405 Orsay Cédex, FRANCE
[†]Département de Mathématiques et d'Informatique, Ecole Normale Supérieure, 45,rue d'Ulm 75230 Paris Cédex 05, FRANCE
[‡]Faculty of Integrated Arts and Sciences, Hiroshima University, Kagami-yama 1-7-1, Higashi-Hiroshima City, Hiroshima 724, JAPAN

asymptotics of reaction-diffusion equations of bistable type and motion by mean curvature. Mean curvature flow apparently does not have stationary patterns.

We now state the main results of this note.

Theorem 1 *(i) Let u_i^ϵ be the solution of problem $(P_i^\epsilon), i = 1,2$. Then, there exist a time T_i^*, a smooth function $S_i(t), t \in [0, T_i^*)$ and a function $u_i(x,t)$ in $L^\infty(\Omega \times (0, T_i^*))$ such that*

$$\lim_{\epsilon \to 0} u_i^\epsilon(x,t) = u_i(x,t) = \begin{cases} -1 & \text{if } r_1 \leq |x| < S_i(t) \\ +1 & \text{if } S_i(t) < |x| \leq r_2 \end{cases}$$

for $0 < t < T_i^*$. T_i^* is either $+\infty$ or the first finite time such that $S_i(T_i^*) = r_1$ or $S_i(T_i^*) = r_2$.

(ii) The function S_1 is the solution of the initial-value problem

$$(M_1) \begin{cases} \dot{S}(t) & = \gamma \dfrac{1}{|\Omega|} \displaystyle\int_\Omega u & = \gamma \dfrac{(r_2^n + r_1^n - 2S(t)^n)}{(r_2^n - r_1^n)}, \qquad t \in [0, T_1^*) \\ S(0) & = r_0, \end{cases}$$

with $T_1^* = +\infty$ and the function S_2 satisfies

$$(M_2) \begin{cases} \dot{S}(t) & = -\dfrac{n-1}{S(t)} + \gamma \dfrac{1}{|\Omega|} \displaystyle\int_\Omega u & = -\dfrac{n-1}{S(t)} + \gamma \dfrac{(r_2^n + r_1^n - 2S(t)^n)}{(r_2^n - r_1^n)}, \\ S(0) & = r_0. \end{cases}$$

$t \in [0, T_2^*)$

In section 2, we study the qualitative behavior of the functions $S_i, i = 1,2$.

For the sake of simplicity, we prove Theorem 1 in the case of Problem (P_1^ϵ). We refer to [8] for more complete proofs. It follows from the maximum principle that $-1 < u_1^\epsilon < 1$. This allows us to perform the change of function

$$u_1^\epsilon(r,t) = \tanh(\frac{Z_1^\epsilon(r,t)}{\epsilon}).$$

In section 3, we give a priori estimates for Z_1^ϵ and prove Theorem 1 (i).

In section 4, we recall some properties of the travelling wave solutions of a related problem.

The proof of Theorem 1 (ii) is given in section 5. We start with an a priori estimate for $(\int_\Omega u^\epsilon)/|\Omega|$ and then construct sub- and supersolutions based on the travelling wave solutions of section 4.

The methods of proof are inspired upon those of Caginalp and X. Chen [3] in their study of the phase-field model. We also refer to X.Chen [5] and to Barles, Soner and Souganidis [1] for the study of related problems.

2 Qualitative behavior of the solutions of the limit problems

2.1 Motion driven by a nonlocal term

For the purpose of simplicity, we omit the index 1 from the notation. Problem (M_1) can be rewritten as

$$(M_1) \begin{cases} \dot{S}(t) &= \phi(S(t)), \qquad t > 0 \\ S(0) &= r_0, \end{cases}$$

where

$$\phi(r) = \gamma \frac{(r_2^n + r_1^n - 2r^n)}{(r_2^n - r_1^n)}.$$

Then the function ϕ is strictly decreasing for $r > 0$ and System (M_1) has a unique stationary state $\tilde{S} = (\frac{r_2^n + r_1^n}{2})^{\frac{1}{n}}$. Thus, if $r_1 < r_0 < \tilde{S}$, S(t) is strictly increasing and converges to $\tilde{S}$ as $t \to +\infty$ and if $\tilde{S} < r_0 < r_2$, S(t) is strictly decreasing and converges as well to $\tilde{S}$.

2.2 Modified motion by mean curvature

Here again we omit the index from the notation. Problem (M_2) can be rewritten as

$$(M_2) \begin{cases} \dot{S}(t) &= \phi(S(t)), \qquad t > 0 \\ S(0) &= r_0, \end{cases}$$

where

$$\phi(r) = -\frac{n-1}{r} + \gamma \frac{(r_2^n + r_1^n - 2r^n)}{(r_2^n - r_1^n)}.$$

Then ϕ' is given by

$$\phi'(r) = \frac{n-1}{r^2} - 2\gamma n \frac{r^{n-1}}{(r_2^n - r_1^n)},$$

and we have that $\phi''(r) < 0$, for all $r > 0$. It turns out that the function ϕ is strictly increasing for $r \in (0, S_0)$ and strictly decreasing for $r \in (S_0, +\infty)$ where

$$S_0 = S_0(\gamma) = [\frac{(n-1)(r_2^n - r_1^n)}{2\gamma n}]^{\frac{1}{n+1}}.$$

The computation of $\phi(S_0(\gamma))$ shows that

$$\phi(S_0(\gamma)) > 0 \text{ if and only if } \gamma > \tilde{\gamma} = \frac{(n^2-1)(r_2^n - r_1^n)}{n(r_2^n + r_1^n)}[\frac{2(n+1)}{r_2^n + r_1^n}]^{\frac{1}{n}}.$$

Thus, we are led to consider 3 cases.

1st case.$\gamma < \tilde{\gamma}$. Then we have $\phi(S_0(\gamma)) < 0$ so that $\phi(r) < 0$ for all $r > 0$. In this case, S is strictly decreasing and there exists $T^* \in (0, \infty)$ such that $S(T^*) = r_1$.

2nd case. $\gamma = \tilde{\gamma}$. We have then $\phi(S_0(\tilde{\gamma})) = 0$ and $\phi(r) < 0$ for all $r \in \mathbb{R}_+^* \backslash \{S_0(\tilde{\gamma})\}$. Moreover, an easy computation shows that

$$\forall\, 0 < r_1 < r_2, \forall n \geq 2, S_0(\tilde{\gamma}) < r_2$$

and

$$S_0(\tilde{\gamma}) > r_1 \Leftrightarrow \frac{r_2}{r_1} > (2n+1)^{\frac{1}{n}}.$$

Then

1. either $r_0 = S_0(\tilde{\gamma}) = [\frac{r_2^n + r_1^n}{2(n+1)}]^{\frac{1}{n}}$. Then $S(t) = r_0$ for all $t > 0$;
2. or $r_0 < S_0(\tilde{\gamma})$. Then $\phi(r_0) < 0$ and $S(t)$ is strictly decreasing and reaches r_1 in finite time;
3. or $r_0 > S_0(\tilde{\gamma})$. Then $\phi(r_0) < 0$ and $S(t)$ is strictly decreasing. If $r_1 < S_0(\tilde{\gamma})$, then $S(t)$ converges to $S_0(\tilde{\gamma})$ as $t \to +\infty$, otherwise $S(t)$ reaches r_1 in finite time.

3rd case. $\gamma > \tilde{\gamma}$. We then have $\phi(S_0(\gamma)) > 0$, which implies that there exist 2 points $0 < S_1(\gamma) < S_0(\gamma) < S_2(\gamma)$ such that $\phi(S_1(\gamma)) = \phi(S_2(\gamma)) = 0$. Then $\phi(r) < 0$ for all $r \in (0, S_1(\gamma)) \cup (S_2(\gamma), +\infty)$ and $\phi(r) > 0$ for all $r \in (S_1(\gamma), S_2(\gamma))$. Since $\phi(r_2) = -\frac{n-1}{r_2} - \gamma < 0$ and $S_0(\gamma) < S_0(\tilde{\gamma}) < r_2$, we always have $S_2(\gamma) < r_2$. Then

1. either $r_0 < S_1(\gamma)$. Then $S(t)$ is strictly decreasing and reaches r_1 in finite time;
2. or $S_1(\gamma) < r_0 < S_2(\gamma)$. Then $S(t)$ is strictly increasing and converges to $S_2(\gamma)$ as $t \to +\infty$.
3. or $S_2(\gamma) < r_0$. Then $S(t)$ is strictly decreasing. If $r_1 < S_2(\gamma)$, then $S(t)$ converges to $S_2(\gamma)$ as $t \to +\infty$, otherwise $S(t)$ reaches r_1 in finite time.

Hence, $S_1(\gamma)$ is an unstable stationary state and $S_2(\gamma)$ is a stable stationary state.

3 Convergence to the moving boundary problem

From now on, we only consider Problem (P_1^ϵ), assume that $\gamma \in (0, 2)$ and omit the index 1 from the notation. The purpose of this section is to prove Theorem 1 (i).

Since by the maximum principle $-1 < u^\epsilon < 1$ in $Q = \Omega \times \mathbb{R}_+$, we may define Z^ϵ by

$$u^\epsilon(r,t) = \tanh(\frac{Z^\epsilon(r,t)}{\epsilon}).$$

Then writing u^ϵ and its derivatives in terms of Z^ϵ, one can show that Z^ϵ satisfies Problem $(\tilde{P}^\epsilon)$

$$(\tilde{P}^\epsilon)\begin{cases} Z_t = \epsilon\Delta Z + 2\tanh(\frac{Z}{\epsilon})(1 - Z_r^2) - \gamma\frac{1}{|\Omega|}\int_\Omega \tanh(\frac{Z}{\epsilon}) & \text{in } Q \\ Z_r(r_1,t) = Z_r(r_2,t) = 1 & t > 0 \\ Z(r,0) = r - r_0 & r \in (r_1, r_2) \end{cases}$$

To begin with, we derive some bounds on Z^ϵ and Z_r^ϵ.

Lemma 1 *Let Z^ϵ be the solution of Problem $(\tilde{P}^\epsilon)$. Then for all $\epsilon > 0$ small enough,*

(i) $r - r_0 - \gamma t \le Z^\epsilon(r,t) \le r + (\gamma+1)t$ *for all* $(r,t) \in \bar{Q}$.

(ii) $1 - \epsilon\frac{(n-1)}{r_1^2}t \le Z_r^\epsilon(r,t) \le 1$ *for all* $(r,t) \in [r_1, r_2] \times [0, \frac{r_1^2}{\epsilon(n-1)}]$.

(iii) For all $T > 0$ and ϵ small enough there exists a positive constant $C(T) > 0$ such that $\|Z_t^\epsilon(r,t)\|_{L^2(Q_T)} \le C(T)$, where $Q_T = \Omega \times (0,T)$.

Proof.

(i) One can check that the functions $Z_-(r,t) = r - r_0 - \gamma t$ and $Z_+(r,t) = r + (\gamma+1)t$ are respectively lower and upper solutions for Problem $(\tilde{P}^\epsilon)$.

(ii) The function $w = Z_r^\epsilon$ satisfies the problem

$$\begin{cases} w_t = \epsilon w_{rr} + \epsilon\frac{n-1}{r}w_r - \epsilon\frac{n-1}{r^2}w + \frac{2}{\epsilon}(1 - \tanh(\frac{Z^\epsilon}{\epsilon})^2)w(1-w^2) & \\ \quad -4\tanh(\frac{Z^\epsilon}{\epsilon})ww_r & \text{in } Q \\ w(r_1,t) = w(r_2,t) = 1 & t > 0 \\ w(r,0) = 1 & r \in (r_1, r_2) \end{cases}$$

One can easily check that the functions $w_-(r,t) = 1 - \epsilon\frac{n-1}{r_1^2}t$ and $w_+(r,t) = 1$ are respectively lower and upper solutions for this problem for all $t \in (0, \frac{r_1^2}{\epsilon(n-1)})$.

(iii) We set $Y_\epsilon = Z^\epsilon - r$, multiply the differential equation for Y_ϵ by $Y_{\epsilon t}$ and integrate by parts.

We deduce from Lemma 1 that one can define level lines $\tilde{R}^\epsilon$ of Z^ϵ by

$$Z^\epsilon(\tilde{R}^\epsilon(z,t),t) = z \tag{3.1}$$

for $t \in [0, \frac{r_1^2}{\epsilon(n-1)}]$ and $z \in [Z^\epsilon(r_1,t), Z^\epsilon(r_2,t)]$. It is convenient to extend $\tilde{R}^\epsilon$ on $\mathbb{R} \times [0, \frac{r_1^2}{\epsilon(n-1)}]$ by

$$R^\epsilon(z,t) = \begin{cases} \tilde{R}^\epsilon(z,t) & z \in [Z^\epsilon(r_1,t), Z^\epsilon(r_2,t)] \\ r_1 & z < Z^\epsilon(r_1,t) \\ r_2 & z > Z^\epsilon(r_2,t) \end{cases}$$

for $t \in [0, \frac{r_1^2}{\epsilon(n-1)}]$. Differentiating (3.1) it easily follows that

$$|R_z^\epsilon| \leq C \text{ and } ||R_t^\epsilon||_{L^2((-M,+M)\times(0,T))} \leq C(M,T)$$

for all positive constants M and T and ϵ small enough.

We are now in a position to prove the first part of Theorem 1.

Proof of Theorem 1(i). Let $T > 0$ be arbitrary. We deduce [7] that there exists a sequence $\epsilon_n \to 0$ and functions $Z \in H^1(Q_T) \cap C^{1/2,1/4}(\overline{Q_T})$ and $R \in H^1_{loc}(\mathbb{R} \times [0,T]) \cap C^{1/2,1/4}(\mathbb{R} \times [0,T])$ such that

$$\begin{array}{ll} Z_{\epsilon_n} \to Z & \text{uniformly in } \overline{Q_T} \\ R_{\epsilon_n} \to R & \text{uniformly in } [-M,+M] \times [0,T] \end{array}$$

as $n \to +\infty$, for all M and T positive. Moreover, it follows from Lemma 1 (ii) that $Z_r = 1$. Hence $Z(r,t) = r - S(t)$, where $S \in H^1((0,T)) \subset C^{1/2}([0,T])$. Moreover, $S(0) = r_0$. Since S is Hölder continuous, it follows that there exists a positive $\tilde{T}$ defined by

$$\tilde{T} = \inf\{t > 0, S(t) = r_1 \text{ or } S(t) = r_2\}$$

Note that the line $r = S(t)$ is the level line where $Z = 0$ and $R(0,t) = S(t)$, for $t \in (0, \tilde{T})$. This implies the pointwise convergence of $u^{\epsilon_n} = \tanh(\frac{Z^{\epsilon_n}}{\epsilon_n})$ to the function u defined by

$$u(x,t) = \begin{cases} -1 & \text{if } r_1 \leq |x| < S(t), \quad t \in [0,\tilde{T}) \\ +1 & \text{if } S(t) < |x| \leq r_2, \quad t \in [0,\tilde{T}). \end{cases}$$

Since we will identify S with the solution S_1 of Problem (M_1) in section 5 below, this result will in fact imply that $\lim_{\epsilon \to 0} u^\epsilon(x,t) = u(x,t)$.

4 The travelling wave solutions

We consider the cubic function

$$f(u,v,\mu) = (1-u^2)(2u-v) + \mu$$

where v and μ are real parameters. There exists a constant $\mu_0 > 0$ such that for all $v \in [-\gamma, +\gamma]$ and $\mu \in [0, \mu_0]$ the equation for u, $f(u,v,\mu) = 0$, has exactly 3 solutions h_-, h_0 and h_+ which satisfy

$$\begin{array}{l} h_-(v,\mu) < h_0(v,\mu) < h_+(v,\mu) \\ h_-(v,\mu) < 0 < h_+(v,\mu) \\ h_\pm(v,\mu) \geq -1 + b\mu \end{array}$$

for some $b > 0$. Note that $h_-(v,0) = -1$, $h_0(v,0) = \frac{v}{2}$, $h_+(v,0) = 1$. We search for a travelling wave solution of the equation

$$u_t = u_{xx} + f(u,v,\mu)$$

which converges to $h_{\pm}(v,\mu)$ as $x \to \pm\infty$, i.e. a pair $(U(.,v,\mu), W(v,\mu))$ such that $u(x,t) = U(x - Wt)$ satisfies the parabolic equation. Set $z = x - Wt$. It comes to search for $U = U(z,v,\mu)$ and $W = W(v,\mu)$ such that

$$\begin{cases} U'' + W(v,\mu)U' + f(U,v,\mu) = 0 & \\ U(-\infty,v,\mu) = h_-(v,\mu) & U(+\infty,v,\mu) = h_+(v,\mu) \\ U(0,v,\mu) = 0 & \end{cases}$$

Next we summarize the properties of $U(z,v,\mu)$ and $W(v,\mu)$ that we will need in the sequel.

Lemma 2 *(i) There exists $A > 0$ and $\alpha > 0$ such that for all $v \in [-\gamma, +\gamma]$ and $\mu \in [0, \mu_0]$, we have*

$$\begin{array}{ll} U_z > 0 & z \in \mathbb{R} \\ |U_z| + |zU_z| + |U_{zz}| + |U_v| \leq A & z \in \mathbb{R} \\ |U(z,v,\mu) - h_+(v,\mu)| \leq A\exp(-\alpha z) & z \in \mathbb{R}_+ \\ |U(z,v,\mu) - h_-(v,\mu)| \leq A\exp(+\alpha z) & z \in \mathbb{R}_- \end{array}$$

(ii) The travelling wave velocity is given by

$$W(v,\mu) = (2h_0 - h_+ - h_-)(v,\mu)$$

In particular, $W(v,0) = v$ and $|W(v,\mu) - v| \leq C\mu$, for some positive constant $C > 0$.

5 Derivation of the interface equation

In this section, we prove that the function S_1 satisfies the initial value problem (M_1). We first give an a priori estimate for $(\int_\Omega u_1^\epsilon)/|\Omega|$ and then construct sub- and supersolutions of Problem (P_1^ϵ). In what follows we omit again the index 1 from the notation.

Lemma 3 *There exists a positive constant C such that*

$$\left|\frac{d}{dt}\frac{1}{|\Omega|}\int_\Omega u^\epsilon\right| \leq C.$$

Proof. We integrate on Ω the equation for u^ϵ to obtain

$$\frac{1}{|\Omega|}\int_\Omega u_t^\epsilon = \frac{\epsilon}{|\Omega|}\int_{\partial\Omega}\frac{\partial u^\epsilon}{\partial n} + \frac{1}{\epsilon}\frac{1}{|\Omega|}\int_\Omega (1 - (u^\epsilon)^2)(2u^\epsilon - \gamma\frac{1}{|\Omega|}\int_\Omega u^\epsilon) \tag{5.1}$$

We deduce from the boundary condition for u^ϵ that the first term on the right hand side of (5.1) is bounded. Next, we show that the second term on the right hand side of (5.1) is bounded as well. We have that

$$\frac{1}{|\Omega|}\int_\Omega (1 - (u^\epsilon)^2) = \frac{n}{r_2^n - r_1^n}\int_{r_1}^{r_2}(1 - \tanh^2(\frac{Z^\epsilon(r,t)}{\epsilon}))r^{n-1}dr.$$

Since $Z_r^\epsilon > 0$, we can make the change of variable $X = \dfrac{Z^\epsilon(r,t)}{\epsilon}$. This gives

$$\frac{1}{|\Omega|}\int_\Omega (1-(u^\epsilon)^2) = \frac{n}{r_2^n - r_1^n}\int_{\frac{Z^\epsilon(r_1,t)}{\epsilon}}^{\frac{Z^\epsilon(r_2,t)}{\epsilon}} (1-\tanh^2 X)\frac{\epsilon(R^\epsilon(\epsilon X,t))^{n-1}}{Z_r^\epsilon(R^\epsilon(\epsilon X,t),t)}dX$$

so that finally

$$\frac{1}{|\Omega|}\int_\Omega (1-(u^\epsilon)^2) \le C\epsilon \int_{\mathbb{R}} (1-\tanh^2 X)dX \le 4C\epsilon.$$

This completes the proof of Lemma 3.

In what follows we construct sub- and supersolutions of Problem (P_1^ϵ). We define the function $S^{\epsilon,\delta}$ by

$$\begin{cases} \dot{S^{\epsilon,\delta}}(t) &= \gamma \dfrac{1}{|\Omega|}\displaystyle\int_\Omega u^\epsilon - \delta, \quad t>0 \\ S^{\epsilon,\delta}(0) &= r_0 - 2\delta. \end{cases}$$

Let $a > 0$ be a small enough constant and let $T_a \in (0,\tilde{T})$ be such that $r_1 + a \le S(t), S^{\epsilon,\delta}(t) \le r_2 - a$ for all $t \in [0,T_a]$. Since S is Hölder continuous and since $\{S^{\epsilon,\delta}\}$ is uniformly continuous on compact subsets of $\mathbb{R}_+$, we know that such a T_a exists. We set $\bar{u}^{\epsilon,\delta} = U(\dfrac{r - S^{\epsilon,\delta}(t)}{\epsilon}, \gamma\dfrac{1}{|\Omega|}\displaystyle\int_\Omega u^\epsilon, \sqrt{\epsilon})$.

Lemma 4 *For all $\delta > 0$ small enough, there exists a constant $\epsilon_\delta > 0$ such that for all $\epsilon < \epsilon_\delta$,*

$$u^\epsilon(r,t) \le \bar{u}^{\epsilon,\delta}(r,t) \textit{ for all } (r,t) \in [r_1,r_2]\times[0,T_a].$$

Consequently

$$R^\epsilon(0,t) \ge S^{\epsilon,\delta}(t), \quad t\in[0,T_a]. \tag{5.2}$$

Proof. The proof of this lemma relies on the comparison principle for Problem (P_1^ϵ). We define the operator $\mathcal{L}$ by

$$\mathcal{L}\phi = \phi_t - \epsilon\Delta\phi - \frac{1}{\epsilon}(1-\phi^2)(2\phi - \gamma\frac{1}{|\Omega|}\int_\Omega u^\epsilon)$$

and successively check that

(i) $u^\epsilon(r,0) \le \bar{u}^{\epsilon,\delta}(r,0)$, for all $r \in [r_1,r_2]$;

(ii) $u^\epsilon(r_i,t) \le \bar{u}^{\epsilon,\delta}(r_i,t)$, for all $t\in[0,T_a], i=1,2$;

(iii) $\mathcal{L}\bar{u}^{\epsilon,\delta}(r,t) \ge 0$ for all $(r,t)\in(r_1,r_2)\times(0,T_a)$.

Next, we sketch the proof of (*iii*). We have that

$$\mathcal{L}\bar{u}^{\epsilon,\delta} = \frac{U_z}{\epsilon}(-\gamma\frac{1}{|\Omega|}\int_\Omega u^\epsilon + \delta + W(\gamma\frac{1}{|\Omega|}\int_\Omega u^\epsilon, \sqrt{\epsilon})) + U_v\gamma\frac{1}{|\Omega|}\int_\Omega u_{\epsilon_t} - \frac{n-1}{r}U_z + \frac{1}{\sqrt{\epsilon}} \tag{5.3}$$

By Lemma 2(ii), the first term on the right hand side of (5.3) is positive if we choose $\epsilon < \epsilon_\delta = \frac{\delta^2}{C^2}$, whereas the second term is positive by Lemma 3 for ϵ small enough. This completes the proof of Lemma 4.

We are now in a position to complete the proof of Theorem 1.

Proof of Theorem 1(ii) . The family of functions $\{S^{\epsilon,\delta}(t), 0 < \epsilon \leq \epsilon_\delta, 0 < \delta < \delta_0\}$ is bounded in $C^{1+\alpha}([0,T_a])$ for all $\alpha \in (0,1)$. Therefore, there exists a function $\tilde{S}$ in $C^{1+\alpha}([0,T_a])$ such that

$$\tilde{S}(t) = \overline{\lim_{\delta\to 0}}\,\overline{\lim_{\epsilon\to 0}}\ S^{\epsilon,\delta}(t), \quad t \in [0,T_a].$$

Since $S^{\epsilon,\delta}$ satisfies

$$S^{\epsilon,\delta}(t) = \gamma \frac{1}{|\Omega|}\int_0^t \int_\Omega u^\epsilon - \delta t + r_0 - 2\delta,$$

we have that

$$\tilde{S}(t) = \gamma\, \frac{1}{|\Omega|}\int_0^t \int_\Omega u + r_0.$$

Thus, it follows from (5.2) that

$$S(t) \geq \gamma\, \frac{1}{|\Omega|}\int_0^t \int_\Omega u + r_0.$$

Similarly, we can show the reverse inequality so that

$$S(t) = \gamma\, \frac{1}{|\Omega|}\int_0^t \int_\Omega u + r_0.$$

Hence, S satisfies the system

$$\begin{cases} \dot{S}\,(t) &= \gamma\, \dfrac{1}{|\Omega|}\displaystyle\int_\Omega u \;= \gamma \dfrac{(r_2^n + r_1^n - 2S(t)^n)}{(r_2^n - r_1^n)}, \qquad t \in (0,T_a) \\ S(0) &= r_0, \end{cases} \tag{5.4}$$

so that S coincides with the solution of Problem (M_1) studied in section 2 on $[0,T_a]$. One then iterates the procedure for $t \geq T_a$ and finds that S satisfies (5.4) on $(0,\tilde{T}) = (0,\infty)$.

6 Derivation of Problem ($\mathbf{P}_\epsilon$) from a reaction-diffusion system

In this section, we show how one can derive Problem (P_1^ϵ) from a system of reaction-diffusion equations which has been extensively studied in literature [10] [5]. We consider

Problem (P_0)

$$(P_0)\begin{cases} u_t = \epsilon\Delta u + \dfrac{1}{\epsilon}(1-u^2)(2u-v) & \text{in } \Omega\times\mathbb{R}_+ \\ \tau v_t = \dfrac{1}{\sigma}\Delta v + u - \dfrac{1}{\gamma}v & \text{in } \Omega\times\mathbb{R}_+ \\ u_r = \dfrac{1}{\epsilon}(1-u^2) & \text{on } \partial\Omega\times\mathbb{R}_+ \\ v_r = 0 & \text{on } \partial\Omega\times\mathbb{R}_+ \\ u(x,0) = u_0(x) & \text{in } \Omega \\ v(x,0) = v_0(x) & \text{in } \Omega \end{cases}$$

where the positive constants σ and τ are small parameters. We suppose that the initial data satisfy the hypothesis,

$$u_0 \text{ and } v_0 \in H^1(\Omega), -1 \le u_0 \le 1 \text{ and } -1/\gamma \le v_0 \le 1/\gamma.$$

One can show that Problem (P_0) has a unique solution (u,v) which satisfies $-1\le u\le 1$ and $-1/\gamma \le v \le 1/\gamma$.

Next we study the limiting behavior of the solution (u,v) of Problem (P_0) as $\sigma\to 0$ and show that $(u,v)=(u_\sigma,v_\sigma)$ converges to the solution of Problem (P_1), namely

$$(P_1)\begin{cases} u_t = \epsilon\Delta u + \dfrac{1}{\epsilon}(1-u^2)(2u-\zeta(t)) & \text{in } \Omega\times\mathbb{R}_+ \\ \tau\zeta'(t) = \dfrac{1}{|\Omega|}\displaystyle\int_\Omega (u(x,t)-\dfrac{1}{\gamma}\zeta(t))dx = \dfrac{1}{|\Omega|}\int_\Omega u(x,t)dx - \dfrac{1}{\gamma}\zeta(t) & \text{in } \Omega\times\mathbb{R}_+ \\ u_r = \dfrac{1}{\epsilon}(1-u^2) & \text{on } \partial\Omega\times\mathbb{R}_+ \\ u(x,0) = u_0(x) & \text{in } \Omega \\ \zeta(0) = \zeta_0 = \dfrac{1}{|\Omega|}\displaystyle\int_\Omega v_0(x)dx & \text{in } \Omega \end{cases}$$

We first give a priori estimates for u_σ.

Lemma 5 *Let $T>0$ be arbitrary. There exists a constant $C=C(T)$ such that*

$$||u_\sigma||_{L^\infty(0,T;H^1(\Omega))} \le C \text{ and } ||u_{\sigma t}||_{L^2(Q_T)} \le C \qquad \text{for all } \sigma>0 \text{ and } \tau>0.$$

To obtain these estimates, we multiply the equation for u_σ by $u_{\sigma t}$ and integrate by parts on Q_T. Similarly, one can show the following estimates for v_σ.

Lemma 6 *Let $T>0$ be arbitrary. There exists a constant $C=C(T)$ such that for all $\sigma>0$ and $\tau>0$,*

(i) $\displaystyle\int\int_{Q_T} (\nabla v_\sigma)^2 \le C\sigma.$

(ii) $\displaystyle\int_\delta^T \int_\Omega v_{\sigma t}^2 \le \frac{C}{\tau\delta}$ *for all* $\delta>0$ *and* $\displaystyle\int_\Omega (\nabla v_\sigma(t))^2 \le \frac{C\sigma}{t}$ *for all* $t\in(0,T)$.

(iii) $\displaystyle\left|\frac{d(\int_\Omega v_\sigma(x)dx)}{dt}\right| \le \frac{C}{\tau}$ *for all* $t\in(0,T)$.

Proof. We multiply the equation for v_σ by v_σ and integrate by parts on Q_T; this gives (i). In order to prove the results in (ii) we multiply this equation by $tv_{\sigma t}$ and integrate by parts on Q_T , whereas (iii) follows directly from integrating the equation for v_σ on Q_T.

Next we prove the following results.

Theorem 2 *Let (u_σ, v_σ) denote the solution of Problem (P_0). There exists a solution pair (u, ζ) of Problem (P_1) and a sequence $(u_m, v_m) = (u_{\sigma_m}, v_{\sigma_m})$ such that, as $m \to +\infty$,*

(i) $\lim_{m\to+\infty} u_m = u$ *weakly in $H^1(Q_T)$, strongly in $L^2(Q_T)$, in $C([0,T], L^2(\Omega))$ and a.e. in Q_T;*

(ii) $\lim_{m\to+\infty} v_m = \zeta$ *weakly in $L^2(0,T,H^1(\Omega))$,*
$\lim_{m\to+\infty} v_m = \zeta$ *weakly in $H^1(\Omega\times(\delta,T))$, strongly in $L^2(\Omega\times(\delta,T))$, in $C([\delta,T];L^2(\Omega))$ and a.e. in $\Omega\times(\delta,T)$, for all $\delta\in(0,T)$;*

(iii) $\zeta = \zeta(t) \in C^{0,1}([0,T])$ *and* $\frac{1}{|\Omega|}\int_\Omega v_m(x,t)dx \to \zeta$ *in $C^{0,\alpha}([0,T])$ for all $\alpha \in (0,1)$.*

Proof. (i) and (ii) are an immediate consequence of the lemmas 5 and 6. Next we give the proof of (iii). It follows from the weakly lower semi-continuity of the functional $w \to \int_\Omega w^2$ that

$$\int_\Omega (\nabla\zeta(x,t))^2 dx \le \liminf_{\sigma\to 0}\int_\Omega (\nabla v_\sigma(x,t))^2 \le \lim_{\sigma\to 0}\frac{C\sigma}{t}$$

so that

$$\nabla\zeta = 0$$

and thus

$$\zeta = \zeta(t).$$

The convergence result in (iii) and the regularity of ζ easily follow from (ii) and from Lemma 6(iii).

Corollary 1 *The pair (u,ζ) is the unique solution of Problem (P_1) and in fact $(u_\sigma, \frac{1}{|\Omega|}\int_\Omega v_\sigma(x,t)dx) \to (u,\zeta)$ as $\sigma \to 0$.*

Next, we prove the last result of this section.

Theorem 3 *Let $\tau \to 0$. Then the pair (u_τ, ζ_τ) converges to the pair $(u^\epsilon, \gamma\frac{1}{|\Omega|}\int_\Omega u^\epsilon)$, where u^ϵ satisfies Problem (P_1^ϵ).*

Proof. The starting point is the remark that the bounds for $||u_\sigma||$ in Lemma 5 do not depend on τ and that $-1/\gamma \le \zeta(t) \le 1/\gamma$. There exist a pair of functions $(u^\epsilon, \zeta^\epsilon)$ and a sequence $(u_m, \zeta_m) = (u_{\tau_m}, \zeta_{\tau_m})$ such that, as $m \to \infty$, u_m converges to u and

$$\zeta_m \rightharpoonup \zeta^\epsilon \text{ weakly in } L^2(0,T).$$

Integrating the differential equation for ζ_m between 0 and t, with $t \in (0,T)$, we deduce the inequality

$$|\int_0^t (\frac{1}{|\Omega|}\int_\Omega u_m - \gamma\zeta_m)| \leq \frac{2}{\gamma}\tau_m \text{ for all } t \in (0,T),$$

which finally implies that

$$\zeta^\epsilon = \frac{1}{\gamma|\Omega|}\int_\Omega u^\epsilon \text{ in } L^2(0,T).$$

Finally, one can show that u^ϵ satisfies Problem (P_1^ϵ).

References

[1] G. BARLES, H. M. SONER, P. E. SOUGANIDIS, *Front propagation and phase field theory*, to appear.

[2] L. BRONSARD, R. KOHN, *Motion by mean curvature as the singular limit of Ginzburg-landau model*, J. Diff. Eq., **90**, 1991, 211-237.

[3] G. CAGINALP, X. CHEN, *Phase field equations in the singular limit of sharp interface problems*, IMA Volumes in Mathematics and its Applications, **Vol.43**, 1-28, 1990.

[4] X. CHEN, *Generation and propagation of interfaces for reaction-diffusion equations*, J. Diff. Eq., **96**, 1992, 116-141.

[5] X. CHEN, *Generation and propagation of interfaces for reaction-diffusion systems*, IMA Preprint n^o708, University of Minnesota, 1990.

[6] L. C. EVANS, H. M. SONER, P. E. SOUGANIDIS, *Phase transitions and generalized motion by mean curvature*, to appear in Comm. Pure Appl. Math.

[7] D. HILHORST, J. HULSHOF, *An elliptic-parabolic problem in combustion theory: Convergence to travelling waves*, Nonlinear Analysis, TMA, **17**, 1991, 519-546.

[8] D. HILHORST, E. LOGAK, Y. NISHIURA, to appear.

[9] P. de MOTTONI, M. SCHATZMAN, *Development of interfaces in $\mathbb{R}^n$*, Proc. Royal Soc. Edin., **116A**, 1990, 207-220.

[10] Y. NISHIURA, M. MIMURA, *Layer oscillations in reaction-diffusion systems*, SIAM J. Appl. Math., **49**, 1989, 481-514.

Axiomatization of shape analysis and application to texture hyperdiscrimination

C. Lopez, J.M. Morel

Abstract. In this conference, we first prove that, under four simple axioms, the multiscale analysis of shapes is given by a curvature motion equation. In a second part, we illustrate the advantages of such an axiomatic analysis to discuss a psychophysical theory of early vision due to B. Julesz: the texture preattentive discrimination theory. The result was unexpected: We prove that the Julesz axiomatic is *too good* for human vision and leads to a hyperdiscrimination algorithm.

1. Introduction

Shape analysis methods try to extract from digital shapes local geometric "features" like holes, corners, orientation, inflexion points ... on which a further "intelligent" analysis could be based. In order to extract such features, the local analysis implies intuitively that one should use differential operators. However, because of perturbations due to the nature of the picture as well as to the recording conditions, pure differential operators are completely unreliable. Therefore, the vision research has defined as a paradigm the concept of *multiscale analysis*. Multiscale analysis aims at a smoothing of the shape. This smoothing allows the reliable extraction of local and semilocal features, but introduces a scale parameter, t, which measures the loss of locality (and the win in globality) of the extracted features, see [Koe], [Wit], [MaMo], ... We shall prove that, under a simple list of obvious axioms, the multiscale shape analysis is given by a curvature motion equation. In a second part of this conference, we shall show numerical experiments illustrating the advantages of such an analysis to discuss a theory of early vision due to B. Julesz : *the texture preattentive discrimination theory.* We shall see that this theory is formalized by a curvature motion and can be discussed in this framework. In particular, we shall see that the Julesz *texton theory* is "too good" as a model for the human perception. Indeed, we prove that a formal multiscale analysis based on this model must be a curvature motion, and achieves "hyperdiscrimination" of textures which should be undiscriminable for Julesz theory.

In section 2, we state the four principles of shape analysis. In section 3, we prove that they lead to a single special curvature motion. Section 4 is devoted to the generalization of the model for image processing. In section 5, we present the Julesz axiomatic of texture discrimination. Section 6 gives a translation of Julesz axioms into a mathematical theory and section 7 gives numerical results on Julesz textures.

2. The four principles of Multiscale Shape Analysis

Define a shape or silhouette as a closed set X whose boundary is a Jordan curve of $\Re^2$. We denote by $T_t(X)$ the *shape analysed at scale t*. X is identified with its characteristic function $X(x) = 1$ if $x \in X$ and 0 else. We call multiscale analysis any family operators $(T_t)_{t\geq 0}$ acting on shapes and we set $X(t) = T_t(X)$.

We first consider a principle which is called in vision theory the *pyramidal architecture* property. According to this assumption, T_t can be computed from T_s for any $s \leq t$ and T_0 is of course the identity. This is natural, since a coarser analysis of the original picture is likely to be deduced from a finer one without any dependence upon the original picture. Of course the finest picture analysis is the identity. In order to formalize the pyramidal architecture we include $T_t = T_{t,0}$ into a family of *transition operators* $T_{t,s}$ indexed by $0 \leq s \leq t$ and satisfying

[Pyramidal architecture] $T_{t+s} = T_{t+s,s}T_s$ for all $0 \leq s$, $0 \leq t$

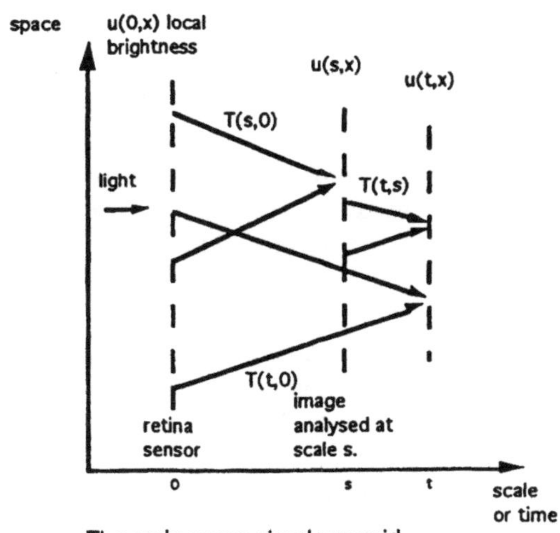

The scale-space visual pyramid.

The shape analysis process is thought of as a flow of the shape through different filters, each associated with a scale t. If we discretize such a process, it yields a serie of operators $S_n = T_{(n+1)h,nh}$ applied successively.

Assume that X and Y are two silhouettes and that for some $x \in \partial Y$ and some $r > 0$, one has $X \cap B(x, 2r) \subset Y \cap B(x, 2r)$. Assume further that the inclusion is strict in the sense that ∂X and ∂Y only meet eventually at x. Then we shall say that *the shape X is included in shape Y around x.*

[Shape local inclusion] If X is included in Y around x, then for h small enough, $T_{t+h,t}(X) \cap B(x,r) \subset T_{t+h,t}(Y) \cap B(x,r)$.

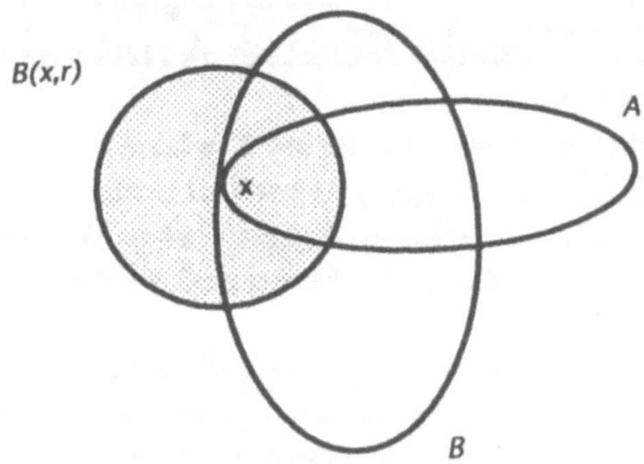

This last axiom implies that the value of $T_{t+h,t}(X)$ for h small, at any point x, is determined by the behaviour of X near x. We are allowed to take r infinite. Therefore the shape local inclusion principle also implies that if a shape is globally contained in another, this order is preserved for every scale [MaMo]. Notice that the combination of both pyramidal and shape local inclusion principles is generally called in Image Processing *causality* [Koe], [Wit], [Kim], [SaTa], [KTZ]. Indeed, the causality assumption in Image Processing states that when the scale increases, no new structure appears in the image.

Both preceding principles allow, as we shall see, to localize the shape analysis in space and time and we therefore only need to state what the multiscale analysis makes of very simple shapes in order to specify it. So we add a basic "principle" which will state what happens to disks. As we shall see, disks are somehow the "basis" on which every shape can be decomposed.

[Basic principle] Let $D = D(x, 1/r)$ be a disk with curvature $1/r$ and center x. Then $T_{t+h,t}(D)$ is a disk with radius $\rho(t, h, 1/r)$ and center x. Moreover, the function $h \to \rho(t, h, 1/r)$ is differentiable with respect to h at $h = 0$ and the differential is continuous with respect to $1/r$.

The "basic principle" implies that the multiscale analysis behaves in a smooth and isotropic way.

In the following, we set

$$g(t, 1/r) = \frac{\partial \rho}{\partial h}(t, 0, 1/r)$$

Note that $g(t,s)$ is defined for $t \geq 0$ and $s \in \Re$, now to define $g(t,0)$ we must assume that $lim\ g(t,1/r)$ exists when r tends to $+\infty$ or $-\infty$. The radius r may be positive or negative, according to the orientation of the normal $n(x)$. In the case where the curve is a Jordan curve enclosing a set X, we take as a *convention that* $\vec{n}(x)$ is *pointing outside* X and *the curvature is negative if* X *is convex at* x, *positive else.* It may seem natural to assume therefore that $g(t,\kappa)$ is odd with respect to κ. Indeed, this corresponds to the assumption that a black disk on grey background and a white disk behave in a same way.

Lemma 1. *The function* $g(t,\kappa) = \frac{\partial \rho}{\partial h}(t,0,\kappa)$ *is nondecreasing with respect to* κ.

Proof. Consider two disks D and D' with respective radii $|r|$ and $|r'| > |r|$ and keep in mind the above convention about the orientation. So we have algebraically: $r' < r$ and $\kappa < \kappa'$. Let D be included in D' and tangent to D' at some point x. By the shape local inclusion principle $T_{t+h,t}(D)$ is contained in $T_{t+h,t}(D')$. By the basic principle both sets are disks concentric to D and D' respectively and with radii $\rho(t,h,\kappa)$ and $\rho(t,h,\kappa')$. The inclusion of both disks therefore yields

$$\rho(t,h,\kappa) - \rho(t,0,\kappa) \leq \rho(t,h,\kappa') - \rho(t,0,\kappa')$$

Dividing by h and passing to the limit when h tends to 0 yields the result.

The next optional shape preserving principle is the affine invariance of shape analysis. Set $AX(x) = X(Ax)$ for any linear map A. The affine invariance can be stated as:

[Affine inv]. For any A and $t \geq 0$, there exists a C^1 function $t'(t,A) \geq 0$ such that $AT_{t'(t,A),t'(s,A)} = T_{t,s}A$. The function $t'(t,\lambda Id)$ is differentiable with respect to λ at $\lambda = 1$, and the function $\phi(t) = \frac{\partial t'}{\partial \lambda}(t,Id)$ is positive for $t > 0$.

This relation means that the result of the multiscale analysis T_t is independent of the size and position in space of the analyzed features : this is very important in the world where we live, since the same plane object X can be seen at very different distances and orientations and therefore with any affine change $X(x) \to X(Ax)$. A linear map corresponds to the anamorphosis of a plane image when it is presented to the eye at an arbitrary distance and with an arbitrary orientation in space.

Lemma 2.[**AGLM**]*(Normalization of scale) Assume that* $t \to T_t$ *is a one to one family of operators verifying [affine invariance].*

Then the funtion $t'(t,B)$ only depends on t and $|detB|$: $t'(t,B) = t'(t, |detB|^{1/2})$ and is increasing with respect to t.
Moreover, there exists an increasing differentiable rescaling function $h : [0,\infty] \rightarrow [0,\infty]$, such that $t'(t,B) = h^{-1}(h(t)|detB|^{1/2})$ and if we set $S_t = T_{h^{-1}(t)}$ we have $t'(t,B) = t|detB|^{1/2}$ for the rescaled analysis.

Remarks.
a) Thus for any multiscale analysis, we can normalize the scale-space relation and give a spatial interpretation to t under a simple rescaling.
b) The assumption on t', $\phi(t) = \frac{\partial t'}{\partial \lambda}(t, Id) > 0$, can be interpreted by looking at the relation $AT_{t'} = T_t A$ when the scale $\lambda = |detB|^{1/2}$ increases, i.e when the image is shrinked before analysis by T_t. Then, the corresponding analysis time before shrinking is increased. In more informal terms we can say that the analysis time increases with the size of the picture.

The proof of the lemma 2 is given in [AGLM].

3. The fundamental equation of Shape Analysis

When a point x belongs to an evolving curve, we denote by $\dot{x}$ the time derivative of x, which is a vector of $\Re^2$. By $curv(x)$ we denote the curvature of a curve which is C^2 at x. Recall that the curvature is defined as the inverse of the radius of the osculatory circle to the curve at x. The curvature is zero if the radius is infinite. By $\vec{n}(x)$ we denote the vector normal to the curve. In the case where the curve is a Jordan curve enclosing a set X, we take as a *convention that $\vec{n}(x)$ is pointing outside X* and *the curvature is negative if X is convex at x, positive else.*

Theorem 1. *(i) Under the three principles (pyramidal, local shape inclusion, "basic"), the multiscale analysis of shapes is governed by the curvature motion equation*

$$\dot{x} = g(t, curv(x))\vec{n}(x) \tag{1}$$

where g is defined in lemma 1.
(ii) If we assume [Affine inv.], then the equation of the multiscale analysis is, up to rescaling:

$$\dot{x} = \gamma(t.curv(x))\vec{n}(x) \tag{2}$$

where γ is defined as following: $\gamma(x) = a.x^{1/3}$ if $x \geq 0$ and $\gamma(x) = b.x^{1/3}$ if $x \leq 0$ and a, b are two nonnegative values.
(iii) If we add that $T_t(X^c) = T_t(X)^c$ [Reverse contrast invariance] then the function g in (i) is odd and $a = b$ in (ii).

By the expression "governed by", we mean that the equation must be verified at any point (t, x) of $X(t)$ where the silhouette is smooth enough to give a classical sense to both terms of the equation.

Remark: It has been proved [GaHa], [Gra], [Ang] that the equation (1) with $g(t, s) = s$ has smooth solutions. This equation has been introduced in picture processing by [KTZ], [MaMo] and [AGLM] in different contexts. An early version of an algorithm leading to equation (1) in [KoDo].

The equation (2) has been introduced and axiomatically justified (with a more complicated axiomatic however) in [AGLM]. It has also been proposed in [SaTa]. Existence and regularity of the solution are proved in [SaTa].

Proof. (i) Let X be a silhouette and assume that $T_t(X)$ has a boundary which is a C^2 manifold in a neighbourhood of a point x of ∂X. Then it has a curvature κ at point x and we consider a subosculatory and a surosculatory disk, that is, a disk D with curvature $\kappa - \epsilon$ and a disk D' with curvature $\kappa + \epsilon$, both tangent to the silhouette at x. Applying the same two principles as in the lemma, we see that

$$T_{t+h,t}(D) \cap B(x,r) \subset T_{t+h,t}(X) \cap B(x,r) \subset T_{t+h,t}(D') \cap B(x,r)$$

Thus, denoting by $x(t+h)$ the point of $\partial T_{t+h}(X)$ such that $x(t+h) - x(t)$ is parallel to $\vec{n}(x)$, we obtain

$$\rho(t,h,\kappa-\epsilon) - \rho(t,0,\kappa-\epsilon) \leq (x(t+h) - x(t)).\vec{n}(x) \leq \rho(t,h,\kappa+\epsilon) - \rho(t,0,\kappa+\epsilon)$$

Dividing by h and passing to the limit when h tends to 0 yields

$$\frac{\partial \rho}{\partial h}(t,0,\kappa-\epsilon) \leq \liminf \frac{x(t+h) - x(t)}{h}.\vec{n}(x)$$

$$\limsup \frac{x(t+h) - x(t)}{h}.\vec{n}(x) \leq \frac{\partial \rho}{\partial h}(t,0,\kappa+\epsilon)$$

We obtain equation (1) by passing to the limit when ϵ tends to 0 and using the fact that $\kappa \longrightarrow \frac{\partial \rho}{\partial h}(t,0,\kappa)$ is continuous.

(ii) After renormalization, we can use the identity

$$T_{t+h,t}D_\lambda = D_\lambda T_{(t+h)\lambda,t\lambda}$$

(where $D_\lambda = \lambda Id$) and so, we can deduce that the function ρ of the basic principle must verifye $\rho(t,h,\frac{\lambda}{r}) = \frac{1}{\lambda}\rho(\lambda t, \lambda h, 1/r)$ Therefore, we obtain the relation (after differentiation with respect to h at 0)

$$g(t, \lambda s) = g(\lambda t, s)$$

for any $t > 0$, $\lambda > 0$ and $s \in \Re$ Changing t in t/λ and taking $\lambda = 1/t$ we get $g(t,s) = g(1,ts) = \beta(ts)$ for the function β defined as $\beta(x) = g(1,x)$.

(iii) After renormalization, we can use the identity $T_{t+h,t}A = AT_{t+h,t}$, where A is the linear transform whose determinant is one

$$(x,y) \to (\lambda x, (1/\lambda)y) \;\; \lambda > 0$$

Let us apply this identity to the unit disk Δ. Look at the point $x_0 = (1,0)$ on the boundary of Δ. Then the velocity of x_0 is $\beta(-t)$, and this velocity is transformed into $\lambda\beta(-t)$. Now, look at $A\Delta$. Since $A\Delta$ is an ellipse with curvature $-\lambda^3$ at point Ax_0, the velocity of Ax_0 is $\beta(-t.\lambda^3)$. Using the first identity, we obtain $\beta(-t.\lambda^3) = \lambda\beta(-t)$. Taking $t = 1$, we get $\beta(x) = b.x^{1/3}$ for $x < 0$ ($b = \beta(-1)$). Now, apply the same technique to Δ^c and we get the result $\beta(x) = a.x^{1/3}$ for $x > 0$ ($a = \beta(1)$).

(iii) With the same technique as above we obtain that the function β is odd.

4. The fundamental equation of Image Processing

Since the multiscale analysis verifies the obvious inclusion principle that if $A \subset B$ then $T_{t+h,t}(A) \subset T_{t+h,t}(B)$, then we can, as well known in the "mathematical morphology" school [Ser], [Mara], associate with a picture u the set of its level sets

$$X_a u = \{(x,y), u(x,y) \geq a\}$$

Then, assuming that each level set is a union of silhouettes, we can simply define $T_t(u)$ as

[morphological principle] For any a, t, h and u: $X_a T_{t+h,t}(u) = T_{t+h,t}(X_a u)$

The corresponding multiscale analysis of images is governed by

$$\frac{\partial u}{\partial t} = g(t, curv(u))|Du| \tag{3}$$

We say that T_t is governed by equation (3) if for any u C^2 and every value a such that $u(x) = a$ implies $Du(x) \neq 0$, the equation is true. However, a generalized sense can be given by using the concept of "viscosity solution" [AGLM]. Conversely, one can find in [CGG] a proof of the existence of viscosity solutions for this equation. We shall only prove here that the equation (3) is verified when the image u is smooth enough.

Let us prove (3) at any (x,a) such that $u(t,x) = a$ and ($u(t,y) = a$ implies $Du(t,y) \neq 0$). The last condition implies by the implicit function theorem that x belongs to a Jordan curve Γ enclosing two regions Γ^+ and Γ^-, with $u(t,x) < a$ on Γ^- , $u(t,x) > a$ on Γ^+ and $u(t,x) = a$ on Γ. Set $u(t) = u(t,.)$. The relation $X_a T_{t+h,t}u(t) = T_{t+h,t}X_a u(t)$ implies that $T_{t+h,t}\Gamma^+$ is equal to a connected component of $X_a u(t+h)$. Therefore, the point $x(t+h)$ defined by equation (1) with initial value $x(t) = x$ belongs to the boundary of $X_a u(t+h)$. So we obtain $u(t+h, x(t+h)) - u(t, x(t)) = 0$. Dividing by h and passing to the limit yields

$$\frac{\partial u}{\partial t}(t,x) + Du(t,x)\dot{x}(t) = 0$$

and using equation (1) and $\vec{n}(x) = -\frac{Du}{|Du|}$ we obtain equation (3).

5. The "Texton Density" Doctrine of Julesz

According to Julesz theory [Jul1], texture discrimination is based on what he calls "local features" (See figure 4 for an example of a pair of discriminable textures and figure 5 for an example of undiscriminable one). He adds that "*those local features that might yield texture discrimination should be very special ones. They should be regarded as the* **basic** *elements of pre-attentive perception, and I have named such hypothetical elements "textons". (...) It seems that elongated blobs of different width, length, and orientation, and perhaps of different first order statistics in small, local neighborhoods are the local conspicuous features yielding texture discrimination. (...) Besides colour, there are two additional texton classes: elongated blobs (of given orientation, width and aspect ratio) and their terminators. Pre-attentive texture discrimination is based either on the difference in the textons or the difference in the first-order statistics (or some parameter defined by the first-order statistic, such as density or standard deviation) of the textons." (...) One can raise (...) the possibility that the pre-attentive textural system (...) can evaluate only first-order statistics of textons. (...) Whatever the final number of these texton classes will be, it is evident that for texture discrimination, the Gestaltist's view has not been corroborated : the global percept of texture was found to be decomposable into more basic elements."*

According to Julesz [Jul2], the axioms of the texton theory are as follows :

Axiom 1: *Human vision operates in two distinct modes:*
Preattentive vision. A system that is parallel (independent of number of elements), acting almost instantaneously, without scrutiny, and covering a large visual field, as in texture discrimination. Basically a preconscious system that knows where texton gradients occur, but is unaware of what these textons are.
Attentive vision. Serial search (scrutiny) by a small aperture of focal attention in steps lasting 15-50 ms each (much faster than eye-movement scanning), as in form perception. Only in this attentional aperture is it consciously known what the textons are.

Axiom 2: *Definition of textons (Restricted to line drawings)*
-Nonoverlapping line segments (black on white background, or white on black background, with specific length, orientation, width, velocity, binocular disparity, and flicker rate).

-Terminators (ends-of-lines) of line segments.
-Crossings of line segments.
-Blobs, whose shape is unimportant, except for average length, width, and orientation.

Axiom 3: *Texture formation, texton density, texton gradients*
-The area called the Δ-neighborhood, in which differences in texton densities (numbers) are determined and texton-gradients computed, is rather limited, and proportional to the size of the texture elements. Thus, textures are formed only if the textons are not to sparse, that is, if the adjacent textons lie within the Δ-neighborhood.
-The Δ-neighborhood and the aperture of focal attention are scaled by the average extent of the texture elements. So the only global operation postulated by the texton theory is the determination of the average element size needed to scale the range Δ of texton interactions.

Axiom 4: *The role of textons*
-Preattentive vision directs attentive vision to the location where texton-gradients (i.e. differences in textons, or in the densities (number of textons)) occur, but ignores the positional relationships between adjacent textons (i.e. the coupling ("glue") between textons is missing).

To summarize, preattentive vision first determines the scale from an aggregate of similar elements, i.e. the average element size, and then computes the *texton gradients* at the boundary of textures, according to the axioms given above. Thus, in preattentive vision, *precise positional relationships between "textons" are not important; only densities matter.* The figure proposed by Julesz [Jul1] illustrates this fact by showing two undiscriminable textures (fig. 5) wich, for Julesz, have the same density of textons but, however, made of different patterns.

6 How to define Texton Densities. An Axiomatic Approach

The texton doctrine, which is based only on synthetic images, neglects to consider an important issue : how can such textons be reliably extracted from images of natural scenes ? Julesz [Jul2] left the question open by accepting a priori any plausible system proposed by biologists or vision researchers : "Are line segments just thin, elongated blobs of white and black (or vice versa), or do they have the luminance profile of the "Mexican-hat-function" or a Gabor-function? These questions can be decided only by doing the psychophysics."

We can give a rigorous answer to this question by noting that the texton characteristics indicated by Julesz are shape elements based on curvature and orientation. Furthermore, these elements are *multiscale* because of the a priori unknown size of the textons. *Therefore texton densities are nothing but multiscale curvature and orientation densities. Now, we have proved in theorem 1 that there is only one way to compute multiscale curvature and orientation.* In order to completely formalize the Julesz theory, we can consider the following equivalences between formal and intuitive concepts:

"*preattentive*" $\Longleftrightarrow$ [*pyramidal*]
(because no time is given to a feedback in the perceptual system, which therefore acts as a serie of filters applied successively.)

"*line drawings*" $\Longleftrightarrow$ [*morphological*]
Indeed, the fact that Julesz admits that the performance of human vision can be characterized by its behavior on line drawings is equivalent to the assumption that the structure of a picture is contained in its level sets. Notice that the Julesz line drawings have some fitness and their boundaries are "silhouettes" or finite union of silhouettes as discussed above.

"*local features*" $\Longleftrightarrow$ [*Basic principle*], [*local inclusion of shape*]
(because the analysis of shape is local for the filters mentionned above. Furthermore, the inclusion preservation is due to the fact that the eye is able to give the right order of sharpness (i.e. curvature) between shapes observed locally).

"Black on white or white on black background" $\Longleftrightarrow$ [Reverse contrast inv.]

"only texton densities matter" $\Longleftrightarrow$ local mean values of multiscale curvature and orientation are computed.

We add the principle [Affine invariance] because it is well known that the discriminability of plane textures is little affected by distance and change of position of the plane observer.

Notice that this formalization of Julesz theory simplifies a lot the kind of texton to be considered: we reduced the blobs, terminators,... to multiscale curvature and orientation. Now, this reduction is obtained from theorem 1 since *no other local feature can be computed multiscale.* We shall now see that, in any case, terminators and elongated blobs are characterized by their multiscale evolution.

7 Experimental verification of Julesz Theory

The preceding equivalences lead to define directly texton densities as local means of multiscale curvatures and orientations. Thus, we do not need to really decompose a picture into "blobs" of several sizes and orientations [MaPe]. However,we almost do it in the following sense: Grayson and Gage-Hamilton proved that any silhouette becomes convex before disappearing when its evolution is governed by equation

(1) with $g(t,s) = s$. In the affine invariant case $g(t,s) = s^{1/3}$, a similar result has been announced by Sapiro and Tannenbaum: any silhouette becomes convex and behaves asymptotically like an ellipse. Thus the "blobs" characterizing any shape in the theory of early vision can be quite well associated with those asymptotic ellipses. This gives a strong confirmation of intuitions contained in the book of D. Marr [Marr]. According to D. Marr, early vision obeys to what he calls *the raw primal sketch* and this sketch is essentially a decomposition of the image into a family of elliptical blobs of several shapes and orientations.

Formalization and computation of textons densities Let $u_0(x)$ be an image and $u(t,x)$ its multiscale analysis by equation (3). Then the curvature and orientation at scale t and location x are given respectively by $curv(u)(x)$ and $\frac{Du}{|Du|}(x)$ which are both well defined and computed by equation (3).

Notice that "blobs" and textons can be white on black or white on black, so we define for any $\theta \in [0,\pi]$ and $t > 0$ the "density on a ball $B(x,\Delta)$ of textons at scale t and orientation θ" by:

$$G_{B_\Delta}(x) * (1_{\Theta_{\theta,\delta\theta}} curv^+(u)(t,x))$$

$$G_{B_\Delta}(x) * (1_{\Theta_{\theta,\delta\theta}} curv^-(u)(t,x))$$

where $\Theta_{\theta,\delta\theta}$ is the set of points where $\frac{Du}{|Du|} = (cos\phi, sin\phi)$ verifies $|\theta - \phi| \leq \delta\theta$. and $G_{B_\Delta}(x)$ is the isotropic gaussian with variance Δ. By a^+ we mean $max(a,0)$ and by a^- $max(-a,0)$.

Then we get a formal definition of texton densities at each scale and orientation. Based on the obtained texton density channels, a segmentation of the image can be achieved. If the Julesz theory is correct, all the pairs of textures which we present as discriminable should be clearly discriminated by at least one of the texton densities.

Conversely, any pair of undiscriminable textures should be undiscriminated by all channels.

The basic facts that we discovered by experimentation are following. The textures proposed by Julesz, and most textures proposed in the litterature as discriminable are easily discriminated by texton densities even rougher than the announced above: Indeed, it is enough to compute global curvature densities, that is, we take $\delta\theta = \pi$ and do not take into account orientation.

Let us now show numerical experiments. First, we analyse basic patterns of Julesz and present their multiscale analysis obtained with the last equation of theorem 1. We store the results until scale 9 in figure 1. From this analysis we extract the multiscale positive curvature (figure 2) and the negative one (figure 3) at the same scales.

Second, we analyse in the same way two pairs of textures. In figure 4, we store the 256X256 original image, the analysed image at scale 4, the positive and negative curvature at scale 4 and the result of a segmentation algorithm applied on the 18 scale-curvature images (the 9 positive curvatures and the 9 negative). We also show the negative curvature at scale 1 to emphasize the need of the scale. In figure 5, we store the original 512X512 image, the corresponding images at scale 7 and we present the result of the segmentation algorithm. It is very surprising because *in that case, we achieve hyperdiscrimination.*

The discrimination is easily achieved and one can even compute the boundary by using the segmentation algorithm commented in [KLM].

N.B. The length unit is the pixel size. In order to make the scale understable in experiments, we normalized the curvature equation in such a way that T_t lets disappear a disk with radius r in time r.

8 Conclusion

The axiomatisation of image processing allows an identification of a single invariant model and subsequent discussion of all psychophysical theories of vision. Moreover, the verification of these theories can face a priori two types of answers: a confirmation or a negation. Now, we have seen a third answer: rationalized vision can be better than the human eye! Psychophysics tends to idealize, as well as mechanics, the behavior of real beings. So the "formalization" of vision goes beyond the experimental expectations and opens the possibility of hyperacuity properties.

References

[AGLM1] L. Alvarez, F. Guichard, P.L. Lions and J.M. Morel. Axiomes et equations fondamentales du traitement d'images. internal Report 9216, CEREMADE. Universite Paris Dauphine, 1992.

[AGLM2] L. Alvarez, F. Guichard, P.L. Lions and J.M. Morel. Axiomatisation et nouveaux operateurs de la morphologie mathematique. C.R. Acad. Sci. Paris, t.315, Serie I, p. 265-268, 1992.

[AhSc] N. Ahuja and B. Schachter. Pattern Models, Wiley, New York 1983.

[ALM] L. Alvarez, P.L. Lions and J.M. Morel. Image selective smoothing and edge detection by nonlinear diffusion (II). SIAM Journal of numerical analysis 29, pp 845-866. 1992.

[Ang] S. Angenent. Parabolic equations for curves on surfaces I, II. University of Wisconsin-Madison Technical Summary Reports, 89-19, 89-24, 1989.

[AsBr] H. Asada and M.Brady. The curvature primal sketch. IEEE Transactions on Pattern analysis and machine intelligence, Vol. 8, No. 1, January. 1986.

[Bar] G. Barles. Remarks on a flame propagation model. Technical Report No 464 INRIA Rapports de Recherche, December 1985.

[BaSo] G. Barles and P.E. Souganidis. Convergence of approximation schemes for fully nonlinear second order equation, Asymp. Anal., to appear.

[BeJu] J. Bergen and B. Julesz. Rapid discrimination of visual patterns. IEEE Trans. on Systems, Man, and Cybernetics, 13 (5), 1983.

[BPR] J. Beck, K. Prazdny and A. Rosenfeld. A theory of textural segmentation, Human and Machine Vision, J.Beck, B. Hope and A. Rosenfeld eds., New York Academic Press, 1983, 1-38.

[Cae] T. Caelli. Three processing characteristics of visual texture segmentation. Spatial Vision, 1(1), 19-30, 1985.

[CGG] Y-G. Chen, Y. Giga and S. Goto. Uniqueness and existence of viscosity solutions of generalized mean curvature flow equations, Preprint, Hokkaido University, 1989.

[Enn] J. Enns. Seeing textons in context. Perception and Psychophysics, 39 (2), 143-147, 1986.

[EvSp] L.C. Evans and J. Spruck. Motion of level sets by mean curvature, I., preprint.

[FMZ] D. Forsyth, J.L. Mundy, A. Zisserman, al. Invariant descriptors for 3-D object recognition and Pose. IEEE Transaction of pattern analysis and machine intelligence. Vol 13 N.10 1991.

[GaHa] M. Gage and R.S. Hamilton. The heat equation shrinking convex plane curves. J. Differential Geometry 23, pp. 69-96, 1986.

[GBS] M. Graham, J. Beck and A. Sutter. Two nonlinearities in texture segregation. Proceedings of the ARVO conference, p161, 1989.

[Gray] M. Grayson. The heat equation shrinks embedded plane curves to round points. J. Differential Geometry 26, pp. 285-314, 1987.

[JGV] B. Julesz, E.N. Gilbert and J. D. Victor. Visual discrimination of textures with identical third order statistics. Biological Cybernetics, 31, 137-140, 1978.

[JuBe] B. Julesz and J.R. Bergen. Textons, the fundamental elements in preattentive vision and perception of textures. Bell System Tech. J. 62(6) 1619-1645, 1983.

[JuKr] B. Julesz and B. Kroese. Features and spatial filters. Nature, 333, 302-303, May 1988.

[Jul1] B. Julesz. Textons, the elements of texture perception, and their interactions. Nature, 290, 12 March 1981.

[Jul2] B. Julesz. Texton gradients: the texton theory revisited. Biol. Cyber. 54, 245-251, 1986.

[Kim] B.B. Kimia. Toward a computational theory of shape. Ph. D. Dissertation, Department of Electrical Engineering, McGill University, Montreal, Canada, August 1990.

[KLM] G. Koepfler, C. Lopez, J.M. Morel. A multiscale algorithm for image segmentation by variational method . Submitted to SIAM journal of Numerical Analysis.

[KoDo] J.J. Koenderink and A.J. van Doorn. Dynamic shape. Biol. Cyber.,53:383-396, 1986.

[Koe] J.J. Koenderink. Solid Shape. MIT Press, cambridge, MA 1990.

[KTZ] B.B. Kimia, A. Tannenbaum, and S.W. Zucker. On the evolution of curves via a function of curvature, 1: the classical case. To appear in J. of Math. Analysis and Applications.

[LSW] Y. Lamdan, J.T. Schwartz and H.J. Wolfson. Object recognition by affine invariant matching. In Proc. CVPR 88, 1988.

[MaMo1] A. Mackworth and F. Mokhtarian. Scale-Based description and recognition of planar curves and two-dimensional shapes. IEEE Transactions on Pattern analysis and machine intelligence, Vol. 8, No. 1, January 1986.

[MaMo2] A. Mackworth and F.Mokhtarian. A theory of multiscale, curvature-based shape representation for planar curves. IEEE Trans. Pattern Anal. Machine Intell. 14, pp 789-805, 1992.

[MaPe] J. Malik and P. Perona. Preattentive texture discrimination with early vision mechanisms. Draft, Department of Electrical Engineering and Computer Science, University of California, Berkeley, CA 94720, June 1989.

[Mara1] P. Maragos. Tutorial on advances in morphological image processing and analysis. Optical engineering, July 1987, Vol. 26 No. 7.

[Marr1] D. Marr. Analysing natural images : a computational theory of texture vision. Cold Spring Harbor Symposium on Quantitative Biology, XL, 647-662, 1976.

[OsSe] S. Osher and J. Sethian. Fronts propagating with curvature dependent speed: algorithms based on the Hamilton-Jacobi formulation. J. Comp. Physics, 79, 12-49, 1988.

[SaTa1] G. Sapiro and A. Tannenbaum. On affine plane curve evolution. Department of electrical engineering. Technion Israel Institute of Technology. Haifa. Israel. EE Pub 821. February 1992, submitted.

[SaTa2] G. Sapiro and A. Tannenbaum. Affine shortening of non-convex plane curves . Department of electrical engineering. Technion Israel Institute of Technology. Haifa. Israel. EE Pub 845. February 1992, submitted.

[Ser] J. Serra. Image analysis and mathematical morphology, Vol. 1. Academic Press 1982.

[Tre] A. Treisman. Preattentive processing in vision. Comput. Vision, Graphics, Image Process., 31(2), 156-177, 1985.

[VoPo1] H. Voorhees and T. Poggio. Detecting textons and texture boundaries in natural images. Proc. Intern. Conf. Comput. Vision, IEEE, 250-258, 1987.

[Wei] I. Weiss. Projective invariants of shapes. In Proc. DARPA IU Workshop, 1989, pp. 1125-1134.

[WeLi] D. Wermser and C.-E. Liedtke. Texture analysis using a model of the visual system. Proc. 6th Intern. Conf. Pattern Recog., 1078-1080, Munich 1982.

[Wit] Witkin, A. P. Scale-space filtering. Proc. of IJCAI, Karlsruhe 1983, 1019-1021.

[YuPo] A. Yuille and T. Poggio. Scaling theorems for zero crossings. IEEE Transactions on Pattern Analysis and Machine Intelligence, 8, January 1986.

Variational approximation of the geometric motion of fronts *

Ricardo H. Nochetto *Maurizio Paolini*
Sergio Rovida and *Claudio Verdi*

Abstract. A double obstacle problem for a singularly perturbed reaction-diffusion equation is used to approximate curvature-dependent evolving interfaces. The solution exhibits a rapid variation from -1 to 1 within a thin transition layer, and coincides with the obstacle ± 1 elsewhere. The singular parameter may depend on space and time, thereby providing an extra degree of freedom for the control of the layer thickness and related accuracy. An adaptive finite element method is introduced for space discretization: it triangulates solely the transition region, and then updates the resulting mesh to follow the layer motion. This results in savings of both CPU time and memory allocation, and gives more flexibility in selecting the local meshsize. Optimal interface error estimates are discussed, which are valid before the onset of singularities, along with convergence results and interface error estimates past singularities. Several numerical experiments for the motion by mean curvature in 2D, 3D, 4D, and 8D, with axial symmetries, illustrate pointwise accuracy, performance and potentials of the proposed method.

1991 Mathematics Subject Classification: 35K57, 35B25, 35A35, 49Q05, 65N15, 65N30.

1. Introduction

The ever increasing interest in the geometric motion of fronts stems from its intrinsic mathematical beauty and difficulty, as well as its applications to phase transitions in material science, flame propagation, combustion theory, crystal growth, etc. In its classical formulation, a front $\Sigma(t) \subset \mathbf{R}^n$ propagates in the normal direction with velocity V depending on the position, the normal vector, and the mean curvature $\frac{1}{n-1}\kappa^m$. The evolution of $\Sigma(t)$ may exhibit singularities and topological changes, such as breaking, merging and extinction. So the classical geometric approach fails to describe the problem [18,22,23,24], whereas the varifold approach leads to nonuniqueness [9], both past singularities.

The level set approach of Osher and Sethian [32,37] provides the basis for a PDE description of such geometric motion. The corresponding weak formulation

* This work was partially supported by: NSF Grant DMS-9008999, MURST (P.N. "Equazioni di Evoluzione e Applicazioni Fisico-Matematiche", "Analisi Numerica e Matematica Computazionale"), CNR (IAN, Contract 91.01312.01).

has been recently studied by Chen, Giga, and Goto [12] and Evans and Spruck [20]. The evolving surface $\Sigma(t)$ is viewed as a zero level set of a function $\omega : \mathbf{R}^n \times [0, +\infty) \to \mathbf{R}$, which satisfies $\partial_t \omega = V|\nabla \omega|$. This becomes the following degenerate parabolic PDE in case $V = \kappa^m$

$$\partial_t \omega = |\nabla \omega| \operatorname{div}\left(\tfrac{\nabla \omega}{|\nabla \omega|}\right),$$

which is interpreted in the viscosity sense [12,14,20]. The resulting generalized motion is defined uniquely, even past singularities, and coincides with the classical one for as long as the latter is smooth. An alternative approach based on the Landau-Ginzburg theory of phase transitions recovers $\Sigma(t)$ as a limit of the zero level set $\Sigma_\varepsilon(t)$ of the solution u_ε to the following singularly perturbed reaction-diffusion PDE [1,5,10,15,16,19,25,35,36,39,40,41]:

$$\varepsilon \partial_t u_\varepsilon - \varepsilon \Delta u_\varepsilon + \tfrac{1}{2\varepsilon} \Psi'(u_\varepsilon) = \tfrac{c_0}{2} g \qquad \text{in } \mathbf{R}^n \times (0, +\infty). \tag{1.1}$$

Here Ψ is a double equal well potential, typically $\Psi(s) = (1 - s^2)^2$, c_0 is a scaling factor given by $c_0 = \int_{-1}^{1} \sqrt{\Psi(s)} ds$, and g is a forcing term. In the singular limit, as $\varepsilon \downarrow 0$, $\Sigma_\varepsilon(t)$ moves with normal velocity $V = \kappa^m + g$.

In this paper we discuss the use of (1.1) as a numerical tool for the actual computation of $\Sigma(t)$. We first note that the asymptotics of (1.1) is insensitive to the details of Ψ, thereby making the choice of Ψ only a matter of convenience [35]. We then propose the following *nonregular* potential [6,7,11,28,29,35]:

$$\Psi(s) = \begin{cases} 1 - s^2 & \text{if } -1 \le s \le 1 \\ +\infty & \text{otherwise.} \end{cases} \tag{1.2}$$

Because of the double obstacle ± 1, u_ε is thus forced to satisfy $|u_\varepsilon(\mathbf{x}, t)| \le 1$ irrespective of g, but more importantly $u_\varepsilon(\cdot, t)$ attains the values ± 1 outside a narrow transition layer $\mathcal{T}_\varepsilon(t)$ of size $\mathcal{O}(\varepsilon)$. Such a crucial property is exploited numerically in that one only has to solve (1.1) within $\mathcal{T}_\varepsilon(t)$, which is a relatively small region. This is in fact essential for the finite element *dynamic mesh algorithm* of §7. In fact it triangulates solely the transition region, and then updates the resulting mesh to follow the layer motion. This results in savings of both CPU time and memory allocation, and gives more flexibility in selecting the local meshsize. For the *regular* potential instead, the asymptotic values for $|\mathbf{x}| \to \infty$ depend on g, and even for $g = 0$ they are never ± 1. Hence any attempt to reduce the computation to a domain comparable with $\mathcal{T}_\varepsilon(t)$ has to deal with boundary conditions, a delicate issue.

We also propose the use of a space-time dependent relaxation parameter $\varepsilon a(\mathbf{x}, t)$ [35]. This results in an extra degree of freedom for the control of the layer thickness and related pointwise accuracy. The underlying finite element mesh is locally refined in accordance with the size of $\varepsilon a(\mathbf{x}, t)$. This guarantees the numerical resolution of $\mathcal{T}_\varepsilon(t)$ and the optimal distribution of spatial degrees of freedom. We implemented an explicit scheme in time with adaptive selection of the time step depending on the smallest triangle size within $\mathcal{T}_\varepsilon(t)$. Consequently, small time steps are only used when singularities develop, whereas relatively large time steps are allowed at the beginning or past singularities, when the evolution is smooth.

Our time stepping procedure thus compares favorably with other explicit marching schemes defined over uniform decompositions of the entire Ω [32,37]. They all can handle singularity formation and topological changes, which, in turn, restrict the applicability of front-tracking techniques to the smooth regime; a typical one is based on discretizing the Laplace-Beltrami operator on the interface [17]. Several numerical experiments for the motion by mean curvature in 2, 3, 4, and 8 dimensions, with axial symmetries, are reported in §8 and corroborate our claims here.

We discuss in §3 the convergence past singularities of $\Sigma_\varepsilon(t)$ to $\Sigma(t)$ as $\varepsilon \downarrow 0$. Our derivation is similar to that in [5,19], but the actual construction of sub and supersolutions is much simpler due to the double obstacle. As a by-product we get a linear rate of convergence under a local nondegeneracy assumption on ω. This improves upon Chen and Elliott [11], who show such a result before singularities develop. In the latter case we prove an optimal quadratic rate of convergence [28,29]. This provides a theoretical support to the reaction-diffusion approach in that ε's of moderate size can still be used to obtain satisfactory numerical results.

The rest of the paper is organized as follows. In §2 we formulate the double obstacle problem and introduce some further notation. Convergence is discussed in §3 and interface error estimates in §4. These results extend to the semidiscrete finite element approximation. Axial symmetries are considered in §5. Fully discrete schemes and implementation details are tackled in §6, whereas the dynamic mesh algorithm is introduced in §7. Flexibility and potentials of the proposed method is fully documented in §8 with several numerical experiments. Final comments conclude the paper in §9.

2. Setting

Let $\Omega \subset \mathbf{R}^n$ be a bounded domain with Lipschitz continuous boundary and set $Q = \Omega \times (0,T)$ for $T > 0$. Let $\Sigma(0) \subset \Omega$ be an oriented $(n-1)$-dimensional manifold and let $\Sigma(t) \subset \Omega$ be the surface that propagates in the normal direction with velocity V given by the sum of the principal curvatures κ^m plus a forcing term g, namely $V = \kappa^m + g$. Let $I(t)$ be the inside and $O(t)$ the outside of $\Sigma(t)$ defined in terms of the viscosity solution ω by

$$I(t) = \{\mathbf{x} \in \Omega : \omega(\mathbf{x},t) < 0\}, \qquad O(t) = \{\mathbf{x} \in \Omega : \omega(\mathbf{x},t) > 0\},$$

whence $\Omega = \Sigma(t) \cup I(t) \cup O(t)$. Let $d(\cdot,t)$ be the signed distance function to the front $\Sigma(t)$ that is positive in $O(t)$. For any $(\mathbf{x},t) \in Q$, let $\mathbf{s}(\mathbf{x},t) \in \Sigma(t)$ be the projection of $\mathbf{x}$ onto $\Sigma(t)$, namely $\operatorname{dist}\big(\mathbf{s}(\mathbf{x},t),\mathbf{x}\big) = |d(\mathbf{x},t)|$.

Let $a \in W^{2,1}_\infty(Q)$ be a density function, which verifies $0 < a_* \le a(\mathbf{x},t)$ for all $(\mathbf{x},t) \in Q$ [35]. We study the following singularly perturbed reaction-diffusion equation with space-time dependent relaxation parameter $\varepsilon a(\mathbf{x},t)$:

$$\varepsilon\partial_t(au_\varepsilon) - \varepsilon \operatorname{div}(a\nabla u_\varepsilon) - \tfrac{1}{\varepsilon a}u_\varepsilon + \beta(u_\varepsilon) \ni \tfrac{\pi}{4}g \qquad \text{in } Q, \tag{2.1}$$

coupled with appropriate initial and boundary conditions, in the singular limit as $\varepsilon \downarrow 0$. Here β is the maximal monotone graph (inverse of the *sign* graph) defined by

$$\beta(s) = \begin{cases} (-\infty, 0] & \text{if } s = -1 \\ 0 & \text{if } -1 < s < 1 \\ [0, +\infty) & \text{if } s = 1 \end{cases} .$$

We then realize that Ψ in (1.2) satisfies $\frac{1}{2}\Psi'(s) = \beta(s) - s$, and (2.1) is a gradient flow for the energy functional

$$\mathcal{F}_\varepsilon(\varphi) = \int_\Omega \varepsilon a |\nabla \varphi|^2 + \tfrac{1}{\varepsilon a}\Psi(\varphi) - \tfrac{\pi}{2} g \varphi,$$

which is known to approximate, in the sense of Γ-convergence as $\varepsilon \downarrow 0$, the *prescribed mean curvature* energy functional at least when $a = 1$ [6] (see also [3,4,26,33] for the quartic potential).

Equation (2.1) is a parabolic double obstacle problem which, on enforcing homogeneous Neumann boundary conditions, reads equivalently: *find* $u_\varepsilon \in H^{1,1}(Q; [-1, 1])$ *such that, for a.e.* $t \in (0, T)$, *the following evolution variational inequality holds for all* $\varphi \in H^1(\Omega; [-1, 1])$:

$$\int_\Omega \varepsilon \partial_t (a u_\varepsilon)(\varphi - u_\varepsilon) + \varepsilon a \nabla u_\varepsilon \cdot \nabla(\varphi - u_\varepsilon) - \tfrac{1}{\varepsilon a} u_\varepsilon(\varphi - u_\varepsilon) - \tfrac{\pi}{4} g(\varphi - u_\varepsilon) \geq 0. \quad (2.2)$$

To determine a standing wave for (2.1), we rescale the space variable as $\frac{\mathbf{x}}{\varepsilon}$ and seek 1D absolute minimizers γ of $\mathcal{F}_\varepsilon$ for $\varepsilon = 1$, $a = 1$, and $g = 0$ that are ± 1 at $\pm\infty$. It turns out that, on imposing $\gamma(0) = 0$, the unique nondecreasing minimizer is given by

$$\gamma(x) = \begin{cases} -1 & \text{if } x < -\frac{\pi}{2} \\ \sin x & \text{if } x \in [-\frac{\pi}{2}, \frac{\pi}{2}] \\ +1 & \text{if } x > \frac{\pi}{2} \end{cases}$$

and satisfies the Euler-Lagrange equation $\mathcal{L}\gamma = \gamma'' + \gamma = 0$ in $(-\frac{\pi}{2}, \frac{\pi}{2})$. The scaling factor multiplying g in (2.1) is related to γ through $\int_{-\pi/2}^{\pi/2} |\gamma'(x)|^2 dx = \int_{-1}^{1} \sqrt{\Psi(s)} ds = \frac{\pi}{2}$. The initial datum for (2.1) is then defined by

$$u_\varepsilon(\mathbf{x}, 0) = \gamma\left(\frac{d(\mathbf{x},0)}{\varepsilon a(\mathbf{s}(\mathbf{x},0))}\right) \qquad \forall\, \mathbf{x} \in \Omega.$$

3. Convergence

In this section we establish, for the simplest case $a = 1$ and $g = 0$, the convergence of $\Sigma_\varepsilon(t)$ to $\Sigma(t)$ as $\varepsilon \downarrow 0$, even past singularities, provided $\Sigma(t)$ does not develop interior. Our result is similar to that of Evans, Soner, and Souganidis for the quartic potential [19], but the present construction of sub and supersolutions is simpler due to the fact that γ attains the values ± 1 outside $[-\frac{\pi}{2}, \frac{\pi}{2}]$. This is illustrated via a very elementary formal proof. We also obtain a linear rate of convergence, which holds whenever the viscosity solution to the motion by mean curvature is nondegenerate, namely its gradient evaluated on the interface does not vanish. This result improves upon Chen and Elliott [11] in that their first order estimate is only valid before the onset of singularities.

The reaction-diffusion PDE (2.1) reads, for $a = 1$ and $g = 0$,

$$\varepsilon \partial_t u_\varepsilon - \mathcal{L}_\varepsilon u_\varepsilon + \beta(u_\varepsilon) \ni 0, \tag{3.1}$$

where $\mathcal{L}_\varepsilon \varphi = \varepsilon \Delta \varphi + \frac{\varphi}{\varepsilon}$. We say that a function v is a *viscosity supersolution* to the double obstacle problem (3.1) if the following happens: (i) $v \geq -1$; (ii) if $v - \phi$ attains a minimum at $(\overline{\mathbf{x}}, \overline{t})$ for some $\phi \in C^\infty(Q)$ and $v(\overline{\mathbf{x}}, \overline{t}) < 1$, then

$$\varepsilon \partial_t \phi - \mathcal{L}_\varepsilon \phi \geq 0 \qquad \text{at } (\overline{\mathbf{x}}, \overline{t}).$$

Similarly we define the concept of *viscosity subsolution*. We say that v is a *viscosity solution* provided it is both viscosity super and subsolution. Consider

$$v_\varepsilon^+(\mathbf{x}, t) = \gamma\left(\tfrac{d(\mathbf{x},t)}{\varepsilon} - \tfrac{\pi}{2}\right) \qquad \forall\, (\mathbf{x}, t) \in Q \tag{3.2}$$

as our candidate for supersolution. As compared to [19] note the absence of any shape correction or vertical shift.

Theorem 3.1. *v_ε^+ is a viscosity supersolution of* (3.1).

The proof of this assertion is quite technical and will be reported elsewhere. We instead discuss now a *formal* derivation. Before extinction time, the signed distance function d is lower semicontinuous (l.s.c.) in $\{d > 0\}$ and satisfies

$$\partial_t d - \Delta d \geq 0 \qquad \text{in } \{d > 0\}$$

in the viscosity sense [14,19,39]. A straightforward calculation shows

$$\varepsilon \partial_t v_\varepsilon^+ - \mathcal{L}_\varepsilon v_\varepsilon^+ = \gamma'(\partial_t d - \Delta d) - \tfrac{\gamma''}{\varepsilon}|\nabla d|^2 - \tfrac{\gamma}{\varepsilon}.$$

In view of the above definition, we only have to examine the case $-1 \leq v_\varepsilon^+(\overline{\mathbf{x}}, \overline{t}) < 1$. If $d > 0$, then $\gamma'' + \gamma = 0$ and $\partial_t d - \Delta d \geq 0$. Since $\nabla d = 1$ and $\gamma' \geq 0$, this yields $\varepsilon \partial_t v_\varepsilon^+ - \mathcal{L}_\varepsilon v_\varepsilon^+ \geq 0$ at $(\overline{\mathbf{x}}, \overline{t})$. If instead $d \leq 0$, then $\gamma = -1$ and $\gamma' = \gamma'' = 0$. Hence $\varepsilon \partial_t v_\varepsilon^+ - \mathcal{L}_\varepsilon v_\varepsilon^+ = \frac{1}{\varepsilon} > 0$ at $(\overline{\mathbf{x}}, \overline{t})$.

We now state without proof another key ingredient for lower and upper semicontinuous (l.s.c. and u.s.c.) functions.

Comparison Lemma 3.1. *Let v^+ be a l.s.c. supersolution and v^- be an u.s.c. subsolution to* (3.1). *If $v^+ \geq v^-$ at $t = 0$ and on $\partial\Omega \times (0, T)$, then $v^+ \geq v^-$ in Q.*

We are now in a position to demonstrate convergence of the zero level sets $\Sigma_\varepsilon(t)$ of u_ε to the generalized motion by mean curvature $\Sigma(t)$. Consider

$$\Sigma^+(0) = \{\mathbf{x} \in \Omega : d(\mathbf{x}, 0) = -\tfrac{\pi}{2}\varepsilon\},$$

which is well-defined and smooth for small ε, and the corresponding generalized motion by mean curvature $\Sigma^+(t)$ originating from $\Sigma^+(0)$. Let ω be the (unique) continuous viscosity solution to the mean curvature flow satisfying $\omega(\mathbf{x},0) = d(\mathbf{x},0)$ [12,20]; hence

$$\Sigma(t) = \{\mathbf{x} \in \Omega : \omega(\mathbf{x},t) = 0\}, \qquad \Sigma^+(t) = \{\mathbf{x} \in \Omega : \omega(\mathbf{x},t) = -\tfrac{\pi}{2}\varepsilon\}.$$

Let v_ε^+ be the supersolution in (3.2) relative to $\Sigma^+(t)$, that is with $d(\cdot,t)$ replaced by the signed distance $d^+(\cdot,t)$ to $\Sigma^+(t)$. Since $v_\varepsilon^+ = u_\varepsilon = \pm 1$ on $\partial\Omega \times (0,T)$ and $d^+(\mathbf{x},0) = d(\mathbf{x},0) + \frac{\pi}{2}\varepsilon$ in the vicinity of $\Sigma(0)$, we see that

$$v_\varepsilon^+(\mathbf{x},0) = \gamma\left(\tfrac{d^+(\mathbf{x},0)}{\varepsilon} - \tfrac{\pi}{2}\right) = \gamma\left(\tfrac{d(\mathbf{x},0)}{\varepsilon}\right) = u_\varepsilon(\mathbf{x},0) \qquad \forall\, \mathbf{x} \in \Omega,$$

and therefore $v_\varepsilon^+ \geq u_\varepsilon$ as a consequence of the Comparison Lemma 3.1. This in turn shows that $u_\varepsilon(\cdot,t)$ attains the values ± 1 outside a narrow transition layer $\mathcal{T}_\varepsilon(t)$ of size $\mathcal{O}(\varepsilon)$. If now $\mathbf{x} \in I(t)$, then $\omega(\mathbf{x},t) < -\frac{\pi}{2}\varepsilon$ for ε sufficiently small, whence $d^+(\mathbf{x},t) < 0$ and $v_\varepsilon^+(\mathbf{x},t) = -1 = u_\varepsilon(\mathbf{x},t)$. This implies the following convergence result for interfaces provided $\Sigma(t)$ does not develop interior. Fattening was not observed in our numerical experiments; see §8.

Theorem 3.2. *Let $\mathbf{x} \in I(t)$ ($\mathbf{x} \in O(t)$). Then for ε sufficiently small, depending on $(\mathbf{x},t)$, we have $u_\varepsilon(\mathbf{x},t) = -1$ ($u_\varepsilon(\mathbf{x},t) = +1$).*

We thus realize that $\Sigma_\varepsilon(t)$ lies between the surfaces $\Sigma^\mp(t) = \{\mathbf{x} \in \Omega : \omega(\mathbf{x},t) = \mp\frac{\pi}{2}\varepsilon\}$. For as long as the evolution is smooth, $\nabla\omega \neq 0$ on $\Sigma(t)$, and thus the distance between $\Sigma^+(t)$ and $\Sigma^-(t)$ is $\mathcal{O}(\varepsilon)$. This linear rate was first shown by Chen and Elliott [11]. As a by-product of our argument, such result is always valid in the vicinity of any point $\mathbf{x} \in \Sigma(t)$ where $\nabla\omega(\mathbf{x},t) \neq 0$, even beyond singularities. Such *nondegeneracy* property is intimately related to interface approximability [27]. Full proofs of the above results, also valid for $g \neq 0$, will appear in [31]. Extensions to variable ε are being investigated.

4. Optimal interface error estimates

We now discuss interface error estimates for the classical, that is before the onset of singularities, motion of fronts $V = \kappa^m + g$. For simplicity we consider $a = 1$. Let $T < t^*$ be fixed, where $t^* > 0$ indicates the first time at which singularities develop. Let dist denote the Hausdorff distance.

Theorem 4.1. *There exist $0 < \mathcal{E} \ll 1$ and a constant $C_T > 0$, depending on T and $\mathcal{E}$, such that for all $0 < \varepsilon \leq \mathcal{E}$ the following optimal interface error estimate holds:*

$$\operatorname{dist}\big(\Sigma(t), \Sigma_\varepsilon(t)\big) \leq C_T\varepsilon^2 \qquad \forall\, t \in [0,T]. \tag{4.1}$$

This error bound is proven in [28] for $g = 0$ and extended in [29] to nonvanishing g. Consider the double obstacle problem

$$\varepsilon\partial_t u_\varepsilon - \mathcal{L}_\varepsilon u_\varepsilon + \beta(u_\varepsilon) \ni \tfrac{\pi}{4}g. \tag{4.2}$$

The proof of Theorem 4.1 is based on constructing suitable sub and supersolutions to (4.2) which contain several corrections to γ dictated by formal asymptotics, and further using a modified distance function. That (4.1) is sharp is shown in [28] for the evolution of a circle in 2D. Besides its intrinsic value, the interest of (4.1) stems from the fact that ε's of moderate size can still produce quite acceptable pointwise accuracy. In this respect, ε plays a role very reminiscent of artificial viscosity in hyperbolic conservation laws, and can be viewed as a numerical, rather than a physical, quantity.

We intend to discuss here the main ideas in deriving (4.1), but without refering to formal asymptotics as in [28,29]. We start out by introducing a *modified distance* function, which combines a shift with a shape correction. For any $(\mathbf{x},t) \in Q$ set

$$d_\varepsilon^+(\mathbf{x},t) = d(\mathbf{x},t) + \varepsilon^2 e^{2Kt}\big(C_1 d^2(\mathbf{x},t) + C_2\big), \tag{4.3}$$

where $K = \|\kappa^s\|_{L^\infty(0,T;\Sigma(\cdot))} + \frac{\pi}{4}\|\nabla g\|_{L^\infty(Q)} > 0$ and $C_1, C_2 \geq 1$ are to be determined later independently of ε. Let $y(\mathbf{x},t) = \frac{d_\varepsilon^+(\mathbf{x},t)}{\varepsilon}$ be the associated modified stretched variable.

In order to define the supersolution v_ε^+, we first introduce several shape corrections to γ as follows:

$$\gamma_\varepsilon^+(x;\alpha,\beta_1,\beta_2) = \min\Big(1, \gamma(x) + \varepsilon\alpha\eta(x) + \varepsilon^2\big(\beta_1\xi_1(x) + \beta_2\xi_2(x)\big)\Big) \qquad \forall\, x \in \mathbf{R}.$$

Here α, β_1, β_2 are parameters and η, ξ_1, ξ_2 are Lipschitz continuous functions which satisfy

$$\mathcal{L}\eta(x) = \gamma'(x) - \tfrac{\pi}{4}, \quad \mathcal{L}\xi_1(x) = x\gamma'(x), \quad \mathcal{L}\xi_2(x) = x, \qquad \forall\, x \in (-\tfrac{\pi}{2}, \tfrac{\pi}{2}), \tag{4.4}$$

and vanish outside $(-\frac{\pi}{2}, \frac{\pi}{2})$. We point out that solvability of $\mathcal{L}\varphi = q$ is solely possible provided q is orthogonal to γ' in $L^2(-\frac{\pi}{2}, \frac{\pi}{2})$, because γ' spans the kernel of $\mathcal{L}$ (Fredholm alternative). This is in fact the case in (4.4). But then η, ξ_1, ξ_2 are determined up to multiples of γ'. This extra degree of freedom is exploited to adjust the first derivatives at $x = -\frac{\pi}{2}$, namely,

$$\eta'(-\tfrac{\pi}{2}) = \xi_1'(-\tfrac{\pi}{2}) = \xi_2'(-\tfrac{\pi}{2}) = 0.$$

Note that $\eta'(\frac{\pi}{2}) = 0$ because η is even, but both ξ_1' and ξ_2' jump at $\frac{\pi}{2}$. In addition, $\gamma_\varepsilon^+ \geq -1$ and η, ξ_1, ξ_2 can be computed explicitly [28,29]; see Figure 4.1. The function v_ε^+ then reads

$$v_\varepsilon^+(\mathbf{x},t) = \gamma_\varepsilon^+\big(y(\mathbf{x},t); \overline{g}(\mathbf{x},t), \overline{\kappa^s}(\mathbf{x},t) + \tfrac{1}{2}\overline{g}^2(\mathbf{x},t), -\tfrac{\pi}{4}\overline{g}'(\mathbf{x},t)\big),$$

where $\overline{g}(\mathbf{x},t) = g(\mathbf{s}(\mathbf{x},t))$, $\overline{g}'(\mathbf{x},t) = (\nabla g)(\mathbf{s}(\mathbf{x},t))\cdot\nabla d(\mathbf{x},t)$ and $\overline{\kappa^s}(\mathbf{x},t) = \kappa^s(\mathbf{s}(\mathbf{x},t))$.

To show that v_ε^+ so defined is a supersolution we resort to a comparison argument, whose elementary proof can be found in [11,28,29].

Comparison Lemma 4.1. *Let $v \in H^{1,1}(Q;[-1,+\infty))$ satisfy*

$$\int_\Omega \varepsilon\partial_t v\varphi + \varepsilon\nabla v\cdot\nabla\varphi - f\varphi \geq 0 \qquad \forall\, \varphi \in H^1(\Omega;[0,+\infty)) \ \text{ and a.e. } t \in (0,T),$$

with $f \geq \frac{1}{\varepsilon}v + \frac{\pi}{4}g$ a.e. in $\{v<1\}$. If in addition $v(\cdot,0) \geq u_\varepsilon(\cdot,0)$, then $v \geq u_\varepsilon$ in Q.

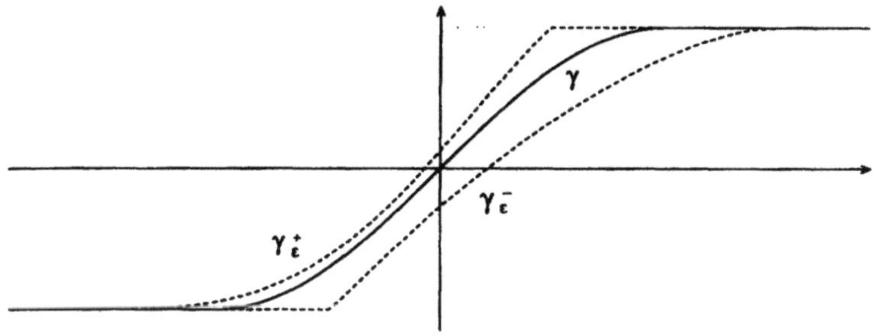

Figure 4.1. Function γ, sub and supersolution $\gamma_\varepsilon^-, \gamma_\varepsilon^+$.

That $v_\varepsilon^+(\cdot,0) \geq u_\varepsilon(\cdot,0)$ is a simple fact. We then set

$$f(\mathbf{x},t) = \begin{cases} \frac{1}{\varepsilon}v_\varepsilon^+(\mathbf{x},t) + \frac{\pi}{4}g(\mathbf{x},t) & \text{if } v_\varepsilon^+(\mathbf{x},t) < 1 \\ 0 & \text{if } v_\varepsilon^+(\mathbf{x},t) = 1 \end{cases}$$

and $\mathcal{T}_\varepsilon(t) = \{\mathbf{x} \in \Omega : |v_\varepsilon^+(\mathbf{x},t)| < 1\}$. Since $f(\mathbf{x},t) < 0$ in $\{v_\varepsilon^+(\mathbf{x},t) = -1\}$ for ε sufficiently small, and $\nabla v_\varepsilon^+ \cdot \nu_\varepsilon \geq 0$ on $\partial\mathcal{T}_\varepsilon(t)$, we only have to demonstrate

$$\mathcal{J}v_\varepsilon^+ = \varepsilon\partial_t v_\varepsilon^+ - \varepsilon\Delta v_\varepsilon^+ - \tfrac{1}{\varepsilon}v_\varepsilon^+ - \tfrac{\pi}{4}g \geq 0 \qquad \text{in } \mathcal{T}_\varepsilon(t). \tag{4.5}$$

The fact that the unit outward vector ν_ε normal to $\partial\mathcal{T}_\varepsilon(t)$ may not exist everywhere must be accounted for with a slightly different argument, that we omit [29]. Proving (4.5) in turn involves a rather lengthly and tedious calculation, whose final outcome is the following expression:

$$\begin{aligned} \mathcal{J}v_\varepsilon^+(y) &= -\tfrac{1}{\varepsilon}\mathcal{L}\gamma(y) - \overline{g}\big(\mathcal{L}\eta(y) - \gamma'(y) + \tfrac{\pi}{4}\big) \\ &\quad - \varepsilon\big(\overline{\kappa^s} + \tfrac{1}{2}\overline{g}^2\big)\big(\mathcal{L}\xi_1(y) - y\gamma'(y)\big) + \varepsilon\tfrac{\pi}{4}\overline{g}'\big(\mathcal{L}\xi_2(y) - y\big) \\ &\quad + \varepsilon^2 e^{2Kt}\big(-4C_1 y\gamma''(y) + C_2\gamma'(y)(2K - \overline{\kappa^s}) + C_2\tfrac{\pi}{4}\overline{g}' - 2C_1\gamma'(y)\big) + \mathcal{O}(\varepsilon^2), \end{aligned}$$

where $\mathcal{O}(\varepsilon^2)$ stands for $C\varepsilon^2 + C_{C_1,C_2}\varepsilon^3$. In light of the definitions of γ, η, ξ_1, and ξ_2, the first four terms on the right hand side vanish. In addition, since $y\gamma(y)+\gamma'(y) \geq 1$ and $K \geq \overline{\kappa^s} + \frac{\pi}{4}|\overline{g}'|$, we get

$$\mathcal{J}v_\varepsilon^+(y) \geq \varepsilon^2 e^{2Kt}\Big(y\gamma(y)(4C_1 - C_2K) + \gamma'(y)(C_2K - 2C_1)\Big) + \mathcal{O}(\varepsilon^2).$$

The choice $C_2K = 3C_1$ then leads to

$$\mathcal{J}v_\varepsilon^+(y) \geq C_1\varepsilon^2 + \mathcal{O}(\varepsilon^2) \geq 0,$$

for C_1 sufficiently large but independent of ε, and $\varepsilon \leq \mathcal{E} \ll 1$ depending on C_1, C_2. This completes the proof of (4.5). A subsolution v_ε^- can be constructed similarly.

To show (4.1) we argue as follows. Set $\Sigma_\varepsilon^\mp(t) = \{\mathbf{x} \in \Omega : v_\varepsilon^\mp(\mathbf{x},t) = 0\}$. Invoking the nondegeneracy property of $v_\varepsilon^\mp$ around the interface, namely, $\nabla v_\varepsilon^\mp \cdot \nabla d = \frac{1}{\varepsilon} + \mathcal{O}(1)$ provided $d_\varepsilon^\mp(\mathbf{x},t) = 0$, in conjunction with the property $v_\varepsilon^\mp(\mathbf{x},t) = \mathcal{O}(\varepsilon)$ for $d_\varepsilon^\mp(\mathbf{x},t) = 0$, we have

$$\mathbf{x} \in \Sigma_\varepsilon^\mp(t) \qquad \text{implies} \qquad d_\varepsilon^\mp(\mathbf{x},t) = \mathcal{O}(\varepsilon^2).$$

Hence, (4.3) shows that $\mathbf{x} \in \Sigma_\varepsilon^{\mp}(t)$ yields $d(\mathbf{x},t) = \mathcal{O}(\varepsilon^2)$, and so

$$\operatorname{dist}\left(\Sigma_\varepsilon^{\mp}(t), \Sigma(t)\right) \le C_T\varepsilon^2 \qquad \forall\, t \in [0,T].$$

The inequalities $v_\varepsilon^-(\mathbf{x},t) \le u_\varepsilon(\mathbf{x},t) \le v_\varepsilon^+(\mathbf{x},t)$ for all $(\mathbf{x},t) \in Q$ then lead to

$$\operatorname{dist}\left(\Sigma_\varepsilon(t), \Sigma_\varepsilon^{\mp}(t)\right) \le \operatorname{dist}\left(\Sigma_\varepsilon^-(t), \Sigma_\varepsilon^+(t)\right) \le C_T\varepsilon^2 \qquad \forall\, t \in [0,T],$$

and the desired error estimate follows.

The above results extend to the case of variable a but their rather technical proofs will be published elsewhere [30]. The construction of sub and supersolutions proceeds again along the same lines but a number of further shape corrections that account for the variation of a must be added. The resulting estimate is *local*, namely

$$\operatorname{dist}\left(\mathbf{x}, \Sigma_\varepsilon(t)\right) \le C_T\varepsilon^2 a^2(\mathbf{x},t) \qquad \forall\, t \in [0,T],\ \ \mathbf{x} \in \Sigma(t).$$

The effect of *anisotropy*, which can be incorporated into (2.1) in various ways, is being investigated.

5. Axial symmetries

We intend to reformulate (2.2) in order to handle axial symmetries. A number of related computational issues, along with concrete examples, will be discussed in §§6 and 8.

Upon splitting $\mathbf{R}^n = \mathbf{R}^{n_1} \times \mathbf{R}^{n_2}$, we denote any point in $\mathbf{R}^n$ by $(\mathbf{x}_1, \mathbf{x}_2)$ with $\mathbf{x}_1 \in \mathbf{R}^{n_1}$, $\mathbf{x}_2 \in \mathbf{R}^{n_2}$. Set $z_1 = |\mathbf{x}_1|, z_2 = |\mathbf{x}_2|, \mathbf{z} = (z_1, z_2)$, and let $Z \subset [0,+\infty)^2$ be so that

$$\Omega = \left\{(\mathbf{x}_1, \mathbf{x}_2) \in \mathbf{R}^n :\ \mathbf{z} \in Z\right\}.$$

Set $Q = Z \times (0,T)$. Assume that $u_\varepsilon(\cdot,0), g, a$ are axially symmetric, that is they just depend on z_1, z_2; then so is $u_\varepsilon(\cdot,t)$ for $t > 0$. To simplify notation, we keep the same symbols for all functions u_ε, a, g and sets $\Sigma_\varepsilon(t)$ involved, even though they now depend only on the variable $\mathbf{z} \in Z$.

We now introduce the usual weight $w(\mathbf{z}) = z_1^{n_1-1} z_2^{n_2-1}$ and corresponding weighted energy spaces

$$H_w^1(Z) = \{\varphi : w^{1/2}\varphi \in L^2(Z), w^{1/2}\nabla\varphi \in [L^2(Z)]^2\},$$
$$H_w^{1,1}(Q) = \{\varphi : w^{1/2}\varphi \in L^2(Q), w^{1/2}\nabla\varphi \in [L^2(Q)]^2, w^{1/2}\partial_t\varphi \in L^2(Q)\}.$$

Homogeneous Neumann boundary conditions are enforced on $\partial Z \backslash \{w(\mathbf{z}) = 0\}$. Let ω_k stand for the surface measure of the k-sphere S^k. The following relations hold for all $\varphi, \chi \in H_w^1(Z), H_w^{1,1}(Q)$:

$$\int_\Omega \varphi(|\mathbf{x}_1|, |\mathbf{x}_2|)d\mathbf{x}_1 d\mathbf{x}_2 = \omega_{n_1-1}\omega_{n_2-1}\int_Z \varphi(\mathbf{z})w(\mathbf{z})d\mathbf{z}, \qquad \nabla\varphi\cdot\nabla\chi = \nabla_{\mathbf{z}}\varphi\cdot\nabla_{\mathbf{z}}\chi.$$

We can then easily write a 2D variational formulation equivalent to (2.2): *find* $u_\varepsilon \in H^{1,1}_w(Q;[-1,1])$ *such that, for a.e.* $t \in (0,T)$ *and all* $\varphi \in H^1_w(Z;[-1,1])$,

$$\int_Z \Big[\varepsilon\partial_t(au_\varepsilon)(\varphi - u_\varepsilon) + \varepsilon a\nabla_{\mathbf{z}} u_\varepsilon \cdot \nabla_{\mathbf{z}}(\varphi - u_\varepsilon) \\ - \tfrac{1}{\varepsilon a}u_\varepsilon(\varphi - u_\varepsilon) - \tfrac{\pi}{4}g(\varphi - u_\varepsilon)\Big] w d\mathbf{z} \geq 0. \tag{5.1}$$

6. Discretization

Continuous piecewise linear finite elements are used for space discretization together with an explicit finite difference method in time. Fully discrete schemes and quadrature issues are discussed below. For simplicity we assume $g = 0$ and a is independent of time.

6.1. Space discretization

Let $\{\mathcal{S}\}$ be a *regular* family of (locally refined) *weakly acute* partitions of Z into triangles S of size h_S that, for simplicity, verify $\overline{Z} = \bigcup_{S\in\mathcal{S}} S$ [13]. If $\mathbf{P}^k(S)$ denotes the set of polynomials of degree $\leq k$ restricted to S, then the discrete spaces and convex set over $\mathcal{S}$ are defined by

$$\mathcal{V}^1_{\mathcal{S}} = \{\varphi \in C^0(\overline{Z}) : \varphi_{|S} \in \mathbf{P}^1(S) \quad \forall\, S \in \mathcal{S}\}, \qquad \mathcal{K}_{\mathcal{S}} = \{\varphi \in \mathcal{V}^1_{\mathcal{S}} : |\varphi| \leq 1\},$$
$$\mathcal{V}^0_{\mathcal{S}} = \{\varphi \in L^\infty(Z) : \varphi_{|S} \in \mathbf{P}^0(S) \quad \forall\, S \in \mathcal{S}\}.$$

Let $\{\mathbf{z}_j\}_{j=1}^J$ and $\{\mathbf{b}_S\}$ be the sets of all vertices and barycenters of $\mathcal{S}$, and $\{\chi_j\}_{j=1}^J$ be the canonical basis functions of $\mathcal{V}^1_{\mathcal{S}}$. Let $\Pi^k_{\mathcal{S}}$ be the Lagrange interpolation operator over $\mathcal{V}^k_{\mathcal{S}}$ $(k = 1, 2)$ defined, for all $\varphi \in C^0(\overline{Z})$, by

$$(\Pi^1_{\mathcal{S}}\varphi)(\mathbf{z}_j) = \varphi(\mathbf{z}_j) \quad \forall\, 1 \leq j \leq J, \qquad (\Pi^0_{\mathcal{S}}\varphi)(\mathbf{b}_S) = \varphi(\mathbf{b}_S) \quad \forall\, S \in \mathcal{S}.$$

The Galerkin discretization of (5.1) then reads as follows: *find* $u_{\mathcal{S}} \in L^2(0,T;\mathcal{K}_{\mathcal{S}}) \cap H^1(0,T;L^2(Z))$ *such that, for a.e.* $t \in (0,T)$ *and all* $\varphi \in \mathcal{K}_{\mathcal{S}}$,

$$\int_Z \Big[\varepsilon\Pi^0_{\mathcal{S}}(a)\Pi^1_{\mathcal{S}}\big(\partial_t u_{\mathcal{S}}(\varphi - u_{\mathcal{S}})\big) + \varepsilon\Pi^0_{\mathcal{S}}(a)\nabla u_{\mathcal{S}} \cdot \nabla(\varphi - u_{\mathcal{S}}) \\ - \tfrac{1}{\varepsilon}\Pi^0_{\mathcal{S}}(\tfrac{1}{a})\Pi^1_{\mathcal{S}}\big(u_{\mathcal{S}}(\varphi - u_{\mathcal{S}})\big)\Big] w d\mathbf{z} \geq 0. \tag{6.1}$$

The resulting (diagonal) mass matrices $\mathbf{M}^a = \{m^a_{jl}\}_{jl}$ and $\mathbf{M}^{1/a} = \{m^{1/a}_{jl}\}_{jl}$ are

$$m^a_{jl} = \int_Z \Pi^0_{\mathcal{S}}(a)\Pi^1_{\mathcal{S}}(\chi_j\chi_l) w d\mathbf{z}, \qquad m^{1/a}_{jl} = \int_Z \Pi^0_{\mathcal{S}}(\tfrac{1}{a})\Pi^1_{\mathcal{S}}(\chi_j\chi_l) w d\mathbf{z},$$

whereas the stifness matrix $\mathbf{K} = \{k_{jl}\}_{jl}$ is given by

$$k_{jl} = \int_Z \Pi_S^0(a)\nabla\chi_j \cdot \nabla\chi_l \; w d\mathbf{z}.$$

Since the weight function w is a polinomial of degree $n-2$, exact computation of the integrals is not practical for a generic value of the dimension n. Stable quadrature formulas, which are exact for linear, quadratic, cubic, and quintic integrants, were used in accordance with n [13,42]. Since w degenerates at the origin as $w(\mathbf{z}) = \mathcal{O}(|\mathbf{z}|^{n-2})$, this causes big relative errors for nonexact quadrature rules. It is also to be stressed that for n_1 and $n_2 \geq 2$ quadrature rules that only use points on interelement boundaries may give rise to a zero diagonal entry at the origin in the mass matrices. This certainly happens if the sole triangle containing the origin has the other two vertices on the coordinate axes. These issues deserve further thought and, perhaps, a complete reformulation of (6.1) in the spirit of [21].

6.2. Time discretization

We have implemented the following *explicit* time stepping of (6.1), which is essential for the dynamic mesh algorithm of §7: *for all* $i = 1, ..., I$ (number of time steps), *find* $U^i \in \mathcal{K}_S$ *such that, for all* $\varphi \in \mathcal{K}_S$,

$$\begin{aligned}\int_Z \Big[\varepsilon\Pi_S^0(a)\Pi_S^1\big((U^i - U^{i-1})(\varphi - U^i)\big) + \tau^i\varepsilon\Pi_S^0(a)\nabla U^{i-1} \cdot \nabla(\varphi - U^i) \\ - \tfrac{\tau^i}{\varepsilon}\Pi_S^0(\tfrac{1}{a})\Pi_S^1\big(U^{i-1}(\varphi - U^i)\big)\Big] w d\mathbf{z} \geq 0.\end{aligned} \tag{6.2}$$

Here τ^i is the time step, thus $t^i = \sum_{k=1}^i \tau^k$, and is selected adaptively to meet the following stability constraint:

$$0 < \sigma \leq ||\tau^i(\mathbf{M}^a)^{-1}\mathbf{K}||_\infty < 1, \tag{6.3}$$

for a fixed σ. This requirement is only necessary, and so verified, in the numerical enlarged transition layer T_ε^{i-1} (see §7.1), which is a small region relative to Z. If τ^i does not satisfy (6.3), then an automatic reduction or increase by a factor σ is performed. This, together with the stability relation $\varepsilon a(\mathbf{b}_S) \geq C h_S$ for all $S \in \mathcal{S}$ [7], in turn eliminates any possible oscillation.

We stress that the choice of an explicit scheme may not be restrictive in this setting. In fact an implicit scheme would require $\tau^i \leq C(\varepsilon a)^2$ in T_ε^{i-1} just for solvability of (6.2) which, together with the relation $h_S = o(\varepsilon a(\mathbf{b}_S))$ necessary for transition layer resolution [6], would anyhow imply $\tau^i \leq C h_S^2$ for a basically linear relation between h_S and $\varepsilon a(\mathbf{b}_S)$.

Let $P : \mathbf{R}^J \to \mathbf{R}^J$ be the componentwise projection on $[-1,1]$ defined by $(P\mathbf{q})_j = \max(-1, \min(1, q_j))$ for $\mathbf{q} = (q_j)_{j=1}^J$. If we identify any function of $\mathcal{V}_S^1$ with the $\mathbf{R}^J$ vector of its nodal values, (6.2) can be written equivalently in matrix form as follows:

$$U^i = P\Big[(\mathbf{M}^a)^{-1}(\mathbf{M}^a - \tau^i\mathbf{K} + \tfrac{\tau^i}{\varepsilon^2}\mathbf{M}^{1/a})U^{i-1}\Big]. \tag{6.4}$$

The actual computation of U^i is thus trivial in that we first have to perform the matrix operations involved on the right hand side of (6.4) (recall that $\mathbf{M}^a$ is diagonal), and then project each component onto $[-1, 1]$.

Remark 6.1. The following semi-explicit scheme can also be used in conjunction with the dynamic mesh algorithm of §7:

$$U^i = P\Big[\big(\mathbf{M}^a - \tfrac{\tau^i}{\varepsilon^2}\mathbf{M}^{1/a}\big)^{-1}(\mathbf{M}^a - \tau^i\mathbf{K})U^{i-1}\Big]. \tag{6.5}$$

This scheme not only retains the advantages of (6.4) but also exhibits better monotonicity properties.

7. Dynamic mesh algorithm

We now describe how to exploit computationally the basic feature of the solution u_ε to (5.1), namely the fact that $u_\varepsilon(\cdot, t)$ attains the values ± 1 outside a narrow transition layer $\mathcal{T}_\varepsilon(t)$ of size $\mathcal{O}(\varepsilon)$ (see §3), which is a small region relative to Z. Our dynamic mesh algorithm triangulates solely T_e^{i-1}, a suitably enlarged discrete transition layer, and then updates the resulting mesh and finite element matrices to follow the layer motion. This results in savings of both CPU time and memory allocation, and gives more flexibility in selecting the local meshsize. For simplicity we assume $g = 0$ and a is independent of time.

7.1. Mesh definition and updating

With the notation of §6, we define the discrete transition region (noncoincidence set) and the coincidence set by

$$\mathcal{T}^i = \{S \in \mathcal{S} : |U^i(\mathbf{b}_S)| < 1\}, \qquad T^i = \textstyle\bigcup_{S \in \mathcal{T}^i} S,$$
$$\mathcal{C}^i = \{S \in \mathcal{S} : |U^i| = 1 \text{ in } S\}, \qquad C^i = (\textstyle\bigcup_{S \in \mathcal{C}^i} S)\backslash T^i,$$

whereas the associated *dynamic mesh* and *enlarged transition layer* are given by

$$\mathcal{T}_e^i = \{S \in \mathcal{S} : S \cap T^i \neq \emptyset\}, \qquad T_e^i = \textstyle\bigcup_{S \in \mathcal{T}_e^i} S.$$

The following *crucial property* holds for both schemes (6.4) and (6.5): *if $\mathbf{z}_j$ is a vertex lying in C^{i-1}, then $U^i(\mathbf{z}_j) = U^{i-1}(\mathbf{z}_j) = \pm 1$*, that is $T^i \subset T_e^{i-1}$. In other words, the discrete transition layer T^{i-1} cannot move faster than one triangle per time step and, consequently, U^i does not have to be computed at vertices lying outside T^{i-1}. We then compute the mass and stiffness matrices $\mathbf{M}_i^a, \mathbf{M}_i^{1/a}$, and $\mathbf{K}_i$ corresponding to the enlarged transition layer T_e^{i-1}, update τ^i on imposing (6.3), and finally solve (6.4) within T_e^{i-1} only. We stress that $\mathcal{T}_e^{i-1}$ will be the only part of $\mathcal{S}$ present at time t^{i-1}, and is minimal in terms of number of triangles. It has to be stressed that $\mathcal{S}$ here is fictitious and just used for definition of $\mathcal{T}_e^{i-1}$.

Once U^i and $T^i \subset \mathcal{T}_e^{i-1}$ have been determined, $\mathcal{T}_e^{i-1}$ is updated as follows. First, all triangles $S \in \mathcal{C}^{i-1}$ satisfying $S \cap T^i = \emptyset$ are removed because their are no longer needed. New triangles are next created at all nodes on the boundary of T^i that are not interior to $\mathcal{T}_e^{i-1}$. The mesh so obtained is $\mathcal{T}_e^i$.

7.2. Mesh generation

We resort to an advancing front algorithm in the spirit of [34]. In order to construct the first mesh, the *moving mesh front* (*mf*) is set initially to be a partition of ∂Z together with the boundary of an equilateral triangle in the interior of the initial transition region $\mathcal{T}_\varepsilon(0)$, both graded according to $ha(\mathbf{z})$, where h is a scaling factor. The isolated triangle serves as a *seed* for the mesh generator and the number of seeds matches the number of connected components of $\mathcal{T}_\varepsilon(0)$. Starting from the seed(s), the mesh generator adds new triangles of size comparable with $ha(\mathbf{z})$ for as long as nodes of *mf* are found to belong to $\mathcal{T}_\varepsilon(0)$.

The dynamic mesh $\mathcal{T}_e^i$, consistent with $ha(\mathbf{z})$, is generated using the *mf* algorithm. Since $\mathcal{T}_e^i$ cannot exit Z and must fit the grid on ∂Z, *mf* comprises also $\partial Z \backslash \partial T^i$. In fact *mf* is always the boundary of the subregion of Z not yet partitioned, but the nodes on $\partial Z \backslash \partial T^i$ are properly flagged, thereby ruling out any computation on them. The *mf* algorithm allows us to either delete or incorporate efficiently an element on *mf*. The deletion of triangles is a new feature with respect to [34]. At this stage we either remove or add the corresponding contributions to the global matrices. It is worth noting that the insertion of new triangles can create new nodes, while the elimination of triangles can lead to removal of nodes. In these cases it is necessary to add or remove rows and columns of matrices. This is not accomplished by a complete renumbering of the nodes, which would be inefficient, but by marking rows/columns which have been removed, so that they can be used later when new nodes are added. This illustrates the strong linkage between mesh generation and matrix maintenance.

Even though $\mathcal{T}_e^{i-1}$ is weakly acute, such property may fail to hold for $\mathcal{T}_e^i$. It is then necessary to postprocess $\mathcal{T}_e^i$ by simply switching the common egde of adjacent triangles whenever the sum of the opposite angles exceeds π. This process is known to converge after finite iterations and lead to a weakly acute triangulation. However mass and stiffness matrices must be modified accordingly. This process introduces interpolation errors on U^{i-1} as well; this issue deserves further investigation.

8. Numerical simulations

We present a number of numerical experiments for the motion by mean curvature in 2, 3, 4, and 8 dimensions, with axial simmetries (see §5). We always indicate the topology of both the initial surface $\Sigma(0)$ and its interior $I(0)$, using the notation S^k and D^k for any k-sphere and k-disk, respectively, as topological manifolds. Hence a torus of dimension $n = n_1 + n_2$ is written as $S^{n_1-1} \times S^{n_2} = \partial(S^{n_1-1} \times D^{n_2+1})$. Asymptotic accuracy of the dynamic mesh algorithm is documented for the case of a shrinking circle for which the exact solution is known.

In all our experiments below the initial datum for (5.1) is defined by

$$u_\varepsilon(\mathbf{z}, 0) = \gamma\left(\tfrac{d(\mathbf{z},0)}{\varepsilon a((\mathbf{z}+\mathbf{s}(\mathbf{z},0))/2)}\right) \qquad \forall\, \mathbf{z} \in Z,$$

and satisfies $\Sigma_\varepsilon(0) = \Sigma(0)$. The further correction in a is used to approximate the first two terms of the inner asymptotic expansion of u_ε [35, §6], namely,

$$u_\varepsilon(\mathbf{z}, 0) = \gamma\left(\tfrac{d(\mathbf{z},0)}{\varepsilon a(\mathbf{s}(\mathbf{z},0))}\right) - \varepsilon(\nabla a \cdot \nabla d)(\mathbf{s}(\mathbf{z},0))\mu\left(\tfrac{d(\mathbf{z},0)}{\varepsilon a(\mathbf{s}(\mathbf{z},0))}\right) + \mathcal{O}(\varepsilon^2),$$

where $\mu(x) = \frac{1}{2}x^2\gamma'(x)$. This in turn alleviates the initial transient regime. The relaxation and spatial discretization scaling factors ε and h are chosen so as to satisfy the stability constraint $\varepsilon \geq Ch$ [7]. The discrete interface Σ^i is defined by $\Sigma^i = \{\mathbf{z} \in Z : U^i(\mathbf{z}) = 0\}$ and the extinction time denoted by $t^\dagger$.

8.1. Two-dimensional experiments

Example I (Circle). The initial circle $\Sigma(0) = \{\mathbf{z} \in [0,2)^2 : |\mathbf{z}| = 1\}$ ($S^1 = \partial(D^2)$, $n_1 = n_2 = 1$) shrinks to a point at $t^\dagger = 0.5$. For $0 \leq t \leq t^\dagger$ the radius is given by $r(t) = \sqrt{1-2t}$.

Table 8.1 summarizes the experiments performed with the dynamic mesh algorithm using the density function $a(\mathbf{z}) = \max(0.1, |\mathbf{z}|)$ and the relation $\varepsilon = 1.36h|\log h|$. We use the notation: J = average number of spatial degrees of freedom, I = number of time steps, CPU= computing time in seconds on a SUN Sparc II workstation, $t^\dagger$ = computed extinction time, $E_t = L^\infty(0, 0.49)$ error for the expected extinction time, $E_d = \max_{t^i \in (0,0.49)} \operatorname{dist}(\Sigma(t^i), \Sigma^i)$.

The full mesh strategy is obviously more expensive, both in terms of CPU time and memory allocation, than the dynamic mesh approach. Striking comparisons with the case $a = 1$ over a quasi-uniform mesh are reported in [35].

h	ε	J	I	CPU	$t^\dagger$	$10^2 E_t$	$10^2 E_d$
.1414	.3762	120	4512	44	.5331	3.02	14.2
.1000	.3132	168	7002	100	.5225	2.16	11.0
.0707	.2548	246	10611	232	.5133	1.31	7.35
.0500	.2037	385	19836	717	.5078	0.79	4.78
.03536	.1607	594	40096	2331	.5046	0.47	3.02
.02500	.1254	877	69556	6367	.5012	0.13	0.92

Table 8.1. Shrinking circle of initial radius $r = 1$ (dynamic mesh algorithm).

Example II (Cross-shape). Let

$$\Sigma_\lambda(0) = \{|\mathbf{z}| < 1 : z_2 = z_1 - \lambda\} \tag{8.1}$$

($n_1 = n_2 = 1$) be the initial (nonregular) curve which, by symmetry considerations, exhibits a nonunique evolution by mean curvature for $\lambda = 0$: fattening of the interface occurs (the zero level set of the viscosity solution becomes fat). The density function is given by $a(\mathbf{z}) = \max(0.05, |\mathbf{z}|)$. Let $h = 0.05$ and $\varepsilon = 0.1513$. Note that even if $\lambda = 0$, the discrete interface selects one of the two possible evolutions because of mesh asymmetries.

Evolutions for $\lambda = 0$ and $\lambda = -0.01$ are shown in Figure 8.1. The positions of $z_1^i = \Sigma^i \cap \{z_2 = 0\}$ versus time for $\lambda = 0$ and $\lambda = 0.01$ are depicted in Figure 8.2. It is apparent that the initial velocity of z_1^i is $+\infty$.

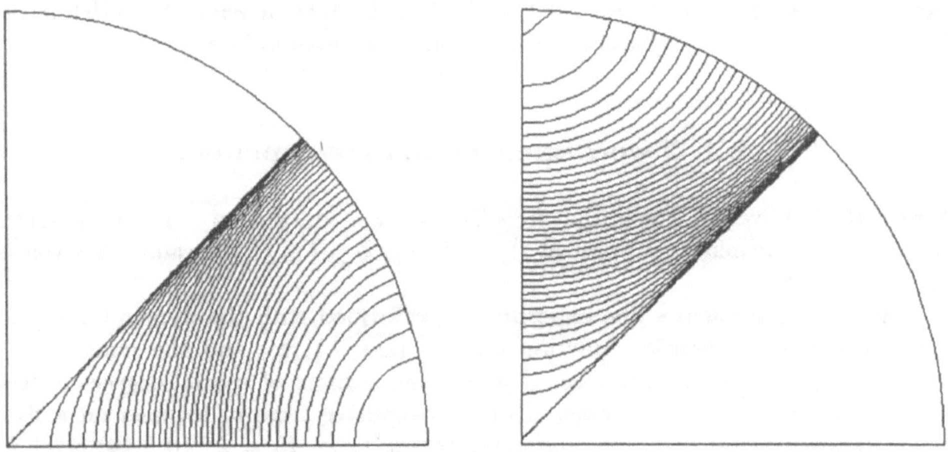

Figure 8.1. Evolution of the cross-shape (Example II): $t = 0.01i$;
a) $\lambda = 0$ ($t_\lambda^\dagger = 0.3801$); b) $\lambda = -0.01$ ($t_\lambda^\dagger = 0.3709$).

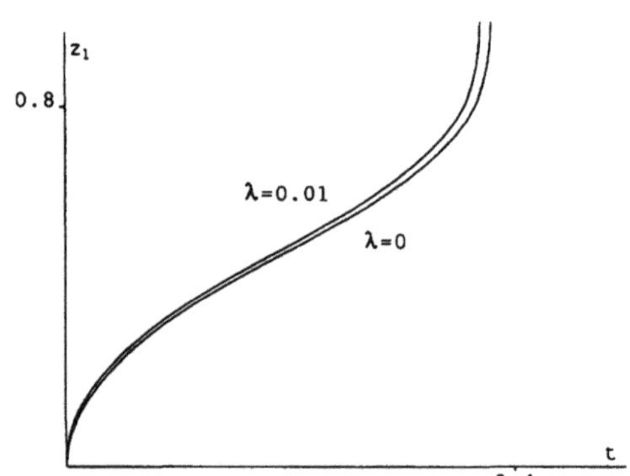

Figure 8.2. Profile of $z_1^i = \Sigma^i \cap \{z_2 = 0\}$ for the cross-shape evolution (Example II).

8.2. Three-dimensional experiments

Example III (Dumbbell). The inside of $\Sigma(0)$ ($S^2 = \partial(D^3)$, $n_1 = 1$, $n_2 = 2$) is defined by $I_\lambda(0) = I_\lambda^1 \cup I^2$, where

$$I_\lambda^1 = \{\mathbf{z} \in Z : z_1 < 1, z_2 < \lambda\}, \quad I^2 = \{\mathbf{z} \in Z : (z_1 - 1)^2 + z_2^2 < 0.25\},$$

$Z = [0,4) \times [0,3)$ and $0 < \lambda < 0.5$ is a parameter. Let $a(\mathbf{z}) = \min(1, 0.05 + \operatorname{dist}(\mathbf{z}, R))$, where $R = [0,0.5] \times [0,0.1]$, $h = 0.1$, and $\varepsilon = 0.2841$.

A critical value λ^* of λ separates smooth evolutions, with the surface shrinking to the origin at $t_\lambda^\dagger$ ($\lambda > \lambda^*$), from nonsmooth evolutions. For $\lambda < \lambda^*$ the front exhibits a singularity at the origin, at $t_\lambda^* < t_\lambda^\dagger$, then splits into two disconnected S^2, and eventually shrinks to two points at $t_\lambda^\dagger$. Uniqueness of dumbbell evolutions have been recently obtained by Huisken [24] and Altschuler, Angenent, and Giga [2] (see also Barles, Soner, and Souganidis [5, Thm. 4.2]). Figure 8.3 illustrates the evolution for two values $\lambda_1 < \lambda^* < \lambda_2$.

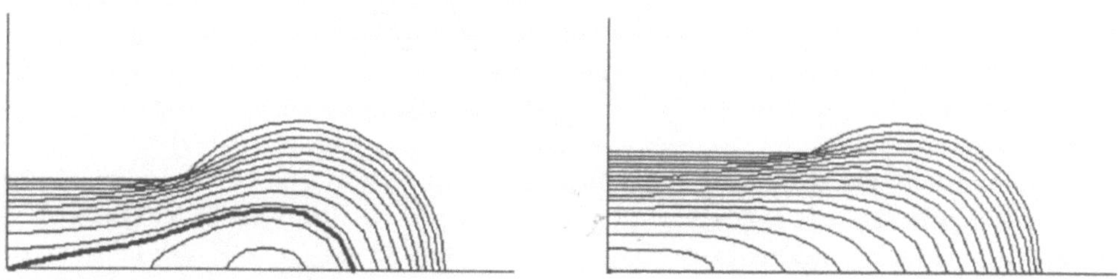

Figure 8.3. Evolution of the dumbbell (Example III): $t = 0.005i$ and $t_{\lambda_1}^*$; a) $\lambda_1 = 0.3$ ($t_{\lambda_1}^* = 0.0472$, $t_{\lambda_1}^\dagger = 0.0569$); b) $\lambda_2 = 0.4$ ($t_{\lambda_2}^\dagger = 0.0826$).

8.3. Four-dimensional experiments

Example IV (4D-cone). The initial surface is defined by (8.1) with $n_1 = n_2 = 2$. We use the same density function and scaling factors as those of Example II.

Evolutions for $\lambda = 0$ and $\lambda = -0.01$ are shown in Figure 8.4. The positions of $z_1^i = \Sigma^i \cap \{z_2 = 0\}$ versus time for $\lambda = 0$ and $\lambda = 0.01$ are depicted in Figure 8.5. The initial velocity seems to be $+\infty$ as happens in Example II. No theoretical results are known to us concerning fattening for this example, although numerical experiments seem to indicate nonuniqueness.

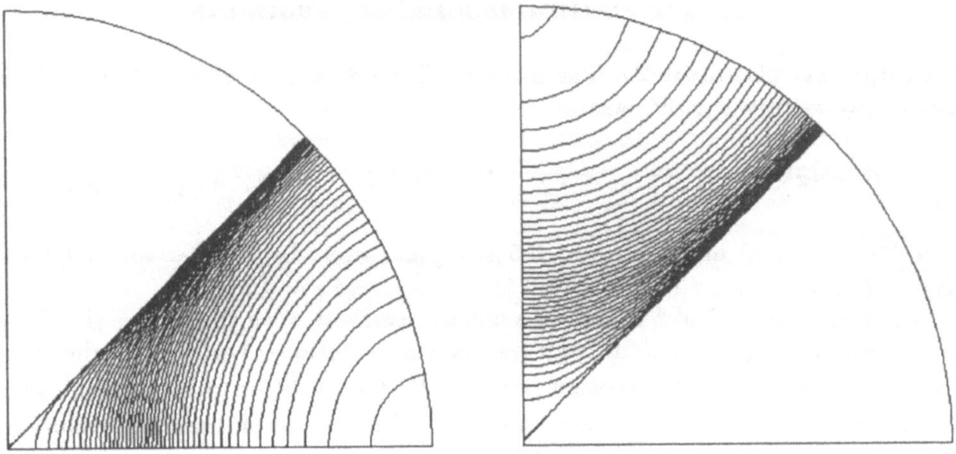

Figure 8.4. Evolution of the 4D-cone (Example IV): $t = 0.01i$;
a) $\lambda = 0$ ($t_\lambda^\dagger = 0.4549$); b) $\lambda = -0.01$ ($t_\lambda^\dagger = 0.3811$).

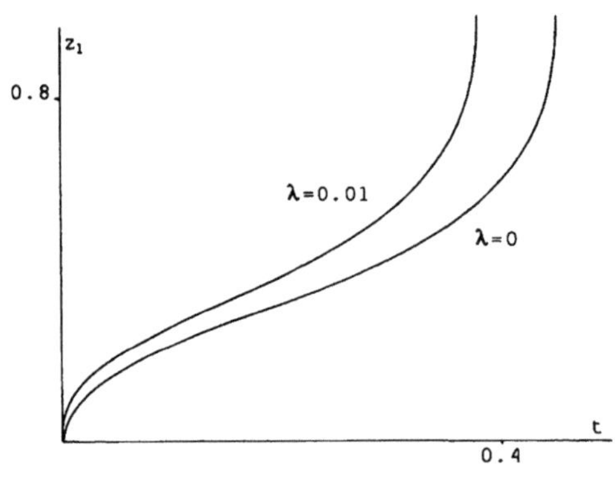

Figure 8.5. Profile of $z_1^i = \Sigma^i \cap \{z_2 = 0\}$ for the 4D-cone evolution (Example IV).

Example V (4D-torus). We consider $S^2 \times S^1 = \partial(S^2 \times D^2)$, $n_1 = 3$, $n_2 = 1$ with an initial surface defined by

$$\Sigma_r(0) = \{\mathbf{z} \in [0,4) \times [0,3) : (z_1 - 1)^2 + z_2^2 = r^2\}. \tag{8.2}$$

The density function is $a(\mathbf{z}) = \min\left(1, \max(0.05, 0.07z_1 + 0.07z_2^2)\right)$, whereas $h = 0.1$ and $\varepsilon = 0.2841$.

The evolution is very similar to the 3D-torus ($S^1 \times S^1 = \partial(S^1 \times D^2)$) which is not reported here (see [35]). Numerical experiments suggest indeed that there exists a critical radius r^* such that for all $r < r^*$ the surface shrinks to $S^2 \subset \{z_2 = 0\}$ and then suddenly disappears at $t_r^\dagger$, whereas for $r > r^*$ the surface focuses at the origin at some time t_r^*, when a singularity occurs, then changes its topology to S^3 and eventually disappears at $t_r^\dagger$.

Uniqueness of the evolution has been proven by Soner and Souganidis for $n \geq 3$, together with the fact $t^\dagger_{r^*} = t^*_{r^*}$ [40]. Our numerical experiments confirm their results (see Figure 8.6).

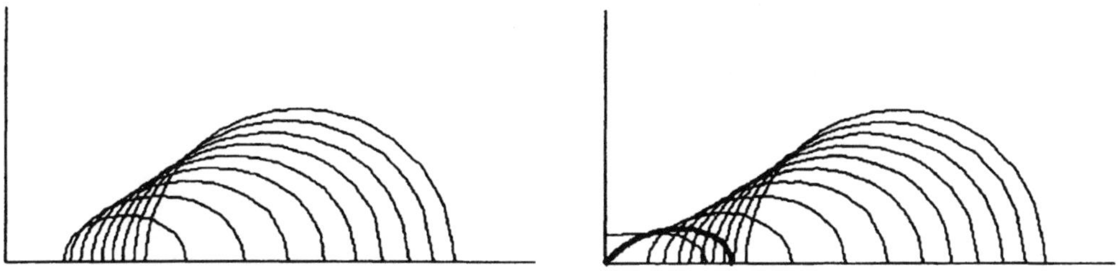

Figure 8.6. Evolution of a 4D-torus (Example V): $t = 0.02i$ and t^*_r; a) $r = 0.520$ ($t^\dagger_r = 0.1740$); b) $r = 0.521$ ($t^*_r = 0.1751$, $t^\dagger_r = 0.1907$).

Example VI (4D-torus). We consider $S^1 \times S^2 = \partial(S^1 \times D^3)$, $n_1 = n_2 = 2$, the initial surface defined by (8.2), and all the other data as in Example V. The only difference then stems from the values of n_1 and n_2, which means different topologies of $\Sigma_r(0)$ and is so significant that leads to a quite distinct evolution.

In contrast to Example V, the numerical experiments show now the existence of two critical radii $0 < r^* < r^{**}$. For $r < r^*$ the surface shrinks at $t^\dagger_r$ to $S^1 \subset \{z_2 = 0\}$ and suddenly disappears without ever touching the origin. For $r > r^{**}$ a singularity develops at the origin at time t^*_r, and then the surface becomes topologically equivalent to S^3 until it shrinks to the origin. For $r^* < r < r^{**}$ the evolution is for a while similar to the case $r > r^{**}$, but a second singularity forms at the origin at a later time $t^{**}_r > t^*_r$ after which the surface becomes again a 4D-torus and shrinks at $t^\dagger_r$ to $S^1 \subset \{z_2 = 0\}$ and disappears. Our simulations suggest that (see Figure 8.7):

$$\lim_{r \downarrow r^*} t^*_r = \lim_{r \downarrow r^*} t^{**}_r < \lim_{r \downarrow r^*} t^\dagger_r.$$

This would imply that the evolution is unique for the critical value r^*. No theoretical results are known to us.

Another interesting feature that distinguishes the 4D and 3D cases is the shape of the surface at the singular time t^*_r, which seems to be tangent to a 4D-cone at the origin with opening α_r with respect to the z_2-axis and verifying $\lim_{r \downarrow r^*} \alpha_r = \frac{\pi}{4}$.

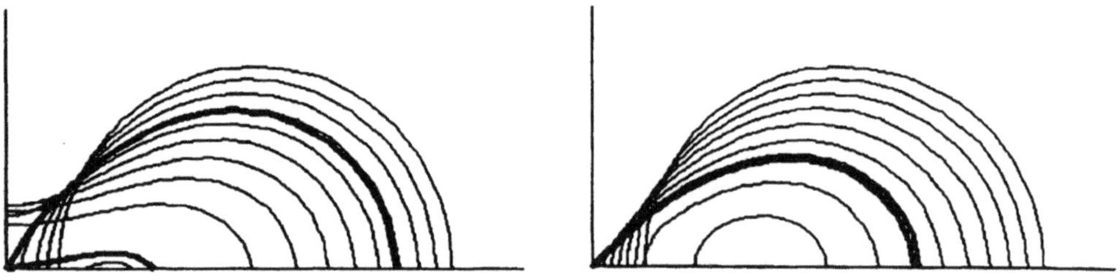

Figure 8.7. Evolution of a 4D-torus (Example VI): $t = 0.02i$, t_r^*, and t_r^{**};
a) $r = 0.795$ ($t_r^* = 0.0643$, $t_r^{**} = 0.1792$, $t_r^\dagger = 0.1803$);
b) $r = 0.78943558$ ($t_r^* = 0.1158$, $t_r^{**} = 0.1200$, $t_r^\dagger = 0.1691$).

Example VII (4D-torus). We consider $S^1 \times S^1 \times S^1 = \partial(S^1 \times S^1 \times D^2)$, $n_1 = n_2 = 2$, with the following initial front:

$$\Sigma_r(0) = \{\mathbf{z} \in [0,3)^2 : (z_1 - 1)^2 + (z_2 - 1)^2 = r^2\} \qquad (0 < r < 1).$$

The density function is $a(\mathbf{z}) = \max(0.05, \min(z_1, z_2))$ and the scaling factors are $h = 0.1$ and $\varepsilon = 0.2841$. A critical radius r^* separates smooth evolutions until extinction time $t_r^\dagger$, case $r < r^*$, from singularity formation at symmetric positions on the coordinate axes at time $t_r^* < t_r^\dagger$, namely the S^1 manifolds $(z_r^*, 0)$, $(0, z_r^*)$; see Figure 8.8. Our experiments suggest that (see Figure 8.9)

$$\lim_{r \downarrow r^*} z_r^* = 0, \qquad \lim_{r \downarrow r^*} t_r^* = \lim_{r \downarrow r^*} t_r^\dagger,$$

which means that for $r = r^*$ the surface shrinks to a point without change of topology (no fattening!). It is also apparent that mesh asymmetry influences the evolution, as shown in Figure 8.9b.

Mesh evolution is shown in Figure 8.10 for $r = 0.625$ and various time steps. Note that only triangles belonging to the boundary of $\mathcal{T}_e^i$ are depicted. These elements, together with the interior triangles of $\mathcal{T}_e^i$, are the only ones present at the corresponding time steps. The evolving front, as well as the triangulation, are connected before the onset of singularities, as illustrated in Figures 8.10a,b,c. At t_r^* and after, both front and mesh split into two connected components (see Figure 8.10d). This situation remains until the interior front disappears along with the corresponding mesh, at t_r^{**}. Then there are only one front and mesh left, as depicted in Figures 8.10e,f,g.

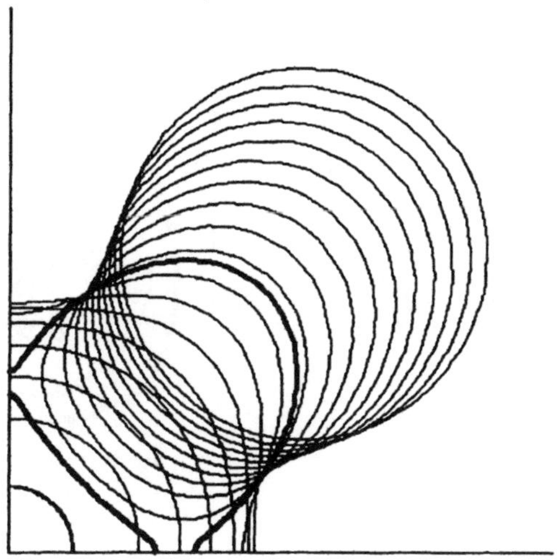

Figure 8.8. Evolution of a 4D-torus (Example VII): $t = 0.025i$ and t_r^*; $r = 0.625$ ($t_r^* = 0.2350$, $t_r^\dagger = 0.4330$).

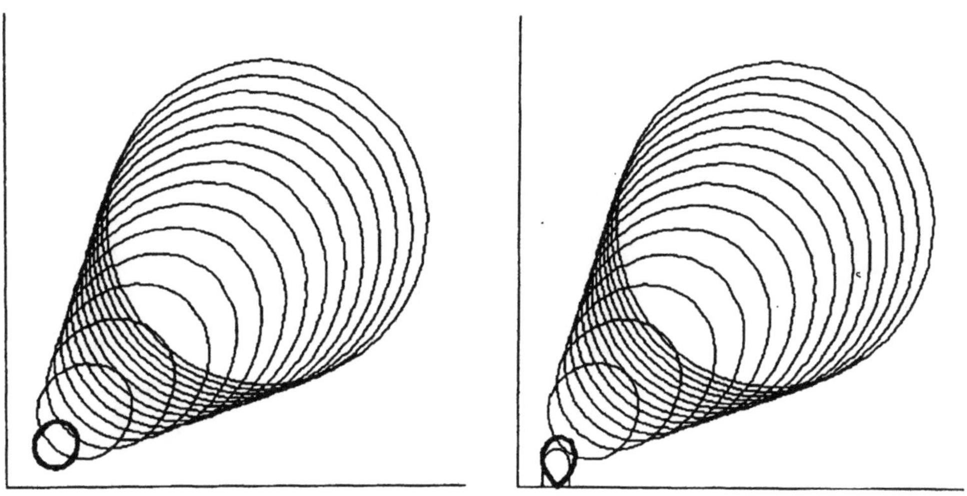

Figure 8.9. Evolution of a 4D-torus (Example VII): $t = 0.025i$ and t_r^*; a) $r = 0.62025$ ($t_r^\dagger = 0.3478$); b) $r = 0.62025125$ ($t_r^* = 0.3485$, $t_r^\dagger = 0.3529$).

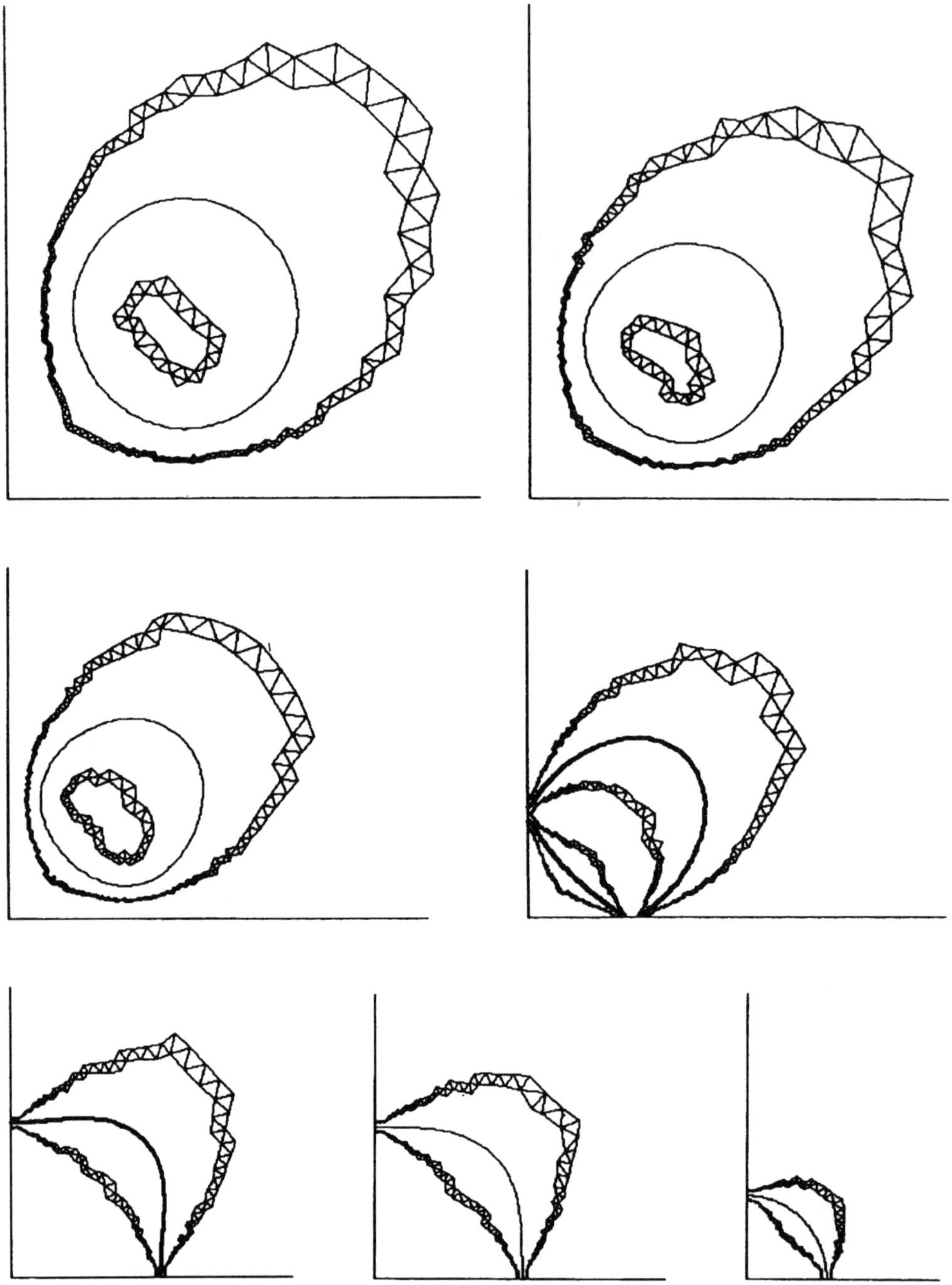

Figure 8.10. Evolution of a 4D-torus (Example VII): Consecutive meshes for $r = 0.625$ and $t = 0,\ 0.1,\ 0.2,\ t_r^*,\ t_r^{**},\ 0.3,\ 0.4$.

Example VIII (4D-toroid). We consider $S^1 \times S^2 = \partial(S^1 \times D^3)$, $n_1 = n_2 = 2$, and the initial surface given by $\Sigma_r(0) = \{\mathbf{z} \in [0,4)^2 : p_r(\mathbf{z}) = 0\}$, where

$$p_r(\mathbf{z}) = \begin{cases} 1 - r - z_1 & \text{if } z_2 < 2 \text{ and } z_1 < 1 \\ z_1 - 1 - r & \text{if } z_2 < 2 \text{ and } z_1 \geq 1 \\ \sqrt{(z_1-1)^2 + (z_2-2)^2} - r & \text{if } z_2 \geq 2. \end{cases}$$

The density function is $a(\mathbf{z}) = \max(0.05, |\mathbf{z}|)$ and the scaling factors are $h = 0.04$ and $\varepsilon = 0.1235$.

Choosing a quite oblong initial front was an attempt to obtain a nonunique evolution for the critical value r^*. The evolution is however very similar to that in Example VI, as depicted in Figure 8.11 for $r_1 < r^* < r_2 < r^{**} < r_3$.

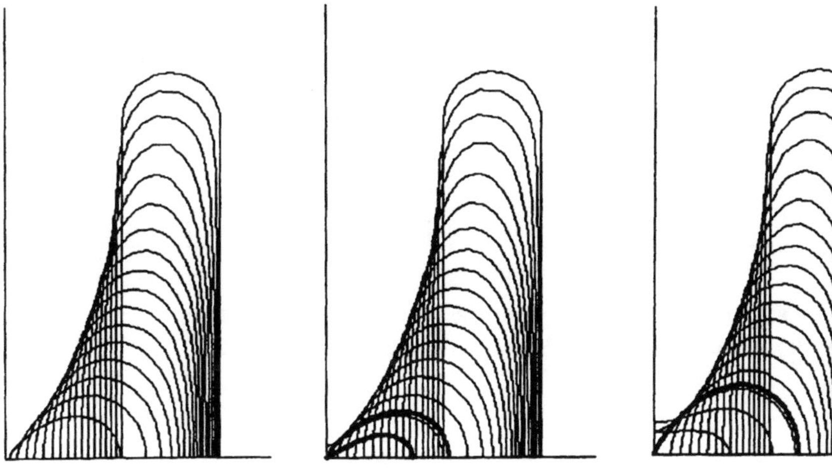

Figure 8.11. Evolution of the 4D-toroid (Example VIII): $t = 0.02i$, t_r^*, and t_r^{**}; a) $r_1 = 0.29393$ ($t_{r_1}^\dagger = 0.3380$); b) $r_2 = 0.29394$ ($t_{r_2}^* = 0.3162$, $t_{r_2}^{**} = 0.3333$, $t_{r_2}^\dagger = 0.3387$); c) $r_3 = 0.295$ ($t_{r_3}^* = 0.2953$, $t_{r_3}^\dagger = 0.3489$).

8.4. Eight-dimensional experiments

Example IX (Simons cone). We consider $n_1 = n_2 = 4$, the initial front given by (8.1), and all the remaining data as in Example II (see Figure 8.12). The purpose of this test is to simulate the evolution of an initial surface related to the minimal cone discovered by Simons [38] and studied by Bombieri, De Giorgi, and Giusti [8], whose existence requires at least dimension 8 of the ambient space. Such cone does not appear to be numerically stable, presumably because of mesh asymmetries, boundary conditions, and the effect of diffusion ($\varepsilon > 0$). However the evolution is now much slower than that of Examples II and IV, as indicated by the very small interface velocity of Figure 8.13 for $\lambda = 0.0015$ (compare with Figures 8.2 and 8.5). We finally stress the remarkable behavior for $\lambda = 0.001$ depicted in Figure 8.12b: the front moves initially very slowly to the right, but then stops and recedes. This behavior is consistent with the stability of the Simons cone and contrasts sharply with our cone simulations in 2D and 4D.

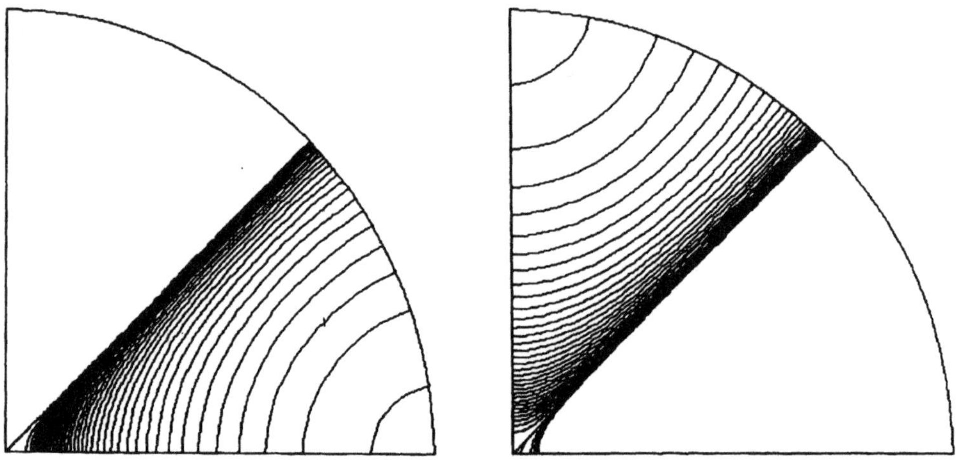

Figure 8.12. Evolution of the Simons cone: $t = 0.01i$; a) $\lambda = 0.0015$ ($t_\lambda^\dagger = 0.5525$); b) $\lambda = 0.001$ ($t_\lambda^\dagger = 0.4237$).

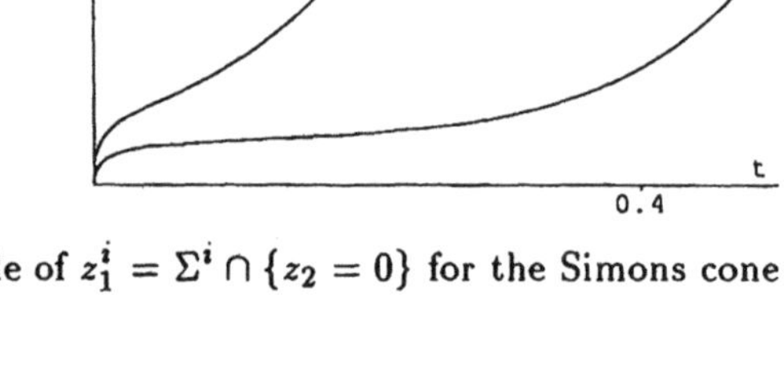

Figure 8.13. Profile of $z_1^i = \Sigma^i \cap \{z_2 = 0\}$ for the Simons cone evolution (Example IX).

9. Conclusions

We have shown how to exploit key features of a double obstacle formulation of a singularly perturbed reaction-diffusion PDE, which is viewed as an approximation of the geometric motion of fronts. Space discretization is based on piecewise linear finite elements over a graded mesh. The so-called *dynamic mesh algorithm* triangulates only the transition region, which is significantly smaller than a full mesh and the sole partition present at any time, and then updates the mesh to follow the layer motion. Time stepping is explicit, but stability constraints force small time steps only when singularities develop, whereas relatively large time steps are allowed at the beginning or past singularities, when the evolution is smooth. Performance is fully documented via a number of numerical simulations for the motion by mean curvature. Such results indicate quite conclusively savings in CPU time for a desired accuracy, together with enhanced singularity resolution through the use of a space-time dependent relaxation parameter. Convergence past singularities, along with optimal interface error estimates before the onset of singularities, provide a sound mathematical basis. These results extend to the semidiscrete finite element approximation. The proposed method exhibits desirable computational properties and potentials that make it appropriate as a building block for adaptive solvers of more complex phase transitions.

References

[1] Allen, S.M., Cahn, J.W., A macroscopic theory for antiphase boundary motion and its application to antiphase domain coarsing. Acta Metall. 27 (1979), 1085–1095.

[2] Altschuler, S., Angenent, S., Giga Y., Mean curvature flow through singularities for surfaces of rotation. Hokkaido Univ. Preprint Series # 130 (1991).

[3] Baldo, S., Minimal interface criterion for phase transitions and mixture of a Cahn Hilliard fluid. Ann. Inst. H. Poincaré Anal. Non Linéaire 7 (1990), 67–90.

[4] Baldo, S., Bellettini, G., Γ-convergence and numerical analysis: an application to the minimal partition problem. Ricerche Mat. (to appear).

[5] Barles, G., Soner, H.M., Souganidis, P.E., Fronts propagations and phase field theory. SIAM J. Control Optim. (to appear).

[6] Bellettini, G., Paolini, M., Verdi, C., Γ-convergence of discrete approximations to interfaces with prescribed mean curvature. Atti Accad. Naz. Lincei Cl. Sci. Fis. Mat. Natur. Rend. (9) Mat. Appl. 1 (1990), 317–328.

[7] —, Numerical minimization of geometrical type problems related to calculus of variations. Calcolo 27 (1991), 251–278.

[8] Bombieri, E., De Giorgi, E., Giusti, E., Minimal cones and the Bernstein problem. Invent. Math. 7 (1969), 243–268.

[9] Brakke, K.A., The Motion of a Surface by its Mean Curvature. Princeton University Press, Princeton 1978.

[10] Bronsard, L., Kohn, R.V., Motion by mean curvature as the singular limit of Ginzburg-Landau dynamics. J. Differential Equations 90 (1991), 211–237.

[11] Chen, X., Elliott, C.M., Asymptotics for a parabolic double obstacle problem. IMA Preprint # 922 (1992), 1–25,

[12] Chen, Y.G., Giga, Y., Goto, S., Uniqueness and existence of viscosity solutions of generalized mean curvature flow equation. J. Differential Geom. 33 (1991), 749–786.

[13] Ciarlet, P.G., The Finite Element Method for Elliptic Problems. North-Holland, Amsterdam 1978.

[14] Crandall, M.G., Ishii, H., Lions, P.L., User's guide to viscosity solutions of second order partial differential equations. Bull. Amer. Math. Soc. (N.S.) 27 (1992), 1–67.

[15] De Giorgi, E., Some conjectures on flow by mean curvature. In: Methods of Real Analysis and Partial Differential Equations (ed. by M.L. Benevento, T. Bruno, C. Sbordone). Liguori, Napoli (to appear).

[16] De Mottoni, P., Schatzman, M., Geometrical evolution of developped interfaces. Trans. Amer. Math. Soc. (to appear).

[17] Dziuk, G., An algorithm for evolutionary surfaces. Numer. Math. 58 (1991), 603–611.

[18] Ecker, K., Huisken, G., Mean curvature evolution of entire graphs. Ann. of Math. (2) 130 (1989), 453–471.

[19] Evans, L.C., Soner, H.M., Souganidis, P.E., Phase transitions and generalized motion by mean curvature. Comm. Pure Appl. Math. 45 (1992), 1097–1123.

[20] Evans, L.C., Spruck, J., Motion of level sets by mean curvature I. J. Differential Geom. 33 (1991), 635–681.

[21] French, D.A., The finite element method for a degenerate elliptic equation. SIAM J. Numer. Anal. 24 (1987), 788–815.

[22] Grayson, M., The heat equation shrinks embedded plane curves to round points. J. Differential Geom. 26 (1987), 285–314.

[23] —, A short note on the evolution of surfaces via mean curvature. Duke Math. J. 58 (1989), 555–558.

[24] Huisken, G., Asymptotic behavior for singularities of the mean curvature flow. J. Differential Geom. 31 (1990), 285–299.

[25] Ilmanen, T., Convergence of the Allen-Cahn equation to Brakke's motion by mean curvature. Preprint (1992).

[26] Modica, L., Mortola, S., Un esempio di Γ-convergenza. Boll. Un. Mat. Ital. B (5) 14 (1977), 285–299.

[27] Nochetto, R.H., A note on the approximation of free boundaries by finite element methods. RAIRO Modél. Math. Anal. Numér. 20 (1986), 355–368.

[28] Nochetto, R.H., Paolini, M., Verdi, C., Optimal interface error estimates for the mean curvature flow. Ann. Scuola Norm. Sup. Pisa Cl. Sci. (4) (to appear).

[29] —, Sharp error analysis for curvature dependent evolving fronts. Math. Models Methods Appl. Sci. (to appear).

[30] —, Variable relaxation parameter for smooth geometric front evolutions: local interface error estimates, preprint.

[31] —, Convergence of parabolic double obstacles to the generalized motion of fronts, in preparation.

[32] Osher, S., Sethian, J.A., Fronts propagating with curvature dependent speed: algorithms based on Hamilton-Jacobi formulations. J. Comput. Phys. 79 (1988), 12–49.

[33] Owen, N.C., Rubinstein, J., Sternberg, P., Minimizer and gradient flows for singularly perturbed bi-stable potentials with a Dirichlet condition. Proc. Roy. Soc. London Ser. A 429 (1990), 505–532.

[34] Paolini, M., Verdi, C., An automatic triangular mesh generator for planar domains. Riv. Inform. 20 (1990), 251–267.

[35] —, Asymptotic and numerical analyses of the mean curvature flow with a space-dependent relaxation parameter. Asymptotic Anal. 5 (1992), 553–574.

[36] Rubinstein, J., Sternberg, P., Keller, J.B., Fast reaction, slow diffusion and curve shortening. SIAM J. Appl. Math. 49 (1989), 116–133.

[37] Sethian, J.A., Recent numerical algorithms for hypersurfaces moving with curvature-dependent speed: Hamilton-Jacobi equations, conservation laws. J. Differential Geom. 31 (1990), 131–162.

[38] Simons, J., Minimal varieties in riemannian manifolds. Ann. of Math. (2) 88 (1968), 62–105.

[39] Soner, H.M., Motion of a set by the curvature of its boundary. J. Differential Equations (to appear).

[40] Soner, H.M., Souganidis, P.E., Singularities and uniqueness of cylindrically symmetric surfaces moving by mean curvature. Carnegie-Mellon Univ. Research Report # 92-NA-010 (1992).

[41] Taylor, J.E., Cahn, J.W., Handwerker, C.A., Geometric models of crystal growth. Acta Metall. 40 (1992), 1443–1474.

[42] Zienkiewicz, O.C., The Finite Element Method in Engineering Science. Mc Graw Hill, London 1971.

On radial growth in crystallization theories of hyperbolic nature

*P. Podio-Guidugli and G. Vergara Caffarelli**

Abstract. The predictions of two recently proposed models of crystal growth, the free-boundary (FB) and the hyperbolic (H) model, are compared. For each model sharp a priori estimates for the time evolution of smooth solutions describing isotropic, radial crystallization in two space dimensions are established. It is shown that smooth radial solutions exist for all time for the H model; for the FB model, until an arbitrarily large, finite horizon is reached, in a finite time. In particular, it is shown that, for both models, crystals melt in finite time if they are initially too small, or else if the liquid phase is stable relative to the solid phase; whereas, if the solid phase is relatively stable, crystal seeds that are initially sufficiently large do grow bigger and bigger, at a rate for which uniform upper and lower bounds are given.

In addition, the linearized equations of certain perturbed motions around radial crystallization motions are deduced, and the evolution of the perturbation with time is studied, in the case of stable, growing crystals. For the H model, radial positional disturbances of the interface that depend on angle are considered; for the FB model, the same type of interface perturbations are coupled with a consistent representation for the perturbed potential that rules the melt's flow. It is found that a curious exchange of roles takes place between positional and velocity disturbances of interface and potential, in the sense that a velocity (positional) disturbance of the interface induces a positional (velocity) disturbance of the potential, and conversely.

1991 Mathematics Subject Classification: 35L70, 35K99, 70K99.

1. Introduction

Melting-freezing waves are oscillations of a solid-liquid interface due to solid→liquid or liquid→solid changes of phase at the interface itself; atoms in the solid phase move only when they leave that phase. For crystals of helium-4 in their melt such waves were predicted by Andreev and Parshin [1] in 1978, and then observed by Keshishev, Parshin, and Babkin [2] in 1979.

* The research here presented was supported by M.U.R.S.T., Progetto Nazionale "Termomeccanica dei Continui Classici e dei Materiali Nuovi" and by C.N.R., Progetto Bilaterale "Modelli Matematici per Materiali Moderni". We are greatly indebted to Mort Gurtin for valuable discussions.

Two models have recently been proposed to capture the peculiar phenomenology of melting-freezing waves and, more generally, to account for the growth of a solid crystal in its liquid bath. Both models are purely mechanical, a simplification judged appropriate when one deals with processes of solidification and liquefaction in superconductors such as solid helium and its melt (cf. Maris and Andreev [3]).

The first model (Gurtin [4]), which we shall refer to as the **FB** (free-boundary) **model**, sees the crystal-interface-bath system as a bounded, n-dimensional rigid body interacting with an infinite environment comprised of an incompressible, inviscid liquid; the interaction takes place through a $(n-1)$-dimensional *massless* material body, the **interface**, for which constitutive equations are postulated; such constitutive equations " ... allow the interfacial energy, the interfacial force, and the interactive dissipation to depend on the orientation and normal velocity of the interface ..." ([4], p. 288); since the crystal is regarded as rigid and quiescent, and the interface has no mass, the inertial effects accounted for by the model arise by the motion of the melt.

The second model (Gurtin and Podio-Guidugli [5]), which we shall call the **H** (hyperbolic) **model**, confines attention to the interface alone: the melt is considered as a source of atoms for the crystallization process, and its motion accounted for only through an *effective inertia* endowed to the interface; moreover, costitutive equations similar to those of the FB model are postulated for the interface.

Both models are based on a framework for continuum theories of crystallization developed by Gurtin and coworkers in [6, 7, 8], but the mathematical problems to which these two models lead are different from each other and from the typical problems of the crystallization theories implemented so far within the same framework. The latter theories are of *parabolic* nature, in that they may be regarded as a broad generalization of the parabolic theory of motion of plane curves by their mean curvature. The H model has instead *hyperbolic* nature, since it leads to an evolution equation for plane interfaces that in its simplest instance appears as a hyperbolic version of the classic curve-shortening equation (vid. Remark 1 in the next section). The nature of the FB model is hard to classify, since that model yields a nonstandard free-boundary problem that not only supports oscillatory solutions interpreted as melting-freezing waves, but also has some parabolic traits (e.g., Rogers [9] argues that, when the problem is linearized about an equilibrium solution that pictures a flat interface, an initial datum having the form of a "square impulse" should propagate at infinite speed).

The H and FB models seem to indicate that the dynamics of crystallization is not necessarily parabolic; no matter how different, they seem capable of mutually matching forecasts that are also physically meaningful. As to melting-freezing waves in helium-4, the predictions of the two models are in fair qualitative agreement [4, 5] both reciprocal and with the available experimental evidences (but only the first model yields the dispersion relation found in experiments by Keshishev, Parshin, and Babkin [2]). As to crystal growth, neither model has been subjected yet to mathematical analysis, let alone numerical computations, in the general case when the constitutive coefficients depend on the place and the orientation of the interface, and the interface itself has no *a priori*, immutable geometrical speciality; however, the few results suitable for comparison [4, 5] agree fairly well on qualitative grounds.

It is the purpose of this paper to compare the predictions of the two models for crystal growth, in the simplest case of an isotropic crystal having initially the shape of a circle and, as to the FB model, of a bath whose flows are both irrotational and, in a sense made precise in [4], slow. The case of spherically symmetric growth of an isotropic crystal has been dealt with briefly in [5] for the H model, and in much more detail in [4] for the FB model; in the framework of the latter, a stability analysis of both spherically symmetric and planar crystallization motions has been performed by Feribach [10].

For each model we here establish sharp a priori estimates for the time evolution of smooth solutions describing **radial crystallization in two space dimensions**. Both the techniques of proof and the resulting estimates are remarkably similar in the two cases. Smooth radial solutions exist for all time for the H model; for the FB model, until an arbitrarily large, finite horizon is reached, in a finite time.[1] In particular, we show that, just as in the three-dimensional case studied in [4, 5], crystals melt in finite time if they are initially too small, or else if the liquid phase is stable relative to the solid phase; whereas, if the solid phase is relatively stable, crystal seeds that are initially sufficiently large grow bigger and bigger, at a rate for which we furnish uniform upper and lower bounds.

In addition, we deduce the linearized equations of certain perturbed motions around such ground crystallization motions; and we determine the evolution of the perturbation with time in the case of stable, growing crystals. Precisely, for the H model we consider radial positional disturbances of the interface that depend on angle, and show that the ensuing perturbations are uniformly controlled by the initial data, so that their ratio to the ground radial solutions decays as the latter grows unboundedly with time. For the FB model, we allow the same types of positional disturbances and accompanying perturbations for the interface, and introduce a consistent representation for the perturbed potential that rules the melt's flow. We find that a curious exchange of roles takes place between positional and velocity disturbances of interface and potential, in the sense that a velocity (positional) disturbance of the interface induces a positional (velocity) disturbance of the potential, and conversely.

Since both models allow for equilibrium solutions when the crystal radius has a critical value, we also look at the evolution of perturbations of such equilibrium solutions. We find that critical-size crystals are inherently unstable under uniform radial perturbations.

1 The finiteness of the horizon is a technical requirement that is dictated by the nature that the velocity potential describing irrotational flows has in dimension two (vid. Remark 2 in the next section).

2. The FB and H models

We refer the reader to [4] and [5], respectively, for motivated presentations of the FB and H models starting from first principles.

The theory of [4], in the so-called *weak-inertia approximation*, leads to the following system of partial differential equations for the evolution of the interfacial curve:

$$\Delta U = 0 \tag{2.1}$$

in the melt and

$$\partial_t U = \beta V - \psi K, \quad \partial_{\mathbf{m}} U = \alpha V \tag{2.2}$$

on the (unknown) interface, subject to suitable initial and far-field conditions. Here $U(\mathbf{x},t)$ is a **velocity potential** for the melt's motion, presumed irrotational; ∂_t denotes partial differentiation with respect to time t, and $\partial_{\mathbf{m}} U := \mathrm{grad} U \cdot \mathbf{m}$ with $\mathbf{m}$ the unit vector of the normal to the interface, oriented toward the melt; V is the **normal velocity**, and K the **curvature**, of the interface; β and ψ, two positive constants, are the **kinetic coefficient** and the **energy** of the interface; and

$$\alpha := \frac{(\rho_c - \rho_m)^2}{\rho_c}, \tag{2.3}$$

with ρ_c, ρ_m the constant **crystal** and **melt mass densities.**

The theory of [5] yields the single *hyperbolic* equation

$$\rho V^\circ + \beta V - \psi K + F = 0, \tag{2.4}$$

where V° is the time derivative of V following a trajectory orthogonal to the interfacial curve, ρ is the **effective inertia density** of the interface, and F is a constant which represents the driving force for crystallization.

Remark 1. To see why it is appropriate to classify the evolution equation (2.4) as a hyperbolic equation, we give it a form directly reminiscent of the standard hyperbolic format. Let $\mathbf{s}(s,t)$ be the position vector of a point on the interfacial curve, with the parameter s interpreted as the arc length. Then,

$$\mathbf{s}^\circ = V\mathbf{m}, \quad \mathbf{s}^\circ \cdot \mathbf{m} = V, \quad \mathbf{s}^{\circ\circ} \cdot \mathbf{m} = V^\circ, \quad \mathbf{s}_s = \mathbf{l}, \quad \mathbf{s}_{ss} = K\mathbf{m},$$

where $\mathbf{l}$ is the unit vector of the tangent to the interface. Consequently, if we let $\mathbf{F} := F\mathbf{m}$, we can write (2.4) as follows:

$$(\psi \mathbf{s}_{ss} - \rho \mathbf{s}^{\circ\circ} - \beta \mathbf{s}^\circ - \mathbf{F}) \cdot \mathbf{m} = 0.^2 \tag{2.5}$$

The "hyperbolicity" of (2.4) is then suggested by the explicit appearance of the "wave operator"

$$\Box_1 \mathbf{s} := \psi \mathbf{s}_{ss} - \rho \mathbf{s}^{\circ\circ}.$$

---

2 For

$$G := (\psi \mathbf{s}_{ss} - \rho \mathbf{s}^{\circ\circ} - \beta \mathbf{s}^\circ - \mathbf{F}) \cdot \mathbf{l},$$

Since it is expressed in terms of the time derivative following orthogonal trajectories, this result may look formal. However, we can recover it after a suitable re-parametrization of the interfacial curve. Precisely, we let the position vector be expressed by

$$\mathbf{p}(p,t) := \mathbf{s}(s(p,t),t), \tag{2.6$_1$}$$

where the mapping $s(\cdot,t)$ has inverse $p(\cdot,t)$, with $p_s(\cdot,t) > 0$, and where

$$\mathbf{p}_p \cdot \mathbf{p}_t = 0, \quad \mathbf{p}_t = V\mathbf{m}. \tag{2.6$_2$}$$

The specialty of (2.6) is that with such a parametrization the curve evolves along its orthogonal trajectories. Since

$$\mathbf{l} = p_s \mathbf{p}_p, \quad \mathbf{s}^{\circ} = \mathbf{p}_t, \quad etc.,$$

(2.5) becomes

$$(\psi(\mathbf{p}_p \cdot \mathbf{p}_p)^{-1}\mathbf{p}_{pp} - \rho\mathbf{p}_{tt} - \beta\mathbf{p}_t - \mathbf{F}) \cdot \mathbf{m} = 0,$$

an equation ruled by the "wave operator"

$$\Box_2 \mathbf{p} := \psi(\mathbf{p}_p \cdot \mathbf{p}_p)^{-1}\mathbf{p}_{pp} - \rho\mathbf{p}_{tt}. \qquad \Box$$

When one restricts attention to radial solutions R(t), the quantities in (2.4) take the form

$$V = \dot{R}, \quad V^{\circ} = \ddot{R}, \quad K = -R^{-1}, \tag{2.7}$$

so that (2.4) becomes

$$(H\ model) \qquad \rho\ddot{R} + \beta\dot{R} + \psi R^{-1} + F = 0. \tag{2.8}$$

On the other hand, in the FB model the following choice of a *harmonic* velocity potential $U(r,t)$ with (arbitrarily large) *finite horizon* R_1 takes care of (2.1) and the far-field conditions formerly left unspecified:

$$U(r,t) = -Ft + B(t)log\frac{r}{R_1}, \; 0 < r \leq R_1 \tag{2.9}$$

---

we see that

$$G = -\rho\mathbf{s}^{\circ\circ} \cdot \mathbf{l} = -\rho V \mathbf{m}^{\circ} \cdot \mathbf{l}.$$

Thus, since $\mathbf{m}^{\circ} \times \mathbf{l} = \mathbf{0}$, the scalar equation (2.5) is equivalent to the vectorial equation

$$\psi\mathbf{s}_{ss} - \rho\mathbf{s}^{\circ\circ} - \beta\mathbf{s}^{\circ} - \mathbf{F} = \mathbf{0}$$

if and only if

$$V\mathbf{m}^{\circ} = \mathbf{0},$$

that is, e.g., for $V \neq 0$, when circles evolve into circles.

(here $B(\cdot)$ is an arbitrary, smooth function of time, and the constant F retains the given interpretation). With (2.9) we have that

$$\partial_t U = -F + \dot{B}(t) log \frac{r}{R_1}, \quad \partial_{\mathbf{m}} U = B(t) r^{-1}; \tag{2.10}$$

thus, equations (2.2) become, respectively,

$$-(log \frac{R}{R_1})\dot{B} + \beta \dot{R} + \psi R^{-1} + F = 0, \quad B(t) R^{-1} = \alpha \dot{R}, \tag{2.11}$$

and may be combined to yield

$$(FB\ model) \qquad -\alpha(log \tfrac{R}{R_1})(R\dot{R})^{\cdot} + \beta \dot{R} + \psi R^{-1} + F = 0. \tag{2.12}$$

For $R \leq R_1$, equations (2.8) and (2.12) describe finite-horizon, radial crystallization in the H and FB models, respectively. The corresponding radial evolution equation in the parabolic theory, namely,

$$(P\ model) \qquad \beta \dot{R} + \psi R^{-1} + F = 0\,,$$

obtains from either (2.8) or (2.12) when the relative "hyperbolicity coefficient" ρ or α is null, i.e., when the interface has no effective inertia or when the crystal and the melt have the same mass density.[3] Here we shall not deal with the parabolic model, but it would not be difficult to show that qualitative results analogous to those to be established in the next section for the H and FB models do hold for the P model as well (and are of course much easier to prove).

Remark 2. Both Gurtin [4] and Feribach [10] considered the problem of radial crystallization within the FB model only in three space dimensions. A difference between the two- and three-dimensional cases comes from the fact that Laplace equation exterior to a n-dimensional sphere has radial solutions $1, logr$ for $n = 2$, and $1, r^{-1}$ for $n = 3$. This leads one to assume the following representations for the velocity potential:

$$U(r,t) = A(t) + B(t) log \frac{r}{R_1}, \ n = 2;$$

$$U(r,t) = A(t) + B(t) r^{-1}, \ n = 3.$$

The far-field requirement that the melt pressure be spatially constant is satisfied by choosing $A(t) = -Ft$ in both cases [4]; for $n = 3$, $A(t)$ is interpreted as the velocity potential at infinity, for $n = 2$, as the velocity potential when r tends to the finite horizon R_1. In both cases, U and $\partial_r U$ have opposite signs within the

3 It would seem appropriate to postulate that the interface has a nonnull effective inertia whenever phase changes induce changes in specific volume, indeed a typical occurrence in the phenomenology of liquefaction-solidification: as to melting-freezing waves, e.g., Feribach reports in [10] that the density of solid helium-4 is roughly 20% greater than that of its melt.

horizon bounds; this property is crucial to make the qualitative predictions of the FB model to agree with those of the H model, as well as to establish, within the framework of the FB model but in the case $n = 2$, a growth theorem that parallels the one given in Section 11 of [4] for $n = 3$ (vid. next section). □

3. Radial crystallization

Under the assumption that both ρ and α are strictly positive, (2.8) and (2.12) may be given a form that will make it a little easier for us to demonstrate various qualitative and quantitative similarities between the two models. Let the constant R_c be defined to be

$$R_c := -\frac{\psi}{F}. \tag{3.1}$$

If we choose $R_1 \geq R_c$, and write R for R/R_1, R_c for R_c/R_1, then (2.8) can be given the form:

$$\ddot{R} + \alpha_1 \dot{R} + \alpha_2 (R^{-1} - R_c^{-1}) = 0, \qquad (3.2)_1$$

with

$$\alpha_1 := \frac{\beta}{\rho}, \quad \alpha_2 := \frac{\psi}{\rho R_1^2}; \qquad (3.2)_2$$

and (2.12) can be written as

$$-(log R)(R\dot{R})^{\cdot} + \bar{\alpha}_1 \dot{R} + \bar{\alpha}_2 (R^{-1} - R_c^{-1}) = 0, \qquad (3.3)_1$$

with

$$\bar{\alpha}_1 := \frac{\beta}{\alpha R_1}, \quad \bar{\alpha}_2 := \frac{\psi}{\alpha R_1^3}. \qquad (3.3)_2$$

The problems of radial crystal growth we wish to study obtain when we supplement either equations (3.2) or equations (3.3) by the initial conditions

$$R(0) = R_o, \quad \dot{R}(0) = 0.^4 \tag{3.4}$$

Whenever in our discussion to follow we do not esplicitly mention the one or the other of these two problems, it will be understood that the circumstance or the property under examination pertains to both problems.

In [4, Section 11] (vid. also [5], Subsection 6.1) Gurtin introduces the following terminology: the crystal is **stable** or **unstable** relative to the melt according as $F < 0$ or $F \geq 0$, i.e., due to (3.1), according as $R_c \geq 0$ or $R_c < 0$. We observe that, if the crystal is stable, then the crystallization problem has the *equilibrium solution*

$$R(t) \equiv R_c, \tag{3.5}$$

4 The case of nonnull initial velocity would not be difficult to handle (vid. [4]).

and R_c is called the **critical radius**; we shall see that the constant R_c enters the description of crystallization motions in many other ways.

Note that, for each time $\tilde{t}$ at which $\dot{R}(\tilde{t}) = 0$, and therefore also for $t = 0$, $\ddot{R}(\tilde{t})$ has the sign of $(R_c^{-1} - R^{-1}(\tilde{t}))$. If $F \geq 0$, i.e., $R_c \leq 0$, then $\ddot{R}(0) < 0$; otherwise, if $F < 0$, and thus $R_c > 0$, we pick $R_o < R_c$, so that once again $\ddot{R}(0)$ is strictly negative. This fact and the initial condition $(3.4)_2$ imply that in both instances $\dot{R}(t) < 0$ for some interval $]0, \bar{t}[$. Now, were not this interval the whole existence interval of the solution $R(t)$, there would be a time $\tilde{t}$ at which $R(t)$ would have a minimum, and thus not only $\dot{R}(\tilde{t}) = 0$ but also $\ddot{R}(\tilde{t}) \geq 0$, a contradiction. Thus, *an unstable crystal and a sufficiently small stable crystal seed both melt*; it remains for us to show that they do it *in a finite time*.

We know from our discussion just above that $\dot{R}(t) < 0$ both for $R_c \leq 0$ and for $0 < R_o < R_c$, so that

$$-R^{-1} < -R_o^{-1}, \tag{3.6$_1$}$$

and, moreover,

$$R_c^{-1} < R_o^{-1}. \tag{3.6$_2$}$$

For the H model, let

$$\gamma := \alpha_2(R_o^{-1} - R_c^{-1}); \tag{3.7}$$

inequalities (3.6) and equation $(3.2)_1$ imply that $\gamma > 0$, and that

$$\ddot{R} + \alpha_1 \dot{R} < -\gamma < 0. \tag{3.8}$$

Two integrations of (3.8) with the initial conditions (3.4) then yield

$$-R_o < R(t) - R_o < -\frac{1}{2}\gamma t^2 + \alpha_1 R_o t; \tag{3.9}$$

since the right side of (3.9) tends to $-\infty$ as $t \to +\infty$, the melting time must be finite (cf. [5], Subsection 6.1).

For the FB model, we use the identity

$$-(log R)(R\dot{R})^{\cdot} = -((log R)R\dot{R})^{\cdot} + \dot{R}^2$$

to give equation $(3.3)_1$ the form

$$-(log R)(R\dot{R})^{\cdot} + \dot{R}^2 + \bar{\alpha}_1 \dot{R} = \bar{\alpha}_2(R_c^{-1} - R^{-1}). \tag{3.10}$$

Next we set

$$\bar{\gamma} := \bar{\alpha}_2(R_o^{-1} - R_c^{-1}), \tag{3.11}$$

so that $\bar{\gamma} > 0$. We then call upon (3.6) to deduce from (3.10) a differential inequality that parallels (3.8), namely,

$$-((log R)(R\dot{R})^{\cdot} + \bar{\alpha}_1 \dot{R} < -\bar{\gamma}. \tag{3.12}$$

Integration of (3.12) yields

$$-(log R(t))R(t)\dot{R}(t) - \bar{\alpha}_1 R_o < -\bar{\gamma} t.$$

On the other hand, it is easy to see that, for e the Napier's number,

$$-(log R)R \leq e^{-1} \quad \text{for all } R > 0, \tag{3.13}$$

so that, recalling once again that $\dot{R}(t) < 0$, we have

$$e^{-1}\dot{R}(t) < -\gamma t + \bar{\alpha} R_o.$$

At this stage, another integration yields

$$-e^{-1}R_o < e^{-1}(R(t) - R_o) < -\frac{1}{2}\bar{\gamma}t^2 + \bar{\alpha}_1 R_o t, \tag{3.14}$$

and we conclude that the melting time must be finite.

We are left with the case of a stable crystal seed whose initial radius is greater than the critical radius.

In this case, for all $t \geq 0$,

$$0 < R_c < R_o \leq R(t) \leq 1;^5 \tag{3.15}$$

we note here for later use that inequalities (3.15) and definitions (3.7), (3.11) imply that, in particular, both γ and $\bar{\gamma}$ are negative. Moreover, a reasoning completely analogous to the one showing that both an unstable crystal and a stable, small crystal seed melt allows us to assert that $\ddot{R}(0)$ is strictly positive, and that, for all $t > 0$,

$$\dot{R}(t) > 0, \tag{3.16}$$

i.e., a stable crystal whose initial size is sufficiently large does grow bigger and bigger. We shall now exploit (3.15), (3.16) to establish *linear* growth estimates both from above and from below, for the two models.

To begin with, we write down the following consequences of $(3.2)_1$ and $(3.3)_1$, respectively:

$$0 \leq \alpha_2 \int_0^t R^{-1}(\tau)d\tau \leq \alpha_2 R_c^{-1} t - \alpha_1(R(t) - R_o) - \dot{R}(t), \tag*{$(3.17)_1$}$$

$$0 \leq \bar{\alpha}_2 \int_0^t R^{-1}(\tau)d\tau \leq \bar{\alpha}_2 R_c^{-1} t - \bar{\alpha}_1(R(t) - R_o) + (log R(t))R(t)\dot{R}(t). \tag*{$(3.17)_2$}$$

Due to (3.15) and (3.16), the last addendum in both of (3.17) is negative, and we have the a priori estimates from above

$$(H\ model) \qquad R(t) - Ro < (\alpha_1 R_c)^{-1}\alpha_2 t, \tag*{$(3.18)_1$}$$

$$(FB\ model) \qquad R(t) - Ro < (\bar{\alpha}_1 R_c)^{-1}\bar{\alpha}_2 t, \tag*{$(3.18)_2$}$$

5 The limitation $R(t) \leq 1$ is mandatory only in the case of the FB model; we stipulate it in both cases because we aim to a comparison of the two models in a common range of validity. However, for the H model, all the estimates we derive (or, on occasions, slightly less sharp versions of those: cf., e.g., $(3.21)_2$ and its consequence listed in footnote 7) hold true in the whole positive half-line.

Passing now to the growth estimates from below, we deal with the easier H model first. As γ is negative, we have from $(3.2)_1$ that

$$0 < -\gamma < \alpha_1 \dot{R}(t) + \ddot{R}(t). \tag{3.19}$$

With the use of the integrating factor $e^{\alpha_1 t}$ we then obtain from $(3.19)_2$ that

$$(H\ model) \qquad R(t) - Ro > max\{0\,, -\gamma\alpha_1^{-1}(t - \alpha_1^{-1})\}. \tag{3.20}$$

The estimates $(3.18)_1$ and (3.20) allow us to conclude that, *in the H model, the radius of a stable crystal seed whose initial size is sufficiently large grows as t for* $t \to +\infty$.[6]

An integrating factor is not in sight for the FB model, but fortunately we can do without it, at the expenses of a somewhat more involved analysis.

Our first step consists in proving the following upper estimates of $\dot{R}(t)$:

$$(H\ model) \qquad \dot{R}(t) \le A = \alpha_1^{-1}\alpha_2(R_c^{-1} - 1), \tag{3.21$_1$}$$

$$(FB\ model) \qquad \dot{R}(t) \le \bar{A} = \bar{\alpha}_1^{-1}\bar{\alpha}_2(R_c^{-1} - 1); \tag{3.21$_2$}$$

these estimates hold uniformly for all times t in the existence interval of the solution $R(t)$ which are smaller than the time necessary to reach the horizon.[7]

Recall that $\ddot{R}(0) > 0$. We have an alternative: either $\ddot{R}(t)$ remains positive for all times or its graph crosses the time axis at a point $\tilde{t}$. In the first instance, since $-R^{-1}(t)$ increases with t, it follows immediately from $(3.2)_1$ that

$$\alpha_1 \dot{R} < \alpha_2(R_c^{-1} - R^{-1}) < \alpha_2(R_c^{-1} - 1) \tag{3.22}$$

an identical chain of inequalities, with overbarred α's, follows from $(3.3)_1$ and the fact that, under the present circumstances,

$$-(logR(t))(R(t)\dot{R}(t))^{\cdot} = -(logR(t))(\dot{R}(t)^2 + R(t)\ddot{R}(t)) > 0. \tag{3.23}$$

In the second instance, $\tilde{t}$ is a critical point of $\dot{R}(t)$, in fact, its unique maximum. To see this it is enough to use the equation $(3.2)_1$ (respectively, $(3.3)_1$ and (3.23)) and deduce that

$$\dot{R}(\tilde{t}) = \psi(R(\tilde{t})), \tag{3.24}$$

where

$$\phi(R) = \alpha_1^{-1}\alpha_2(R_c^{-1} - R^{-1}) \tag{3.25$_1$}$$

6 This result improves the corresponding result in [5], Subsection 6.1

7 For the H model, $(3.21)_1$ trivially yields the estimate

$$\dot{R}(t) \le A = \alpha_1^{-1}\alpha_2 R_c^{-1},$$

which holds for all $t > 0$.

(respectively,

$$\phi(R) = \frac{2\bar{\alpha}_1^{-1}\bar{\alpha}_2(R_c^{-1} - R^{-1})}{1 + \sqrt{1 - 4\bar{\alpha}_1^{-1}\bar{\alpha}_2(R_c^{-1} - R^{-1})logR}}). \tag{3.25$_2$}$$

The desired conclusion follows from the fact that both mappings $\phi(R)$ in (3.25), as well as $R(t)$, are increasing.[8]

We now return to the evolution equations with the estimates (3.21) at hand.

For the H model, an integration of (3.19) yields

$$-\gamma t < \alpha_1(R(t) - R_o) + \dot{R}(t) < \alpha_1(R(t) - R_o) + A; \tag{3.26}$$

consequently,

$$(H\ model) \qquad R(t) - R_o > max\{0, -\gamma\alpha_1^{-1}(t - A(-\gamma)^{-1})\}; \tag{3.27}$$

since $A(-\gamma)^{-1} > \alpha_1^{-1}$, the estimate (3.20) is a little sharper than (3.27).

Similarly, for the FB model we have from (3.10) that

$$-\bar{\gamma} < -((logR)R\dot{R})^{\cdot} + \dot{R}^2 + \bar{\alpha}_1\dot{R} < -((logR)R\dot{R})^{\cdot} + (\bar{\alpha}_1 + \bar{A})\dot{R}, \tag{3.28}$$

so that, after an integration and with the use of (3.13),

$$-\bar{\gamma}t < -(logR)R\dot{R} + (\bar{\alpha}_1 + \bar{A})(R(t) - R_o) < e^{-1}\bar{A} + (\bar{\alpha}_1 + \bar{A})(R(t) - R_o); \tag{3.29}$$

it follows from (3.29) that

$$(FB\ model) \qquad R(t) - R_o > max\{0, -\bar{\gamma}(\bar{\alpha}_1 + \bar{A})^{-1}(t - \bar{A}(-\gamma e)^{-1})\}. \tag{3.30}$$

On the basis of $(3.18)_2$ and (3.30) we see that, provided the crystal is stable relative to the melt and the initial radius is greater than the critical radius, *a solution to the radial crystallization problem exists in the FB model until the horizon is reached* (in a *finite* time that may be estimated both from above and from below).

---

8 As a bonus, the above argument also shows that $(R(t)\dot{R}(t))^{\cdot} > 0$ for all t greater than zero and less than the time necessary to reach the horizon. Indeed, if $\ddot{R}(t) \geq 0$, the announced property is trivially verified. If $\ddot{R}(t < 0$, and hence $t > \tilde{t}$, since both $R(t) > R(\tilde{t})$ and $0 < \dot{R}(t) < \dot{R}(\tilde{t})$, we have from $(3.25)_2$ that

$$\bar{\alpha}_1\dot{R}(t) + \bar{\alpha}_2(R^{-1}(t) - R_c^{-1}) < \bar{\alpha}_1\dot{R}(\tilde{t}) + \bar{\alpha}_2(R^{-1}(\tilde{t}) - R_c^{-1}) < 0;$$

this and equation $(3.3)_1$ imply that $(R(t)\dot{R}(t))^{\cdot}$ must be positive.

4. Perturbation of crystallization motions in the H model

We find it expedient to use polar coordinates (r, θ) in the plane, and represent the radial crystallization motions whose perturbations we wish to study in the form

$$\mathbf{R}(\theta, t) = R(t)\mathbf{n}(\theta), \tag{4.1}$$

where $\mathbf{n}(\theta) \equiv (cos\theta, sin\theta)$, and $R(t)$ is a solution of the evolution equation (3.2) meeting the initial conditions (3.4).

The perturbed motions we consider have the form

$$\mathbf{r}(\theta, t; \epsilon) = (R(t) + \epsilon u(\theta, t))\mathbf{n}(\theta),^{9} \tag{4.2}$$

with $u(\cdot, t)$ a 2π-periodic function for all t, subject to the initial conditions

$$\mathbf{r}(\theta, 0; \epsilon) = (R_o + \epsilon u_o(\theta))\mathbf{n}(\theta), \ \mathbf{r}_t(\theta, 0; \epsilon) = \mathbf{0}, \tag{4.3}$$

where u_o is a given 2π-periodic function.

Remark 1. The form (4.2) may seem special; however, to within second-order terms in the perturbation parameter ϵ, it is just as general as the forms

$$\bar{\mathbf{r}}(\theta, t; \epsilon) = (R(t) + \epsilon u(\theta, t))\mathbf{n}(\theta - \epsilon\phi(\theta, t)), \tag{4.4}$$

where the *radial* perturbation $u(\theta, t)$ is supplemented by a *phase* perturbation $\phi(\theta, t)$; and

$$\tilde{\mathbf{r}}(\theta, t; \epsilon) = (R(t) + \epsilon u(\theta, t))\mathbf{n}(\theta) + \epsilon v(\theta, t)\mathbf{t}(\theta), \tag{4.5}$$

where $\mathbf{t}(\theta) \equiv (sin\theta, -cos\theta)$, and a *tangential* perturbation $v(\theta, t)$ is taken into account. Indeed, if we assume (4.4), (4.2) is recovered by noticing that, for $\psi := \theta - \epsilon\phi(\theta, t)$, $\epsilon u(\theta, t) \cong \epsilon u(\psi, t)$,[10] and thus

$$\bar{\mathbf{r}}(\theta, t; \epsilon) \cong (R(t) + \epsilon u(\psi, t))\mathbf{n}(\psi) = \mathbf{r}(\psi, t; \epsilon).$$

On the other hand, it is obvious that, conversely, (4.2) can be given the form (4.4). Moreover, since $\mathbf{n}(\theta - \epsilon\phi(\theta, t)) \cong \mathbf{n}(\theta) - \epsilon\phi(\theta, t)\mathbf{t}(\theta)$, we see that, provided we choose $v(\theta, t) = -R(t)\phi(\theta, t)$,

$$\bar{\mathbf{r}}(\theta, t; \epsilon) \cong \tilde{\mathbf{r}}(\theta, t; \epsilon). \qquad \square$$

Remark 2. In (4.5) the tangential perturbation $v(\theta, t)$ may be chosen in such a way as to satisfy (to within $O(\epsilon^2)$-terms) the request that the perturbed curve evolve along its orthogonal trajectories (cf. Remark 1, Section 2). Indeed, condition $(2.6)_1$ translates now into

$$\tilde{\mathbf{r}}_\theta \cdot \tilde{\mathbf{r}}_t \cong 0, \tag{4.6}$$

9 Thus, $\mathbf{r}(\theta, t; 0) = \mathbf{R}(\theta, t)$, $\mathbf{r}_\epsilon(\theta, t; 0) = u(\theta, t)\mathbf{n}(\theta)$.

10 We have here introduced the symbol $\cong$ to mean that equality holds to within $O(\epsilon^2)$-terms.

and a straightforward computation shows that (4.6) is met if and only if $v(\theta,t)$ is determined in terms of the basic motion $R(t)$ and the radial perturbation $u(\theta,t)$ by the first-order equation

$$Rv_t - \dot{R}(v+u_\theta) = 0. \qquad \square$$

In view of the preceding remarks, the study of perturbed motions we aim at is essentially the study of the evolution of the radial perturbation $u(\theta,t)$; we now derive the relative governing equation, which is the first-order approximation of (2.4).

It follows from (4.2) that

$$V \cong \dot{R} + \epsilon u_t, \qquad V^\circ \cong \ddot{R} + \epsilon u_{tt}. \tag{4.7}$$

It also follows from (4.2) that

$$\begin{aligned} \mathbf{r}_\theta &\cong -R\mathbf{t} + \epsilon(u_\theta\mathbf{n} - u\mathbf{t}), \\ \mathbf{r}_{\theta\theta} &\cong -R\mathbf{n} + \epsilon((u_{\theta\theta} - u)\mathbf{n} - 2u_\theta\mathbf{t}), \\ \mathbf{m} &\cong \mathbf{n} + \epsilon R^{-1}u_\theta\mathbf{t}, \end{aligned} \tag{4.8}$$

and hence

$$-K = (\mathbf{r}_\theta \cdot \mathbf{r}_\theta)^{-1}\mathbf{r}_{\theta\theta} \cdot \mathbf{m} \cong R^{-1} - \epsilon R^{-2}(u_{\theta\theta} + u). \tag{4.9}$$

If we insert (4.7) and (4.9) into (2.4), and use (2.8) and $(3.2)_2$, we obtain

$$\rho u_{tt} + \beta u_t - \psi R^{-2}(u_{\theta\theta} + u) = 0; \tag{4.10}$$

therefore, if we again write R for R/R_1 and recall $(3.2)_2$, the radial perturbation of a given radial crystallization motion $R(t)$ is a solution $u(\theta,t)$, 2π-periodic with respect to θ for all $t \geq 0$, of the linear, hyperbolic evolution problem

$$u_{tt} + \alpha_1 u_t - \alpha_2 R^{-2}(u_{\theta\theta} + u) = 0, \tag{4.11$_1$}$$

$$u(\theta,0) = u_o(\theta), \; u_t(\theta,0) = 0. \tag{4.11$_2$}$$

If we seek solutions of (4.11) of the form

$$u(\theta,t) = c_0(t) + \sum_{k=1}^{\infty}(\check{c}_k(t)sin(k\theta) + \hat{c}_k(t)cos(k\theta)), \tag{4.12}$$

we find that the typical Fourier coefficient c_k must solve the initial-value problem

$$\ddot{c_k} + \alpha_1\dot{c_k} + \alpha_2(k^2-1)R^{-2}c_k = 0, \tag{4.13$_1$}$$

$$c_k(0) = c_{k_o}, \; \dot{c_k}(0) = 0, \; (k = 0,1,2,\ldots), \tag{4.13$_2$}$$

where the initial data c_{k_o} are the Fourier coefficients of $u_o(\theta)$.

There are three types of problems (4.13). For $k=1$ the solution is $c_1(t) \equiv c_{1_o}$; as (4.11) confirms, *a perturbation of the type*

$$u(\theta,t) = \check{c}_{1_o}sin\theta + \hat{c}_{1_o}cos\theta \tag{4.14}$$

persists unaltered for all time.

For $k = 0$, with the use of the integrating factor $e^{\alpha_1 t}$ we obtain from (4.13) that

$$\dot{c}_0(t) = \alpha_2 \int_0^t R^{-2}(\tau) e^{-\alpha_1(t-\tau)} c_0(\tau) d\tau, \tag{4.15}$$

and another integration yields

$$c_0(t) = c_{0_o} + \alpha_1^{-1}\alpha_2 \int_0^t R^{-2}(\tau)(1 - e^{-\alpha_1(t-\tau)}) c_0(\tau) d\tau. \tag{4.16}$$

We note that, for each time $\tilde{t}$ at which $\dot{c}_0(\tilde{t}) = 0$, and therefore also for $t = 0$, $\ddot{c}_0(\tilde{t})$ has the sign of $c_0(\tilde{t})$; thus,

(i) $c_0(t)$ has neither positive maxima nor negative minima.

But, it follows from (4.16) that

(ii) there is an interval $]0, \bar{t}[$ where $\dot{c}_0(t)$ has the sign of c_{0_o}.

Properties (i) and (ii) together imply that $c_0(t)$ is monotone. Moreover, since (4.16) implies that

$$|c_0(t)| \le |c_{0_o}| + \alpha_1^{-1}\alpha_2 \int_0^t R^{-2}(\tau) |c_0(\tau)| d\tau, \tag{4.17}$$

and since, by (3.20),

$$\int_0^{+\infty} R^{-2}(\tau) d\tau = \delta < +\infty, \tag{4.18}$$

an application of the Gronwall lemma yields

$$|c_0(t)| \le |c_{0_o}| e^{\alpha_1^{-1}\alpha_2\delta}. \tag{4.19}$$

Thus, *a perturbation of the type*

$$u(\theta, t) = c_0(t) \tag{4.20}$$

stays bounded by its initial value for all time.

The third and last type of problems (4.13) obtains when $k \ge 2$; we prove that $|c_k(t)|$ is bounded by $|c_{k_o}|$ for all k and t, provided that the crystal-interface-bath system be **strongly dissipative**, in the sense that

$$\alpha_1^2 \ge \alpha_2 (R_o R_c)^{-1}. \tag{4.21}$$

Firstly, we proceed as in the standard energy estimates, and deduce from (4.13) that

$$(e^{2\alpha_1 t}\dot{c}_k(t))^{\cdot} + \alpha_2(k^2 - 1) R^{-2}(t) e^{2\alpha_1 t} (c_k^2(t))^{\cdot} = 0. \tag{4.22}$$

Next, we introduce the auxiliary mapping

$$d_k(t) := \alpha_2(k^2 - 1) R^{-2}(t) e^{2\alpha_1 t} c_k^2(t), \; d_k(t) \ge 0, \tag{4.23}$$

and integrate (4.22) to arrive at

$$d_k(t) \le d_k(0) + 2\int_0^t (\alpha_1 - R^{-1}(\tau)\dot{R}(\tau))d_k(\tau)d\tau. \tag{4.24}$$

With a view towards applying the Gronwall lemma to the last inequality, we note that, if (4.21) holds, we can use the estimate in footnote 7 to prove that

$$\alpha_1 R(t) \ge \alpha_1 R_o \ge A = \alpha_1^{-1}\alpha_2 R_c^{-1} \ge \dot{R}(t), \tag{4.25}$$

and hence

$$\alpha_1 R(t) - \dot{R}(t) \ge 0.^{11} \tag{4.26}$$

Therefore, (4.24) yields

$$d_k(t) \le d_k(0)R_o^{-2}R^2(t)e^{2\alpha_1 t},$$

and hence, by definition (4.23),

$$c_k^2(t) \le c_{ko}^2. \tag{4.27}$$

We remark that (4.27) guarantees that the properties of (L^2) summability and smoothness stipulated for the initial data are shared by the solution.

Remark 3. If (4.26) holds, one can also prove the following rate estimate:

$$\dot{c}_k^2 \le \alpha_2(k^2-1)c_{ko}^2 R^{-2}(t). \tag{4.28}$$

This follows after integration by parts from

$$R^2(t)(\dot{c}_k^2(t))^{\cdot} + 2\alpha_1 R^2(t)\dot{c}_k^2(t) + 2\alpha_2(k^2-1)(c_k^2(t))^{\cdot} = 0.$$

In particular, for $t \to +\infty$, $\dot{c}_k(t) \to 0$. □

Remark 4. It is of some interest to look at the evolution of perturbations of type (4.2) of the equilibrium solution (3.5). Since equation $(4.13)_1$ has now constant coefficients, it is the matter of a straightforward calculation to verify that, for $t \to +\infty$, (i) $c_k(t) \to 0$ for $k \ge 2$; (ii) once again, $c_1(t) \equiv c_{1o}$; (iii) for $k = 0$, a perturbation of type (4.20) induces melting of a critical-size crystal if $c_{0o} < 0$, whereas if $c_{0o} > 0$ the crystal tends to grow; this points to the inherent instability of critical-size crystals. □

11 This conclusion may be reached under milder hypotheses than (4.24): e.g., one can manipulate certain integral consequences of $(3.2)_1$ to show that (4.26) follows if one assumes that

$$\alpha_1^2 \ge \alpha_2(R_o R_c)^{-1} \log\frac{e}{2}.$$

We conjecture that the conclusion should hold also under the minimal hypothesis that $\alpha_1 \ne 0$.

5. Perturbation of crystallization motions in the FB model

Linearizing a free-boundary problem is a delicate matter; we proceed step by step.

As to the perturbed interface, we stick to the representation (4.2); in particular, then, we can count on the approximations for V, $\mathbf{m}$, and K given in $(4.7)_1$, $(4.8)_3$, and $(4.9)_2$, respectively.

When it comes to perturbing the velocity potential of melt's flows, we choose not to perturb its value at the horizon: this we do for consistency with our previous choice of looking in this paper only to the consequences of a small positional disturbance of the interface (cf. the initial conditions (4.3)). We provisionally set

$$P(r,\theta,t;\epsilon) = U(r,t) + \epsilon w(r,\theta,t), \tag{5.1}$$

with

$$U(r,t) = -Ft + B(t) log \frac{r}{R_1}, \; 0 < r \leq R_1, \tag{5.2}$$

(cf. (2.9)), and with

$$\Delta w(r,\theta,t) = 0, \; w(R_1,\theta,t) = 0, \tag{5.3}$$

for each fixed value of t. It follows from (5.1)-(5.2) that

$$\partial_t P = -F + \dot{B} log \frac{r}{R_1} + \epsilon w_t, \tag{5.4}$$

$$\begin{aligned} \partial_{\mathbf{m}} P &= \nabla P \cdot \mathbf{m} \\ &\cong (Br^{-1}\mathbf{n} + \epsilon(w_r \mathbf{n} - r^{-1} w_\theta \mathbf{t})) \cdot (\mathbf{n} + \epsilon r^{-1} u_\theta \mathbf{t}) \\ &\cong Br^{-1} + \epsilon w_r. \end{aligned} \tag{5.5}$$

Therefore, when we write (2.2) at the perturbed interface, we obtain for $(2.2)_1$

$$\begin{aligned} \partial_t P &\cong -F + \dot{B} log \frac{R}{R_1} + \epsilon(\dot{B} R^{-1} u + w_t), \\ \beta V - \psi K &\cong \beta(\dot{R} + \epsilon u_t) + \psi R^{-1}(1 - \epsilon R^{-1}(u_{\theta\theta} + u)); \end{aligned}$$

for $(2.2)_2$,

$$\begin{aligned} \partial_{\mathbf{m}} P &\cong BR^{-1} + \epsilon(-BR^{-2}u + w_r), \\ \alpha V &\cong \alpha(\dot{R} + \epsilon u_t). \end{aligned}$$

These equations yield (2.12) and the evolution equations for the perturbations $u(\theta,t)$ and $w(R(t),\theta,t)$, namely,

$$w_t(R,\theta,t) = \beta u_t - \psi R^{-2}(u_{\theta\theta} + u) - \alpha R^{-1}(R\dot{R})^{\cdot} u, \tag*{$(5.6)_1$}$$

$$w_r(R,\theta,t) = \alpha R^{-1}(Ru)_t, \tag*{$(5.6)_2$}$$

with $R(t)$ a given solution of (3.2) meeting the initial conditions (3.4), just as was done before for the H model.

Remark 1. Equations (5.6) yield a formula for $\dot{w}(R(t),\theta,t)$ in terms of $u(\theta,t)$ and $R(t)$:

$$\dot{w}(R,\theta,t) = \beta u_t - \psi R^{-2}(u_{\theta\theta} + u) + \alpha(\dot{R}u_t - \ddot{R}u). \qquad \square$$

Our next step is to stipulate, alongside the representation (4.12) for $u(\theta,t)$, the following representation for $w(r,\theta,t)$:

$$w(r,\theta,t) = b_0(t)log\frac{r}{R_1} + \sum_{k=1}^{+\infty}((\frac{r}{R_1})^k - (\frac{r}{R_1})^{-k})(\check{b}_k(t)sin(k\theta) + \hat{b}_k(t)cos(k\theta)). \quad (5.7)$$

A direct consequence of this assumption is that, for each index k in $\{0,1,2,\ldots\}$, the pair $(b_k(t),c_k(t))$ is determined as the solution of a specific initial-value problem. Precisely, we find that

$$-\alpha R log\frac{R}{R_1}\dot{b}_0 = -\beta b_0 + \alpha(\beta\dot{R} + \psi R^{-1} + \alpha(R\dot{R})^{\cdot})c_0, \quad (5.8)_1$$

$$\alpha R\dot{c}_0 = b_0 - \alpha\dot{R}c_0; \quad (5.8)_2$$

and, for $k \geq 1$,

$$-\alpha R((\frac{R}{R_1})^k - (\frac{R}{R_1})^{-k})\dot{b}_k = -\beta k((\frac{R}{R_1})^k + (\frac{R}{R_1})^{-k})b_k + $$
$$+ \alpha(\beta\dot{R} + \psi(1-k^2)R^{-1} + \alpha(R\dot{R})^{\cdot})c_k, \quad (5.9)_1$$

$$\alpha R\dot{c}_k = k((\frac{R}{R_1})^k + (\frac{R}{R_1})^{-k})b_k - \alpha\dot{R}c_k. \quad (5.9)_2$$

These equations are to be thought of as supplemented by assignments of initial values $(b_0(0),c_0(0))$ and $(b_k(0),c_k(0))$, respectively.

Since $R(0) = R_o$, $\dot{R}(0) = 0$, equation $(5.8)_2$ shows that a velocity disturbance $\dot{c}_0(0)$ of the interface induces a positional disturbance $b_0(0)$ of the potential, and conversely; moreover, since $\ddot{R}(0) > 0$, when $\dot{c}_0(0) = 0$ (so that $b_0(0) = 0$) we have from $(5.8)_1$ that a positional disturbance $c_0(0)$ of the interface is equivalent to a velocity disturbance $\dot{b}_0(0)$ of the potential. It is not difficult to see that this exchange of roles between positional and velocity disturbances of interface and potential takes place for each index $k \geq 1$.

Remark 2. Equations (5.8), (5.9) have the common structure of a first-order system with time-dependent coefficients:

$$\dot{\mathbf{y}}_k = \mathbf{A}_k(t)\mathbf{y}_k, \; \mathbf{y}_k(0) = \mathbf{y}_{ko}, \; k = 0,1,2,\ldots, \quad (5.10)$$

where, on writing again R for R/R_1,

$$\mathbf{y} \equiv \begin{pmatrix} \frac{\alpha_1}{\beta}b_k \\ Rc_k \end{pmatrix}, \; k = 0,1,2,\ldots; \quad (5.11)_1$$

$$\mathbf{A}_0 \equiv \begin{pmatrix} \frac{-\bar{\alpha}_1}{-R\log R} & \frac{\bar{\alpha}_1\dot{R}+\bar{\alpha}_2R^{-1}+(R\dot{R})^{\cdot}}{-R^2\log R} \\ 1 & 0 \end{pmatrix}, \tag{5.11$_2$}$$

$$\mathbf{A}_k \equiv \begin{pmatrix} \frac{-\bar{\alpha}_1 k(R^k+R^{-k})}{-R(R^k-R^{-k})} & \frac{\bar{\alpha}_1\dot{R}+\bar{\alpha}_2(1-k^2)R^{-1}+(R\dot{R})^{\cdot}}{-R^2(R^k-R^{-k})} \\ k(R^k+R^{-k}) & 0 \end{pmatrix}, \; k \geq 1. \tag{5.11$_3$}$$

For each ground solution $R(t)$, we see that both the trace of $\mathbf{A}_k(t)$ and the determinant of $sym\mathbf{A}_k(t)$ are strictly negative for each index $k \geq 0$ and all $t \geq 0$. If the equilibrium solution is chosen as ground solution ($R(t) \equiv R_c$), the system (5.10)-(5.11) yields a second-order differential equation for $c_k(t)$. When the initial conditions

$$c_k(0) = c_{ko}\,, \dot{c}_k(0) = \frac{\bar{\alpha}_1}{\beta R_c} b_k(0) = 0$$

are appended to this equation, results completely analogous to those sketched in Remark 4, Section 4, are obtained. □

References

[1] Andreev, A.F., A.Y. Parshin, Equilibrium shape and oscillations of the surface of quantum crystals. Sov. Phys. JETP, 48, (1978), 763-766 .

[2] Keshishev, K.O., A.Y. Parshin, A.V. Babkin, Experimental detection of crystallization waves in He4. JETP Lett., 30, (1979), 56-59.

[3] Maris, H.J., A.F. Andreev, The surface of crystalline helium 4. Phys. Today, Feb. issue, (1987), 25-30.

[4] Gurtin, M.E., A mechanical theory for crystallization of a rigid solid in a liquid melt; melting-freezing waves. Arch. Rational Mech. Anal., 110, (1990), 361-378.

[5] Gurtin, M.E., P. Podio-Guidugli, A hyperbolic theory for the evolution of plane curves. SIAM J. Math. Anal., 22, (1991), 575-586.

[6] Gurtin, M.E., Multiphase thermomechanics with interfacial structure. 1. Heat conduction and the capillary balance law. Arch. Rational Mech. Anal., 104, (1988), 195-221.

[7] Gurtin, M.E., S. Angenent, Multiphase thermomechanics with interfacial structure. 2. Evolution of an isothermal interface. Arch. Rational Mech. Anal., 108, (1989), 323-391.

[8] Gurtin, M.E., A. Struthers, Multiphase thermomechanics with interfacial structure. 3. Evolving phase boundaries in the presence of bulk deformation. Arch. Rational Mech. Anal., 112, (1990), 97-160.

[9] Rogers, R.C., On a linear model of mechanical phase transitions. J. Appl. Math. Phys. (ZAMP), 41, (1990), 728-733.

[10] Fehribach, J.D., Mullins-Sekerka stability analysis for melting-freezing waves in helium. IMA Preprint Series No. 859, (1991), to appear in European J. Appl. Math..

Physics and Geometry of Flow under Curvature: Singularity Formation, Minimal Surfaces, Geodesics, and Surface Tension

James A. Sethian
Department of Mathematics
University of California
Berkeley, California 94720

David L. Chopp
Department of Mathematics
University of California
Los Angeles, California 90024

Introduction

In this paper, we study physical and geometric applications of hypersurfaces whose speed depends at least in part on the local mean curvature. The main tool is a numerical technique, introduced in [25], that accurately follows the evolving hypersurface by embedding it as the zero level set in a family of hypersurfaces. The resulting partial differential equations for the motion of the level set function may be solved by using numerical techniques borrowed from hyperbolic conservation laws. The advantage to this approach is that sharp corners and cusps are accurately tracked, and topological changes in the evolving hypersurface are handled naturally with no special attention.

This technique, known as the level set approach, is used to analyze a collection of problems. In Part One, we give the mathematical and numerical formulation for this technique. In Part Two, we discuss a collection of geometric problems. First, the collapse of a hypersurface under motion by mean curvature is studied. In [29], numerical experiments were performed of the collapse of a dumbbell, and showed that the handle pinches off and splits the single dumbbell into two separate hypersurfaces, each of which collapses to a point. In this paper, we show that an extension of this problem produces an interesting result: a multi-armed dumbbell leaves a separate, residual closed object at the center after the singularity forms. Hypersurfaces propagating under Gaussian curvature are also considered.

Next, the level set approach is used to generate minimal surfaces attached to a one-dimensional wire frame in three space dimensions. Given a wire frame, we construct a surface passing through that 1-D curve and view it as the zero level set of a higher dimensional function. The mean curvature equation for this function is then evolved in time, producing a minimal surface as the final limiting state.

Next, we compute hypersurfaces of constant non-zero mean curvature by adding a hyperbolic component to the flow partial differential equation. As examples, catenoid-like surfaces of a variety of non-zero curvatures are computed. The extension of the level set formulation to the computation of surfaces of any prescribed function of the curvature is given.

Finally, we end our geometric examples with the generalization of the curvature flow algorithm to curves on two-dimensional manifolds in R^3. In this context, the curves flow with speed dependent on the geodesic curvature of the curve. Examples of curves on a cube, sphere, and torus are given.

In Part Three, we summarize some of the work so far in applying these level set approaches to physical problems in which motion by curvature plays a key role. Here, we provide motivation and reference for work in flame propagation, two fluid problems, crystal growth and dendritic solidification, droplet dynamics, and image reconstruction.

Part I: The Level Set Formulation

Equations of Motion

Consider a closed curve $\gamma(t)$ where t is time, $t \in [0, \infty)$, moving with speed F normal to itself. The speed F may depend on local properties of the curve such as the curvature or normal vector. The origin of the work to follow propagating interfaces began in [27,28], where the role of curvature in the speed function F for the propagating front $\gamma(t)$ was shown to be analogous to the role of viscosity in the corresponding hyperbolic conservation law for the evolving slope of $\gamma(t)$. This led to the level set formulation of the propagating interface introduced in [25]. In general terms, let $\gamma(0)$ be a closed, non-intersecting, $(N-1)$ dimensional hypersurface and construct a function $\phi(\bar{x}, t)$ defined from $R^N \times R$ to R such that the level set $\{\phi = 0\}$ is the front $\gamma(t)$, that is

$$\gamma(t) = \{\bar{x} \in R^N : \phi(\bar{x}, t) = 0\} \tag{1}$$

In order to construct such a function $\phi(\bar{x},t)$, appropriate initial conditions $\phi(\bar{x},0)$ and an associated partial differential equation for the time evolution of $\phi(\bar{x},t)$ must be supplied. We initialize ϕ by

$$\phi(\bar{x},0) = \pm d(\bar{x}) \tag{2}$$

where $d(\bar{x})$ is the signed distance from $\bar{x}$ to the initial front $\gamma(t = 0)$. In order to derive the partial differential equation for the time evolution of ϕ, consider the motion of some level set $\{\phi(\bar{x},t) = C\}$. Let $\bar{x}(t)$ be the trajectory of some particle located on this level set, so that, (see [23]),

$$\phi(\bar{x}(t),t) = C \tag{3}$$

The particle velocity $\frac{\partial \bar{x}}{\partial t}$ in the direction $\bar{n}$ normal to the level set C is given by

$$\frac{\partial \bar{x}}{\partial t} \cdot \bar{n} = F. \tag{4}$$

Differentiating Eqn. (3) with respect to t and combining with Eqn. (4) yields

$$\begin{aligned} \phi_t + F\|\nabla\phi\| &= 0 \\ \phi(\bar{x}, t=0) &= \text{given}. \end{aligned} \tag{5}$$

Eqn. (5) yields the motion of the interface $\gamma(t)$ as the level set $\phi = 0$, thus

$$\gamma(t) = \{x : \phi(\bar{x},t) = 0\}. \tag{6}$$

Eqn. (5) is referred to as the *level set formulation*. For certain speed functions F, it reduces to some familiar equations. For example, for $F = 1$, the equation becomes the eikonal equation for a front moving with constant speed. For $F = 1 - \epsilon\kappa$, where κ is the curvature of the front, Eqn. (5) becomes a Hamilton-Jacobi equation with parabolic right-hand-side, similar to those discussed in [7]. For $F = \kappa$, Eqn. (5) reduces to the equation for mean curvature flow. When required, the curvature κ may be determined from the level set function ϕ. For example, in three space dimensions the mean curvature is given by

$$\kappa = \frac{\begin{array}{r}(\phi_{xx})(\phi_y^2+\phi_z^2)+(\phi_{yy})(\phi_x^2+\phi_z^2)+(\phi_{zz})(\phi_x^2+\phi_y^2)\\ -2(\phi_x\phi_y\phi_{xy}+\phi_y\phi_z\phi_{xz}+\phi_x\phi_z\phi_{xz})\end{array}}{2(\phi_x^2+\phi_y^2+\phi_z^2)^{3/2}}. \tag{7}$$

There are several advantages of this approach given in Eqn. (5). First, the fixed coordinate system avoids the numerical stability problems that plague approximation techniques based on a parameterized approach. Second, topological changes are handled naturally, since the level surface $\phi = 0$ need not be simply connected. Third, the formulation clearly applies in any number of space dimensions.

To illustrate, in Figure 1 the motion of a circle in the xy-plane propagating outward with constant speed is shown. Fig. 1a shows the initial circle, while Fig. 1b shows the same circle as the level set $\phi = 0$ of the initial surface

$$\phi(x,y,t=0) = \sqrt{x^2+y^2} - 1.$$

The one-parameter family of moving curves $\gamma(t)$ is then matched with the one-parameter family of moving surfaces in Figs. 1c and 1d. This level set approach to front propagation has been employed in a variety of investigations, as discussed in Part Two. In some particular cases, the reaction diffusion method has also shown some promising results. The theoretical underpinnings of the level set approach have been examined in detail by Evans and Spruck [9,10]; for further theoretical work, see also [4,8,11,15].

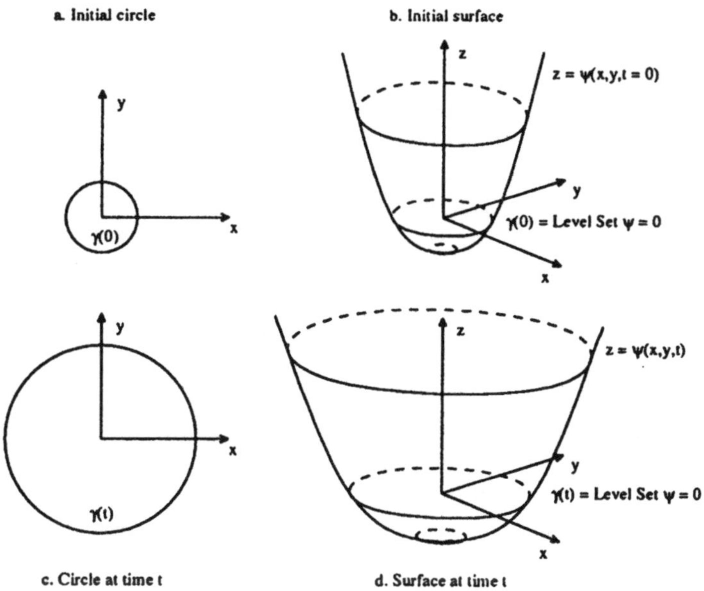

Figure 1: Eulerian formulation of equations of motion

Numerical Approximation

A successful numerical scheme to approximate Eqn. (5) hinges on the link with hyperbolic conservation laws. As motivation, consider the simple case of a moving front in two space dimensions that remains a graph as it evolves, and consider the initial front given by the graph of $f(x)$ with f, f', periodic on $[0,1]$. Let $y(x,t)$ be the height of the propagating function at time t, thus $y(x,0) = f(x)$. The normal at (x,y) is $(1, -y_x)$, and the equation of motion becomes $y_t = F(\kappa)(1+y_x^2)^{1/2}$. Using the speed function $F(\kappa) = 1 - \epsilon\kappa$, where the curvature $\kappa = y_{xx}/(1+y_x^2)^{3/2}$, we get

$$y_t - (1+y_x^2)^{1/2} = \epsilon \frac{y_{xx}}{(1+y_x^2)} \tag{8}$$

To construct an evolution equation for the slope $u = dy/dx$, we differentiate both sides of the above with respect to x and substitute to obtain

$$u_t + \left[-(1+u^2)^{1/2}\right]_x = \epsilon \left[\frac{u_x}{(1+u^2)}\right]_x \tag{9}$$

Thus, the derivative of the Hamilton-Jacobi equation with curvature-dependent right-hand-side for the changing height $y(x,t)$ is a viscous hyperbolic conservation law for the propagating slope u. With this hyperbolic conservation law, an associated entropy condition must be invoked to produce the correct weak solution beyond the development of a singularity in the evolving curvature. Complete details may be found in [29].

Considerable care must be taken in devising numerical schemes to approximate the level set Eqn. (5). Because a central difference approximation to the gradient produces the wrong weak solution, we instead exploit the technology of hyperbolic conservation laws in devising schemes which maintain sharp corners in the evolving hypersurface and choose the correct, entropy-satisfying weak solution. One of the easiest such schemes is a variation of the Engquist-Osher scheme presented in [25]. This scheme is upwind in order to follow the characteristics at boundaries of the computational domain. The scheme is as follows. Decompose the speed function F into $F = F_A + F_B$, where F_A is treated as the hyperbolic component which must be handled through upwind differencing, and the remainder F_B which is to approximated through central differencing. Let ϕ^n_{ijk} be the numerical approximation to the solution ϕ at the point $i\Delta x$, $j\Delta y$, $k\Delta z$, and at time $n\Delta t$, where Δx, Δy, Δz is the grid spacing and Δt is the time step. We can then advance from one time step to the next by means of the numerical scheme

$$\begin{aligned}\phi^{n+1}_{ij} = \phi^n_{ij} + F_A\Delta t\cdot \\ &((\min(D^-_x\phi_{ij},0))^2 + (\max(D^+_x\phi_{ij},0))^2 + (\min(D^-_y\phi_{ij},0))^2 \\ &\quad + (\max(D^+_y\phi_{ij},0))^2 + (\min(D^-_z\phi_{ij},0))^2 + (\max(D^+_z\phi_{ij},0))^2)^{1/2} \\ &+ \Delta t F_B \|\nabla\phi\|\end{aligned} \tag{10}$$

Here, the difference operators D^-_x refers to the backward difference in the x direction. The other difference operators are defined similarly.

Examples

Grayson [16] has proven that any non-intersecting closed curve in R^2 moving with speed $F(\kappa) = -\kappa$ must collapse smoothly to a circle; see also [12,13,14]. In Figure 3, is a demonstration of a closed spiral curve shrinking towards a circle. Note that the calculation follows a family of spirals lying on the higher dimensional surface. The particular front corresponding to the propagating curve vanishes when the evolving surface moves entirely above the xy-plane, that is, when $\phi(x,y,t) > 0$ for all (x,y).

As a different example, let the wound spiral in the previous example represent the boundary of a flame burning with speed $F(\kappa) = 1 - \epsilon\kappa$, $\epsilon = 0.1$. Here, the entropy condition is needed to account for the change in topology as the front burns together. In Fig. 4a, the initial spiral as the boundary of the shaded region is given. In Fig. 4b, the spiral expands, and pinches off due to the outward normal burning and separates into two flames, one propagating outward and one burning in. In Figure 4c, the front is the boundary of the shaded region. The outer front expands and the inner front collapses and disappears. In Fig. 4d all that remains is the outer front which asymptotically approaches a circle.

Part II: Geometric Problems in Curvature Flow

Collapsing Dumbbells under Mean Curvature Flow

In this section, singularity formation of hypersurfaces in three space dimensions propagating under mean curvature is studied. Theoretical discussion of such flows have been made in [3,17,20]. Numerical calculations based on a marker Langrangian approach have been made in [2].

A well-known example is the collapse of a dumbbell, studied numerically in [27], and theoretically in [18,21]. In Fig. 2, the cross-section of the evolution of a dumbbell collapsing under its mean curvature ($F(\kappa) = -\kappa$) is given. In Figure 2a, various time snapshots of the collapsing dumbbell are shown. As can be seen from the evolving shape, the center handle of the dumbbell pinches off, separating the collapsing hypersurface into two pieces.

A more complicated version is shown in Figure 5, which shows the collapse of a four-armed dumbbell. Four different singularities form leaving a residual pillow in the center which collapses smoothly through a spherical shape to a point.

As a final demonstration of this process, Figures 6 and 7 show the collapse of a diagonal lattice of tubes. The lattice shown in Figure 6a (Time = 0.0) has periodic boundary conditions; thus, the figure represents one section of an infinite lattice. As the hypersurface collapses, the pillow emerges at the intersection of the tubes. A wholly different result is shown in Figure 7, where the same tube lattice is shown, only this time with thicker tubes (Figure 7a). In this case, the separate pillows appear in the *holes* of the lattice, as the evolving surface collapses around them.

Next, a three-dimensional version of the spiral collapsing under mean curvature is computed. The three-dimensional spiral hypersurface shown in Figure 8 is actually hollow on the inside; the opening on the right end extends all the way through the object to the leftmost tip. As the hypersurface collapses under its mean curvature, the inner sleeve shrinks faster than the outer sleeve, and withdraws to the rightmost edge before the outer sleeve collapses around it.

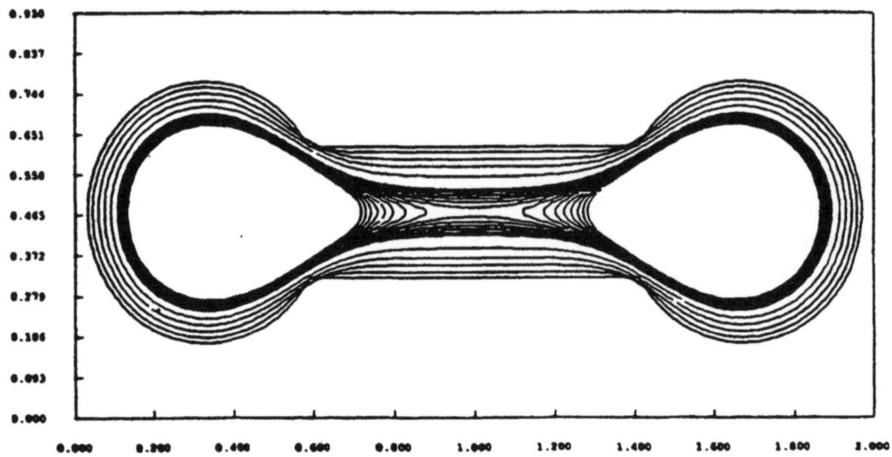

Figure 2: Collapsing Dumbbell

Collapsing Surfaces under Gaussian Curvature Flow

A variation on the above study can be performed using Gaussian curvature instead of mean curvature. Starting with the definition of Gaussian curvature κ_{Gaussian} for a surface (see [19]), an expression for κ_{Gaussian} in terms of the level set function ϕ can be obtained, namely

$$\kappa_{\text{Gaussian}} = \frac{\begin{array}{l}\phi_x^2(\phi_{yy}\phi_{zz} - \phi_{yz}^2) + \phi_y^2(\phi_{xx}\phi_{zz} - \phi_{xz}^2) + \phi_z^2(\phi_{xx}\phi_{yy} - \phi_{xy}^2) \\ \quad - 2[\phi_x\phi_y(\phi_{xz}\phi_{yz} - \phi_{xy}\phi_{zz}) + \phi_y\phi_z(\phi_{xy}\phi_{xz} - \phi_{yz}\phi_{xx}) \\ \quad + \phi_x\phi_z(\phi_{xy}\phi_{yz} - \phi_{xz}\phi_{yy})]\end{array}}{(\phi_z^2 + \phi_y^2 + \phi_x^2)^2}. \tag{11}$$

Suppose we consider flow of surfaces under Gaussian curvature. If the closed hypersurface is convex, the Gaussian curvature will not change sign, and the surface should collapse as it flows, see [24]. In Figure 9, the motion of a flat disk-like surface collapsing under its Gaussian curvature is shown. The sharply curved regions move in quickly, since they are regions of high Gaussian curvature, and the hypersurface moves towards a spheroidal shape.

In the case of non-convex closed hypersurfaces, the problem acts like the backwards heat equation due to the fact that Gaussian curvature is the *product* of the two principle curvatures. Thus, non-convex hypersurfaces go unstable in most cases as shown in Figure 10.

Construction of Minimal Surfaces

In this section, the level set perspective is used to construct minimal surfaces. Consider a closed curve $\Gamma(s)$ in R^3; $\Gamma : [0,1] \to R^3$. The goal is to construct a membrane with boundary Γ and mean curvature zero.

Given the bounding wire frame Γ, consider some initial surface $S(t=0)$ whose boundary is Γ. Let $S(t)$ be the family of surfaces parameterized by t obtained by allowing the initial surface $S(t=0)$ to evolve under mean curvature, with boundary given by Γ. Defining the surface S by $S = \lim_{t\to\infty} S(t)$, one expects that the surface S will be a minimal surface for the boundary Γ. Thus, given an initial surface $S(0)$ passing through Γ, construct a family of neighboring surfaces by viewing $S(0)$ as the zero level set of some function ϕ over all of R^3. Using the level set Eqn. (5), evolve ϕ according to the speed law $F(\kappa) = -\kappa$. Then the minimal surface S will be given by

$$S = \lim_{t\to\infty} \{\bar{x} : \phi(\bar{x}, t) = 0\}. \tag{12}$$

The difficult challenge with the above approach is to ensure that the evolving zero level set always remains attached to the boundary Γ. This is accomplished by creating a set of boundary conditions on those grid points closest to the wire frame and link together the neighboring values of ϕ to force the level set $\phi = 0$ through Γ. The details for this construction can be found in Chopp [5].

As a test example, the minimal surface spanning two rings has an exact solution given by the catenoid

$$r(x) = a\cosh(x/a) \tag{13}$$

where $r(x)$ is the radius of the catenoid at a point x along the x axis, and a is the radius of the catenoid at the center point $x = 0$. In Figure 11, the minimal surface spanning two rings each of radius 0.5 and at positions $x = \pm.277259$ is computed. A cylinder spanning the two rings is taken as the initial level set $\phi = 0$. Next, in Figure 12, this same problem is computed, but the rings are placed far enough apart so that a catenoid solution cannot exist. As the surface evolves, the middle pinches off and the surface splits into two surfaces, each of which quickly collapses into a disk. The final shape of a disk spanning each ring is indeed a minimal surface for this problem.

This example illustrates one of the virtues of the level set approach. No special cutting or *ad hoc* decisions are employed to decide when to break the surface. Instead the characterization of the zero level set as but one member of a family of flowing surfaces allows this smooth transition. More complex examples of minimal surfaces are given in [6].

Surfaces of Constant Mean Curvature

The above technique can be extended to produce surfaces of constant but non-zero mean curvature. In order to construct a surface of constant curvature κ_0, start with any initial surface passing through the initial wire frame and allow it to propagate with speed

$$F(\kappa) = \kappa_0 - \kappa. \tag{14}$$

Here, the "constant advection term" κ_0 is taken as the hyperbolic component F_A, and treated using the entropy-satisfying upwind difference solver, while the parabolic term κ is taken as F_B, and is approximated using central differences.

Using the two ring "catenoid" problem as a guide, in Figure 13 this technique is used to compute the surface of constant curvature spanning the two rings. In each case, the initial shape is the cylinder spanned by the rings. The final computed shapes for a variety of different mean curvatures are shown.

Surfaces of Non-Constant Mean Curvature

Finally, this technique is extended to allow the calculation of surfaces of a prescribed function of the curvature. Suppose we wish to find a surface of curvature $A(\bar{x})$ passing through a given wire frame, where A is some given function of a point $\bar{x}$ in three-dimensional space. Using the above approach, the initial bounded zero level surface is evolved with speed

$$F(\kappa) = A(\bar{x}) - \kappa. \tag{15}$$

As a simple example, the surface spanning two rings with curvature at any point x along the x-axis given by $10\cos(10x)$ is constructed. The obtained wavy surface with prescribed curvature is shown in Figure 14d. The rings are located at ± 0.305, with radius 0.5.

Geodesic Curvature Flow

The curvature flow algorithm can be generalized to other two-dimensional spaces. For example, we may let the level set function ϕ be defined on a two-dimensional differentiable manifold in R^3 with speed depending on geodesic curvature. The fixed boundary condition techniques for minimal surfaces can also be applied here. In this case, a curve with fixed endpoints should flow towards a geodesic of the manifold, i.e. a curve with constant geodesic curvature zero.

We begin the examples of geodesic curvature flow with flow on a sphere. The gap in Figure 15 shows the boundary of the domain, the function is assumed to be periodic across the gap. Figure 15 shows an initial circle just smaller than a great circle shrinking to a point at the top.

Figure 16 shows a single curve flowing on a torus. In order to implement the level set method on this surface, the computations must be done on a collection of coordinate patches. For the full details of how coordinate patches are used and connected, see Chopp [5].

Another example of flow on a submanifold is when the manifold is thaway from the center over a ridge.

Finally, in Figures 18, 19, we show flows on a cube. The cube is constructed with six coordinate patches corresponding to the faces of the cube. Additional examples can be found in [5].

Part III. Physical Applications

In this part, we discuss several physical problems that have been analyzed using the above level set technique.

Flame Propagation

The original motivation for our use of a level set approach arose from problems in flame propagation and combustion, where the propagating flame front can become extremely convoluted, due to cusping and wrinkling. The dynamics of a burning front are complex, depending in part on vorticity released at the front whose strength depends on the local curvature at the interface. A level set approach is well-suited to such problems, since the accurate determination of the curvature can be used to generate the proper amount of flame-induced vorticity. Here, the basic idea is to use an idealized model of combustion and view the flame as an infinitely thin front across which the reactants undergo a single step irreversible on by means of a second-order finite difference projection scheme for fluid mechanics. A random vortex method is coupled to the level set technique in [26] to study the dynamics of a flame attached to a flame holder in confined channel. The full effects of exothermicity and vorticty-production are analyzed, showing wrinkling, cusping, and the development of flame brushes as a function of turbulence levels.

Two Fluid Problems

Another fluid dynamic application of the level set approach was performed in [23]. Here, the level set technique was used to track the boundary between two different fluids in the Rayleigh-Taylor instability. Analysis of two different approaches was done; one in which the level set equation was solved directly, and one in which a conservative form of the equation was incorporated into the other four conservation laws. Results show the development of large convective rolls and the effects of viscosity on the instabilities in the flow.

Crystal Growth and Dendritic Solidification

In [30], the level set approach was applied to computing the motion of dendritic boundaries in unstable solidification. The motivation here was the ability of the level set approach to accurately compute the curvature which is required in the evaluation of the correct jump conditions across the solid/liquid interface.

The physical motivation for the problem is as follows. Begin with a supercooled liquid in a box. Under certain conditions, a seed of frozen material placed in the bath will grow in an unstable manner as the boundary freezes, producing complex shapes such as those produced in snowflakes. One model of this growth assumes the heat equation in both the solid and liquid phases, together with a set of jump conditions across the interface which determine the normal velocity of the interface in terms of the temperature and the local curvature.

The application of the level set approach to this problem begins with a boundary integral formulation of the problem due to Strain. This boundary integral formulation gives the normal velocity of the interface in terms of the current and past positions of the interface. The central idea is to view the boundary as the zero level set of the level set function, and to then extend the boundary integral formulation to provide a velocity field throughout the entire domain. This extension velocity field is then used to move all the level sets and thus update the position of the freezing boundary. For details, see [30].

Using this technique, the motion of dendritic boundaries under such effects as heat release, kinetic effects, crystalline anisotropy, initial shape, and undercooling are considered. Results show a variety of physical phenomena, including tip splitting, side growth, cusping, and morphological changes. A complete description of this algorithm and the results may be found in [30].

Droplet Dynamics

An application currently under way links the level set technique to a projection method in order to study the dynamics of a droplet falling into a bath. Here, the local curvature determines the surface tension which is responsible for the "crown"-like structure that develops in the splash. The two-dimensional results show the change of topology as the falling droplet merges with the still bath, and the resulting symmetric spiking of the fluid into the air. A three-dimensional simulation is underway. For details, see [1].

Image Processing

A different application of a level set approach is in the area of edge detection in bitmap images. Here, the main idea is extract from a pixel map the location of the edges of a potentially complex collection of boundaries. In [22] level set schemes are merged with propagating snake techniques to produce an algorithm which robustly identifies images in tomographic scans and density maps.

Acknowledgments: The first author was supported in part by the Applied Mathematics Subprogram of the Office of Energy Research under contract DE–AC03–76SF00098, and the National Science Foundation and DARPA under grant DMS–8919074. The second author was supported in part by the National Science Foundation under grant CTS–9021021. A video (VHS format) of the evolving surfaces is available from the first author. All calculations were performed at the University of California at Berkeley and the Lawrence Berkeley Laboratory. The figures were computed using the facilities of the Graphics Group at the Lawrence Berkeley Laboratory. We would like to thank L. Craig Evans, F.A. Grunbaum, Ole Hald, and V. Oliker for helpful discussions.

References

[1] A. Bourlioux and J. A. Sethian Two and Three Dimensional Droplet Dynamics. to be submitted for publication, 1993.

[2] K. A. Brakke. *The Motion of a Surface By Its Mean Curvature.* Princeton University Press, Princeton, New Jersey, 1978.

[3] K. A. Brakke. Surface evolver program. Research Report GCG 17, the Geometry Supercomputer Project, 1200 Washington Ave. South, Minneapolis, MN 55455, 1990.

[4] Y. Chen, Y. Giga, and S. Goto. Uniqueness and existence of viscosity solutions of generalized mean curvature flow equations. *Journal of Differential Geometry,* 33:749–786, 1991.

[5] D. L. Chopp. Flow under geodesic curvature. Technical Report CAM 92-23, UCLA, May 1992.

[6] D. L. Chopp. Computing minimal surfaces via level set curvature flow. *Journal of Computational Physics,* to appear.

[7] M. G. Crandall and P. L. Lions. Viscosity solutions of Hamilton-Jacobi equations. *Transactions of the American Mathematical Society,* 277(1), 1983.

[8] L. C. Evans, H. M. Soner, and P. E. Souganidis. Phase transitions and generalized motion by mean curvature. *Communications on Pure and Applied Mathematics,* 45(9):1097–1123, 1992.

[9] L. C. Evans and J. Spruck. Motion of level sets by mean curvature I. *Journal of Differential Geometry,* 33:635–681, 1991.

[10] L. C. Evans and J. Spruck. Motion of level sets by mean curvature II. *Transactions of the American Mathematical Society,* 330:321–332, 1992.

[11] M. Falcone, T. Giorgi, and P. Loretti. Level sets of viscosity solutions and applications. Technical report, Instituto per le Applicazioni del Calcolo, Rome, 1990. preprint.

[12] M. Gage. An isoperimetric inequality with applications to curve shortening. *Duke Math Journal,* 50(1225), 1983.

[13] M. Gage. Curve shortening makes convex curves circular. *Inventiones Mathematica,* 76(357), 1984.

[14] M. Gage and R. S. Hamilton. The equation shrinking convex planes curves. *Journal of Differential Geometry*, 23(69), 1986.

[15] Y. Giga and S. Goto. Motion of hypersurfaces and geometric equations. *Journal of the Mathematical Society*, to appear.

[16] M. Grayson. The heat equation shrinks embedded plane curves to round points. *Journal of Differential Geometry*, 26(285), 1987.

[17] M. Grayson. A short note on the evolution of surfaces via mean curvature. *Duke Mathematical Journal*, 58:555–558, 1989.

[18] M. Grayson. Shortening embedded curves. *Journal of Differential Geometry*, 129:71-111, 1989.

[19] N. J. Hicks. *Notes on Differential Geometry*. Van Nostrand, Princeton, NJ, 1965.

[20] G. Huisken. Flow by mean curvature of convex surfaces into spheres. *Journal of Differential Geometry*, 20(237), 1984.

[21] G. Huisken. Asymptotic behavior for singularities of the mean curvature flow. *To appear, Journal of Differential Geometry*.

[22] R. Malladi and J. A Sethian. Edge detection and Propagating Interface Techniques. Center for Pure and Applied Mathematics Technical Report, 1993.

[23] W. Mulder, S. Osher, and J. A. Sethian. Computing interface motion in compressible gas dynamics. *Journal of Computational Physics*, 100(2):209–228, 1992.

[24] V. Oliker. Evolution of nonparametric surfaces with speed depending on curvature, I. the Gauss curvature case. preprint, 1991.

[25] S. Osher and J. A. Sethian. Fronts propagating with curvature-dependent speed: Algorithms based on Hamilton-Jacobi formulations. *Journal of Computational Physics*, 79(1), November 1988.

[26] C. Rhee, L. Talbot, and J. A, Sethian. A Level Set Approach to Flame Propagation. Submitted for publication, Combustion and Flame, 1993.

[27] J. A. Sethian. Curvature and the evolution of fronts. *Communications in Mathematical Physics*, 101:487–499, 1985.

[28] J. A. Sethian. Numerical methods for propagating fronts. In P. Concus and R. Finn, editors, *Variational Methods for Free Surface Interfaces*. Springer-Verlag, New York, 1987.

[29] J. A. Sethian. A review of recent numerical algorithms for hypersurfaces moving with curvature-dependent speed. *Journal of Differential Geometry*, 31:131–161, 1989.

[30] J. A. Sethian and J. Strain. Crystal growth and dendrite solidification. *Journal of Computational Physics*, 98(2):231–253, 1992.

[31] J. Zhu and J. A. Sethian. Projection methods coupled to level set interface techniques. *Journal of Computational Physics*, 102(1):128–138, 1991.

Figure 3: Collapsing 2-dimensional spiral

Figure 4: Burning spiral

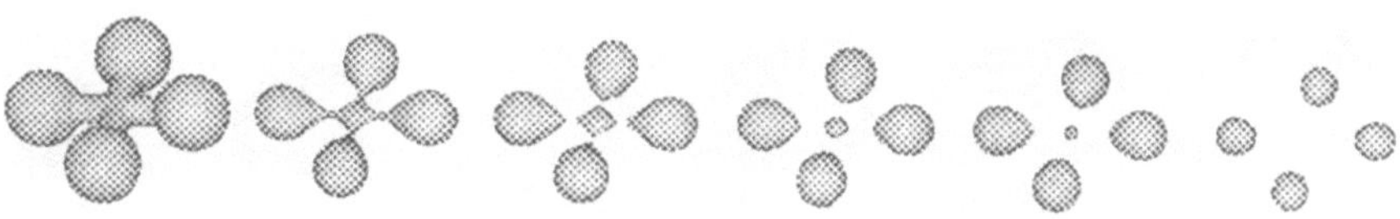

Figure 5: Collapse of a four-armed dumbbell

Figure 6: Collapse of a periodic lattice, small tubes

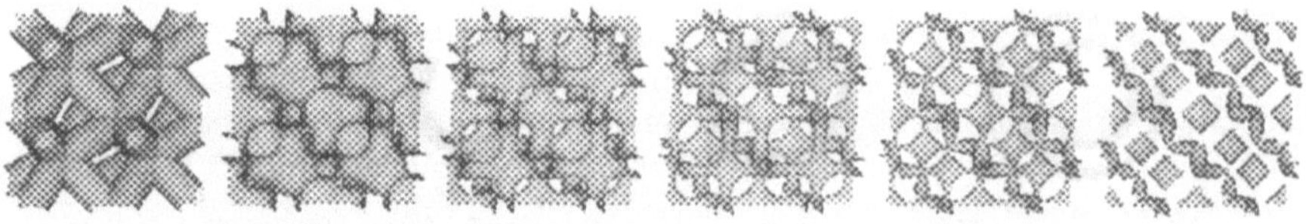

Figure 7: Collapse of a periodic lattice, large tubes

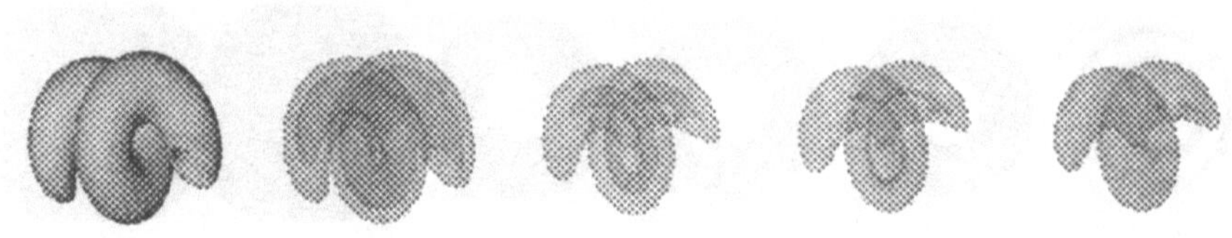

Figure 8: Collapse of a twisted test tube

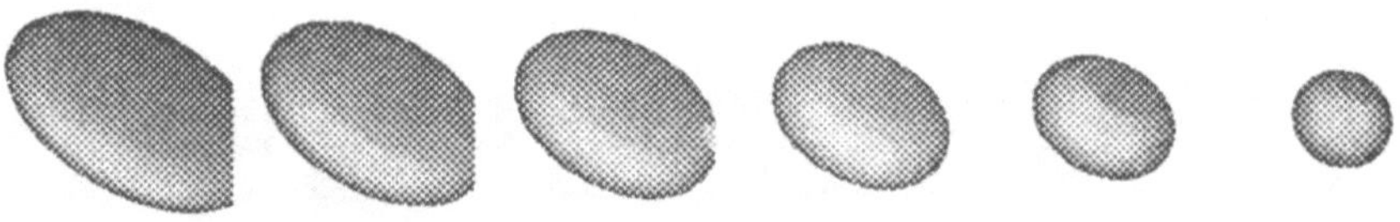

Figure 9: Collapse of a surface under Gaussian curvature

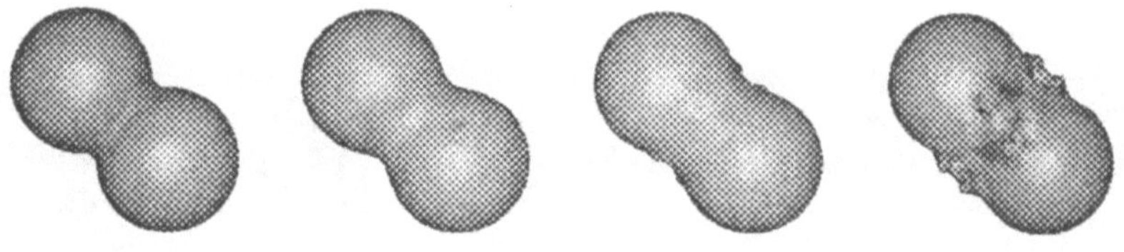

Figure 10: Collapse of a non-convex surface under Gaussian curvature

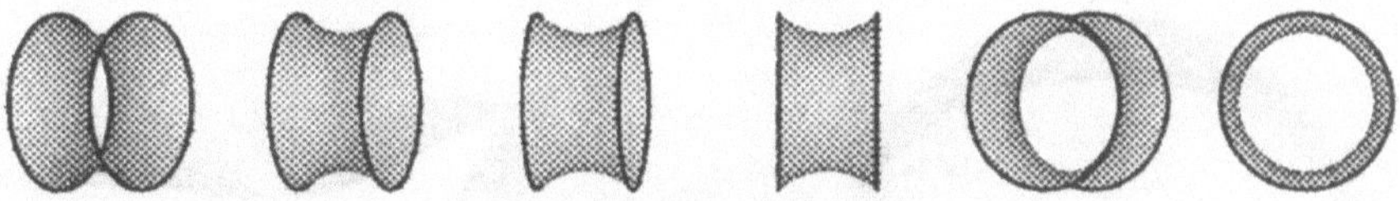

Figure 11: Euler's catenoid minimal surface

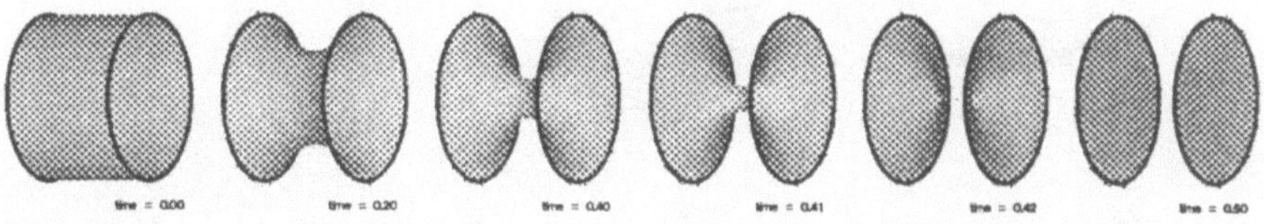

Figure 12: Splitting catenoid evolution

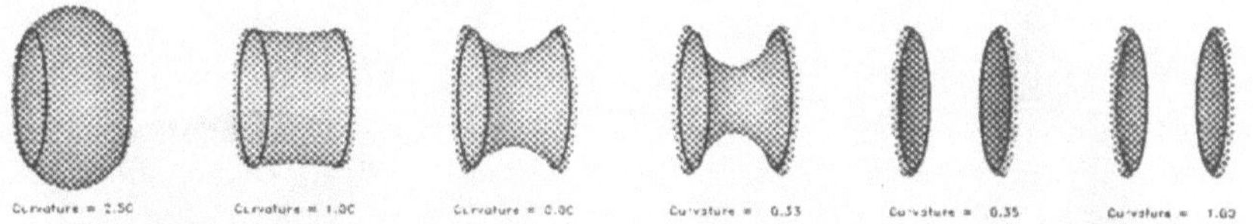

Figure 13: Constant mean curvature surfaces with fixed boundary

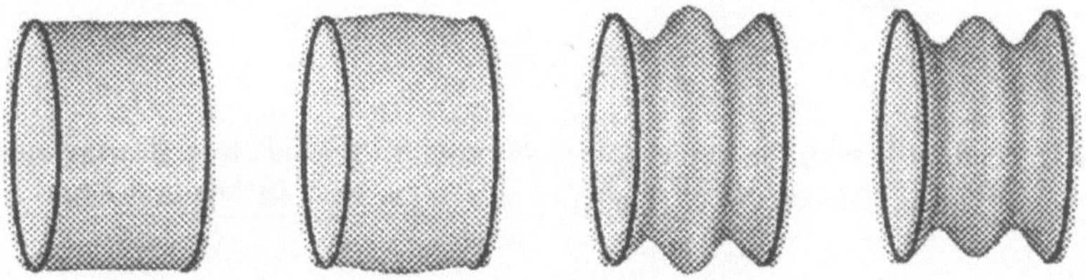

Figure 14: Non-constant prescribed curvature surface

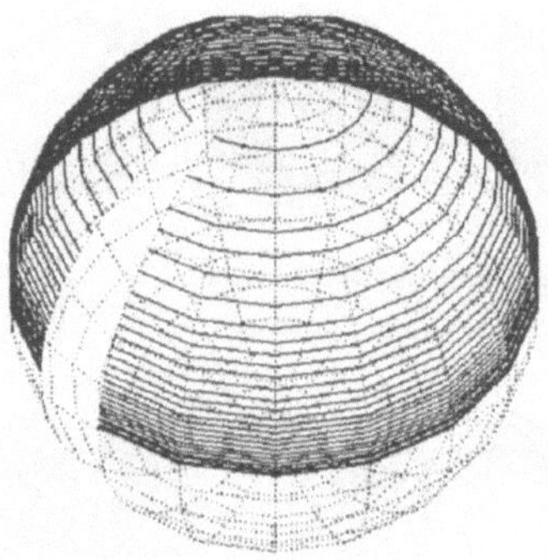

Figure 15: Circle shrinking on a sphere

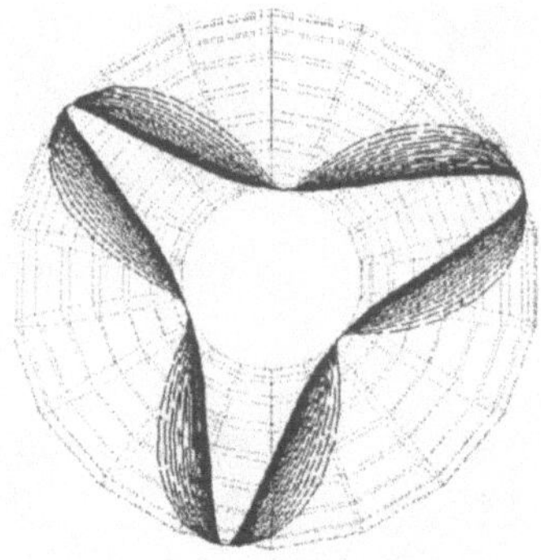

Figure 16: A single curve flowing on a torus

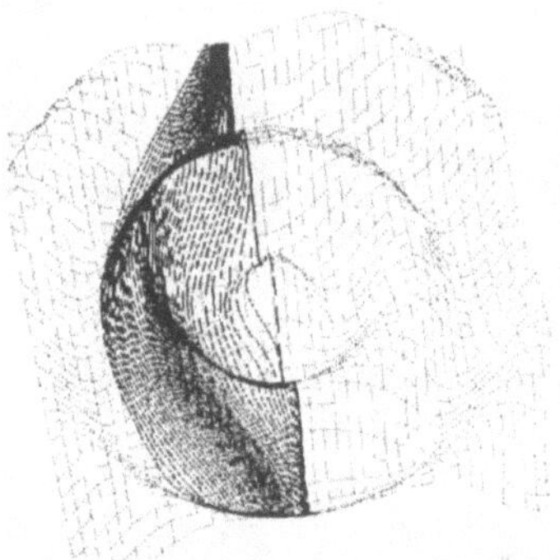

Figure 17: A curve flowing on the graph of $f(x, y) = 2\cos(2\sqrt{x^2 + y^2})$

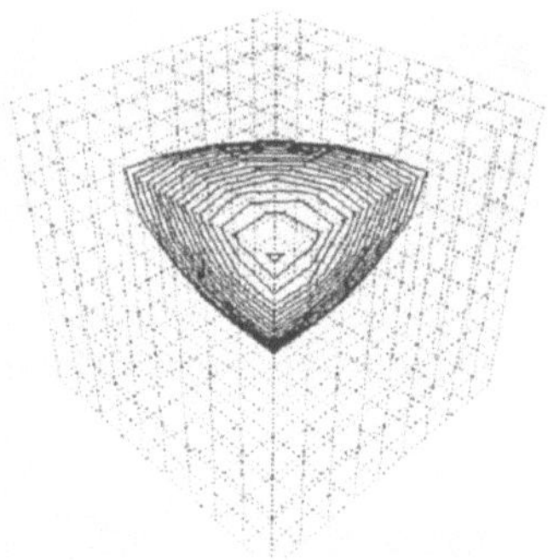

Figure 18: A single loop flowing on a cube with orthogonal edges

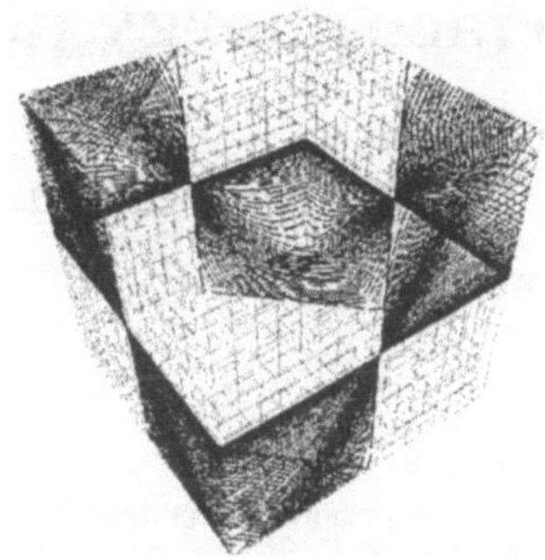

Figure 19: A single loop pulled over alternating corners

Computing hypersurfaces which minimize surface energy plus bulk energy

John M. Sullivan

Abstract. An area-minimizing hypersurface is dual to a vector field which calibrates it. In [S], we used this idea to construct an algorithm to find the minimum-energy surface on any boundary, where the surface energy density can depend on position or normal vector. Here we extend this algorithm to allow not only surface energies, but also bulk energies, integrated over the volume enclosed by the surface. Minimizing such energies is an important step in the definition of flat curvature flows [ATW]. By modifying the bulk energy to include undercooling terms as well as distance terms, these flows are especially useful for computing :rystal growth.

1991 Mathematics Subject Classification: 49Q20, 49M35, 90C08, 53A10.

1. Introduction to prescribed curvature problems

Many physical problems have a mathematical formulation in terms of minimizing the energy of an interface surface, where the energy might be just area, or could have a density dependent on position or orientation. Here we consider problems which have such a surface energy, given by integrating $\Phi(x, n)$, the surface energy density at x for a hypersurface with normal vector n, and which also have a bulk energy, given by integrating a density H over the volume on "one side" of the surface. We assume that the surface is a hypersurface in $\mathbf{R}^n$, but other ambient manifolds can be handled in the same manner.

To find a local minimum of such an energy, we could use some gradient descent method; many such approaches to mean curvature flow, for instance, have been studied, as described elsewhere in this volume. But to find the global minimum, within a homology class, the most fruitful approach is to use convex duality. Dual to a minimum-area surface is a calibration, a divergence-free vector field normal to the surface which proves it has minimum area. In [S], we approximated this pair of dual problems by a finite network-flow problem, a linear program solvable by a special algorithm. Here we extend this approximation and algorithm to the case with bulk energies.

In $\mathbf{R}^n$, the homology class of a hypersurface is given by its list of boundary components, with oriented multiplicities. We will specify such a class by giving one surface S_0 in the class. Then any homologous S can be written as $S = S_0 + \partial T$ for some region T. (We will make this precise by working in the context of currents, as in [ATW] or [S].) Here, if S and S_0 are contained in a bounded region, then we may take T to be supported in the same region.

1.1. Prescribed mean curvature

One energy we might consider is

$$\mathcal{E}_0(S) := \Phi(S) - \int_T H := \int_S \Phi(x, n_S) - \int_T H(x),$$

which is independent (up to an additive constant) of the initial surface S_0 chosen to represent the homology class. Consideration of this energy leads to what is known as the prescribed mean curvature problem [G], because, under certain regularity conditions, any surface S which is critical for $\mathcal{E}$ and passes through a point x will necessarily have Φ-mean curvature equal to $H(x)$ there. (This can be taken to be the definition of the weighted mean curvature associated to the energy density Φ; if $\Phi \equiv 1$ so that the surface energy is just area, then it is the usual mean curvature of a smooth surface S.)

Intuitively, the dual problem asks for a vector field $f(x)$ in space which has speed bounded by the value of Φ in the direction of f (meaning that $f \cdot f \leq \Phi(x, f)$, where by convention we extend Φ to be positive-homogeneous in its second argument) and has divergence $\nabla \cdot f = H(x)$. We try to maximize the flux of such a flow through the initial surface S_0, noting that this flux equals the flux through S minus the integral $\int_T H(x)$. This difference equals the energy $\mathcal{E}_0(S)$ iff f is everywhere normal to S with maximum speed.

Often, the prescribed mean curvature problem does not have a global minimum solution. For instance, one might hope to find a spherical soap bubble by specifying H to be a constant throughout $\mathbf{R}^3$, and taking $\Phi \equiv 1$. Then a sphere of radius $2/H$ is a critical point for the energy $\mathcal{E}$, but it is an unstable critical point, and spheres of greater radius have decreasing energies approaching $-\infty$. Correspondingly, there is no feasible solution to the dual problem: there is no vector field in $\mathbf{R}^n$ of constant divergence with global speed bounds. Some other technique is needed to find constant mean curvature soap bubbles.

In fact, unless the function H is close to zero everywhere, there will be no global minimum (in a homology class) for the prescribed mean curvature problem. If there is any region R with $|\int_R H| > \Phi(\partial R)$, then we can add the boundary ∂R to our surface S with high multiplicity (positive or negative as appropriate) to get a new surface with lower energy. If $\Phi \equiv 1$, this means, for instance, that we cannot have $H > nk$ on a ball of radius $1/k$. However, we should note that many positive results are known [G] if we restrict our attention to a small enough region.

1.2. The absolute curvature problem

Although the prescribed mean curvature problem often fails to have a global minimum, there is a related problem which is guaranteed to have a minimum energy solution. We wrote the difference $\partial S - \partial S_0$ as the boundary of an n-current T; if we separate T into positive and negative pieces, $T = T_+ - T_-$ where T_+ and T_- have disjoint support and positive multiplicities, then our modified energy is

$$\mathcal{E}(S) = \int_S \Phi(x, n_S) + \int_{T_+ + T_-} H(x) = \Phi(S) + \int_{T_+} H + \int_{T_-} H.$$

Here we take H to be nonnegative. We can think of H as specifying the absolute value of the mean curvature of a minimizing surface; the mean curvature vector will point "away from" the initial surface S_0. This modified problem, which we will call an *absolute curvature problem*, does depend on the particular surface S_0, not just on its homology class. All our results would apply to a generalization, where we integrate different nonnegative functions H_+ and H_- over T_+ and T_- respectively. We will not, however, state our results in this generality.

Note that any solution must lie in the convex hull of the initial surface S_0, because intersecting T with a half-space containing S_0 can only decrease $\mathcal{E}(S)$.

This kind of energy was introduced in [ATW], where, to define a generalized flow by mean curvature for an initial surface S_0 through a time Δt, the function $H(x)$ is cleverly prescribed to be the distance from x to S_0, normalized by dividing by Δt. A solution to this particular absolute curvature problem should then be a surface which, if it has moved a distance $d = H\Delta t$ from S_0, has Φ-mean curvature H there. Existence and regularity results are proven in [ATW] for such solutions, which are shown to approximate other known curvature flows.

To find a global minimum for such a problem computationally, we need to make some approximations. In fact, we replace the problem by a discrete version, which asks for the best surface in a cell complex. Given any complex which fills $\mathbf{R}^n$, we can restrict our attention to hypersurfaces made up of facets of these cells, that is, to $(n-1)$-chains in the complex. If we know the surface energy of each such facet, and then bulk energy of each n-cell, then finding the best surface becomes a linear programming problem. In section 2, we describe an algorithm for solving this network flow problem quickly. In section 3, we formulate the absolute curvature problem precisely in the context of currents. In section 4, we prove that for any absolute curvature problem, there is a cell complex in which its solution will have a good approximation; in particular the complex used is a hypergrid, obtained by slicing space with parallel planes in many directions.

2. A discretized problem and its algorithm

We describe a network-flow problem which is dual (in the sense of linear programming) to the problem of minimizing energy among surfaces in a cell complex. This problem has a fast algorithm with a nice geometric interpretation.

2.1. The discrete absolute curvature problem

Suppose we have a cell complex in $\mathbf{R}^n$. We pay special attention to the cells of dimension n and $n-1$, and to the graph or flow network G to which they are dual. Each node v of this graph lies in an n-cell $\hat{v}$ of the complex, and if $\hat{v}$ and $\hat{w}$ are adjacent, there is an edge or pipe $p = (v, w)$ dual to the facet $\hat{p}$ between v and w. We write $-p = (w, v)$ for the oppositely oriented edge.

The energy $\mathcal{E}$ of any hypersurface in the complex will be given in terms of

$$u_p := \Phi(\hat{p}) > 0 \quad \text{and} \quad h_v := \int_{\hat{v}} H.$$

If Φ is symmetric as a function of normal direction, then $u_p = u_{-p}$. Although we keep these interpretations in mind, our linear program works the same for any nonnegative values u_p, h_v, independent of whether these could come from the energies of some cell complex.

We should add a note about finiteness. If we know *a priori* that the surface we are looking for will lie in a certain region of our complex, then we can replace all cells outside that region by a single cell v_∞ with $h_{v_\infty} = \infty$. Then the surface we find in the modified complex will be the same as the one we would find in the original complex. Thus we speak of a cell complex filling out $\mathbf{R}^n$, but we really need to work in just a compact region (say the convex hull of the initial surface) to get the true solution in $\mathbf{R}^n$, so we can assume that the dual graph is finite.

Definition. A *flow network* is a finite graph G with nonnegative functions u_p on the edges and h_v on the nodes. We will talk about cohomology in the graph in terms of homology in a dual complex. Thus, a integer combination $\sum a_p p$ of edges (pipes) will be viewed as a surface $A = \sum a_p \hat{p}$. (Here we take by convention $a_{-p} = -a_p$ and we take sums over just one orientation of each edge.) A linear combination $T = \sum t_v v$ of nodes (vertices) is viewed as a region, with boundary ∂T equal to the set of pipes out of this region. So A and B are homologous surfaces if $B = A + \partial T$ for some region T. (Focusing our attention instead on the graph, we would say that $B - A$ was a cocycle in the edges of G.) The values u_p are called the *capacities* of the pipes p, for reasons we will see later.

Definition. Given a flow network G and an initial surface A, the *discrete absolute curvature problem* asks to find the homologous surface $B = \sum b_p \hat{p}$ of least energy $\mathcal{E}$. If $T = \sum t_v \hat{v}$ is the region satisfying $B = A + \partial T$, then the energy of B is defined to be

$$\mathcal{E}(B) = \sum |b_p| u_{\pm p} + \sum |t_v| h_v.$$

Here, the sign in $u_{\pm p}$ is that of b_p.

The condition that B be homologous to A can be written as a set of constraints, one for each pipe $p = (v, w)$, namely $b_p = a_p + t_v - t_w$. Thus this problem is a linear programming problem: find b_p and t_v satisfying these constraints and minimizing the energy $\mathcal{E}$. Note that T is not quite determined by B; we could add a constant to each t_v. However, we always have in mind the choice minimizing $\sum |t_v| h_v$; for the networks we have described with $h_{v_\infty} = \infty$, this sum is finite only for this choice of T.

2.2. The dual flow problem

The standard duality for linear programs gives us a dual problem, which has variables $f_p = -f_{-p}$ corresponding to the constraints on every pipe p. These can be interpreted as a pseudoflow F on the network G. The new problem asks to maximize the sum $\sum a_p f_p$ which is the flux of F through A, subject to a condition on each pipe p, namely that $|f_p| \leq u_p$, and a condition on each node v. Then the condition says that $|\nabla_v^F| \leq h_v$, where ∇_v^F is the net flow out of v, the divergence of F there.

Proposition. *This linear programming problem dual to the discrete absolute curvature problem is equivalent to the minimum-cost circulation problem in a modified network.*

Remark. The minimum-cost circulation problem on a flow network G asks for a circulation of minimum cost, where a circulation is a pseudoflow with no divergence anywhere, and there is a given cost $-a_p$ for each unit of flow through pipe p. Thus our dual problem as described above is a generalization to what we might call quasicirculations, where the pseudoflow is allowed to have a limited amount of divergence at each point. If all the h_v are zero, then our problem is exactly the minimum-cost circulation problem.

Proof. We construct the modified network as follows: let G^H be the graph with the nodes and edges of G, plus an extra edge $p_\infty(v)$ joining v to v_∞ for each node v. We set the capacity of $p_\infty(v)$ top be h_v (in each direction), and set $a_p = 0$ for these new edges. Then circulations $\bar{F}$ in G^H correspond exactly to quasicirculations F in G, if we let $\bar{F} = F$ on the edges of G, and let it take the values it must on the new edges to eliminate the divergence. Since the new edges have no cost, a minimum-cost circulation in G^H is the same as a minimum-cost quasicirculation for G. □

Many special algorithms are known for network-flow problems, starting with the work of Ford and Fulkerson in the 1950's. A nice summary is given in [GTT]. We will describe an algorithm for our problem that works well for the kinds of data we encounter (where most a_p are zero), and has a clear geometric interpretation. A similar interpretation for a minimum-cost circulation algorithm was given in [S], with more complete details.

2.3. A primal-dual algorithm

Our algorithm is a primal-dual algorithm, which starts with a feasible pair of solutions ($F = 0$ and $B = A$), and at each step improves one or the other, until they are both optimal. As always in linear programming, we know that for any feasible solutions, the primal and dual energies satisfy $\sum f_p a_p \leq \mathcal{E}(B)$. Equality here holds iff the solutions are both optimal, so this gives us the *orthogonality conditions* for optimality. We always know that the flux through B is no more than the surface energy of B; for these to be equal, we must have $f_p = u_p$ whenever $b_p > 0$ (and so $f_p = -u_{-p}$ whenever $b_p < 0$). We call these the *capacity* conditions. The other orthogonality conditions ensure that the difference in flux through B and A equals the bulk energy of T. These *divergence* conditions say that $\nabla_v^F = h_v$ whenever $t_v > 0$, and that $\nabla_v^F = -h_v$ whenever $t_v < 0$.

Initially, $t_v = 0$ since $B = A$, so the divergence conditions are satisfied. But the capacity conditions are not satisfied, since there is not flow to capacity through the surface. Our algorithm repeatedly finds some unsatisfied capacity condition, and attempts to satisfy it by increasing the flow or (if this is not possible) by moving the surface. However, we never let ourselves violate any other orthogonality condition. This implies that divergence conditions are never violated, and only the original capacity conditions are violated: new pieces of surface where $b_p > 0$ but $a_p = 0$ always have full flow through them. After each such step, we have one fewer violated condition, so after a finite number of iterations, bounded by the size $|A| := \sum |a_p|$, we must have an optimal solution.

In particular, we start each iteration by finding a facet $\hat{p}_0$ in the surface (so $a_{p_0}, b_{p_0} > 0$) where the flow is not to capacity ($f_{p_0} < u_{p_0}$). We try to increase f_{p_0} by finding a flow path including p_0, either a loop or a path from a source to a sink. Any node v is available as a source if $\nabla_v^F < h_v$, except that at nodes where $t_v < 0$ (so $\nabla_v^F = -h_v$) we do not allow the divergence to be increased, as this would violate the orthogonality condition. The conditions for sinks are analogous. Each pipe can support additional flow up to its spare capacity, except that we do not allow any flow backwards through the current surface D, as this would violate a capacity condition.

Given a pseudoflow F on a flow network G with capacities u_p, we can consider the residual capacities $u_p - f_p$, the maximum amount that can be added to f_p. The network G with these capacities is called G_F. In particular, we can think of deleting from G the directed edges already used to capacity. Here, we in fact want to delete additionally the edges going backwards through B. Call the resulting network $G_{F,B}$. Our iterative step (for an edge $p = (t, s)$) asks to find a maximum quasiflow from t to s in $G_{F,B}$.

Again, we can reduce this maximum quasiflow problem to Ford and Fulkerson's original maximum flow problem by adding extra edges to infinity. Here we let $G^H_{\bar{F},\bar{B}}$ be the graph G^H with capacities modified to account for the current flow $\bar{F}$, and edges backwards through the current surface ($b_p < 0$) deleted. We can interpret t_v as $b_{p_\infty(v)}$, and we delete also the appropriate added edges where this is nonzero, to avoid violating new divergence conditions.

Thus we have demonstrated the following proposition.

Proposition. *The discrete absolute curvature problem for initial surface A in network G can be solved by at most $|A|$ iterations, each a maximum flow problem on a subgraph of G.* □

2.4. Algorithms for maximum flows

There are many good algorithms known [GTT] for the maximum flow problem, also known as the max-flow/min-cut problem. We will give a quick summary of one that runs in reasonable time. We explain it in terms of the equivalent maximum quasiflow problem encountered above.

The algorithm must try to improve the flow f_{p_0} until it equals u_{p_0}. If this cannot be done, we have a minimum cut, that is, an improved surface demonstrating that f_{p_0} cannot be further increased.

To improve the flow, we start with the node at the head of p_0, and perform a breadth-first search of nodes accessible from there, looking to reach one available as a sink or to reach the tail of p_0. The search may only follow pipes that can support additional flow. If we reach the tail of p_0, we augment the flow in this loop by the minimum spare capacity. If we reach a sink, we start from the tail of p_0 and work backwards, looking for a source node, again by a breadth-first search through allowable pipes. If we find a source and sink, then we augment the flow along the path between them as much as possible (determined by the minimum of the spare capacities and the available source and sink values). In these cases, we have improved the flow f_{p_0}. If it is still below the capacity u_{p_0}, we must repeat the whole process, but the choice at each stage of a shortest possible augmenting path guarantees that the number of repetitions is $\mathcal{O}(|V||E|)$, where $|V|$ is the number of vertices in G, and $|E|$ the number of edges [S, GTT].

If, on the other hand, the search for a sink or source node fails, we improve the surface rather than improving the flow. Assume that it was the search for a sink that failed, and let Z be the set of nodes reached from the head of p_0. (We handle the case of failure to find a sink in a symmetric way.) We want to move one sheet of surface across to the other side of the set Z; this is accomplished by subtracting one from t_v for each node in Z and adding one to b_p for each pipe out of Z. (Thus B changes by the boundary of Z, and we maintain the condition $B = A + \partial T$.)

Since Z was the entire set of reachable nodes, it must be true that no pipe out of Z has any spare capacity (except when $-\hat{p}$ is used in the current surface). Also, every node v in Z was unavailable as a sink, so it has either $\nabla_v^F = -h_v$ or $t_v > 0$. This means exactly that no new orthogonality conditions are violated when we move the surface: the only places where b or t is made nonzero are places where the orthogonality is already met.

Also, a simple calculation shows that the energy $\mathcal{E}(B)$ is decreased by this move, since the net flow out of Z equals the total divergence in Z. If there were no nodes in Z where $t_v > 0$, and no pipes out of Z where $b_p < 0$, then there would be no change in the energy. But each such pipe or node gives us a decrease in energy, and we know there is at least one such pipe, the original p_0.

Modified algorithms let us find maximum flows (or quasiflows) in time $\mathcal{O}(V^3)$ or even less [GTT]. Thus we can summarize the results of section 2 in the following theorem.

Theorem. *The discrete absolute curvature problem for initial surface A in network G can be solved by the network-flow algorithm described above, taking total time* $\mathcal{O}(|A||V|^3)$. □

3. Minimum energy currents

Here we will give a precise statement of the absolute curvature problem in the context of integral currents, and show that a minimum-energy solution exists. Currents are objects over which differential forms can be integrated; they form the basis for much of geometric measure theory. See [M] for a nice introduction; [S] and [ATW] also give more background than we give here.

Definition. A *rectifiable set* of dimension k is a measurable set of finite Hausdorff k-measure in $\mathbf{R}^n$ which is a countable union of Lipschitz images of k-planes. Since we know how to integrate a differential form over a rectifiable set, we can view it as a current, and in fact a *rectifiable k-current* T in $\mathbf{R}^n$ is a countable sum of rectifiable sets with integer multiplicities. Its *mass* $M(T)$ is the supremum of $T(\phi)$ over all ϕ pointwise bounded by 1 on unit simple vectors, which equals the Hausdorff measure of the underlying set, counted with multiplicities; this is required to be finite for a rectifiable current. Its *boundary* ∂T of a current T is defined by its action on $(k-1)$-forms, to make Stokes' Theorem true: $\partial T(\phi) = T(d\phi)$. An *integral current* is a rectifiable current whose boundary is also rectifiable (and thus in particular has finite mass).

Much of the usefulness of currents comes from the following compactness theorem, which proves the existence of minimizers for many problems.

Theorem [M 5.5]. *Given a compact set* $K \in \mathbf{R}^n$ *and a mass bound* $M_0 > 0$, *the set of all integral currents* T *supported in* K *with* $M(T) < M_0$ *and* $M(\partial T) <$

M_0 is compact with respect to the flat norm $\mathcal{F}$. This norm is defined by setting $\mathcal{F}(T,S) = \inf M(A) + M(B)$, where the infimum is over all ways of writing $T - S$ as $\partial A + B$. □

Note that $\mathcal{F}(\partial T, \partial S) \leq \mathcal{F}(T,S)$, so that if currents converge in the flat norm, then so do their boundaries. For n-currents, the flat norm $\mathcal{F}$ coincides with the mass norm M.

The structure of top-dimensional currents is relatively simple: any integral current T of dimension n can be written as a sum of positive and negative parts

$$T = T_+ - T_- = \sum_i T_{+i} - \sum_i T_{-i},$$

where each T_{+i} and T_{-i} is associated to a Lebesgue-measurable set with multiplicity one, $T_{+1} \cap T_{-1} = \emptyset$, and $T_{+i} \subset T_{+(i-1)}$, $T_{-i} \subset T_{-(i-1)}$. If we decompose two currents T and S in this manner, then

$$\mathcal{F}(T,S) = \mathcal{F}(T_+, S_+) + \mathcal{F}(T_-, S_-) = \sum_i \mathcal{F}(T_{+i}, S_{+i}) + \sum_i \mathcal{F}(T_{-i}, S_{-i}),$$

where, in each case, the flat distance between two n-currents equals the mass of their difference. Thus convergence of n-currents implies convergence of their positive and negative parts separately.

A parametric integrand Φ for hypersurfaces is a positive continuous function of position and unit normal vector. We assume that Φ is convex at each point, meaning that if it is extended to be positive-homogeneous on normal vectors at that point, it is a convex function of these. Such an integrand Φ can be integrated over an integral $(n-1)$-current T to give $\Phi(T) \geq 0$. For instance the mass $M(T)$ of a current can be obtained from the integrand $\Phi \equiv 1$. We assume here that our integrands Φ are independent of position, but all the results extend to integrands with Lipschitz dependence on position. Since Φ is positive, it has a positive lower bound $\Phi(T) \geq \Phi_0 M(T)$.

A parametric integrand for n-currents is simply a continuous nonnegative density function $H(x)$ on $\mathbf{R}^n$. We may assume that H has an upper bound H_0, since we will be working in a compact region.

It is not hard to check that $M(T)$ is lower semicontinuous with respect to the flat norm $\mathcal{F}$, and the same is true [ATW] for any convex parametric integrand Φ.

We are now ready to state the absolute curvature problem and prove the existence of solutions as integral currents. We are give an initial surface, an integral $(n-1)$-current S_0 supported in some compact convex set $K \subset \mathbf{R}^n$, and two energy functions, a surface energy given by a convex parametric integrand Φ, and a bulk energy, given by a nonnegative continuous function $H(x)$ on K.

Then the absolute curvature problem asks for a n-current T such that $S = S_0 + \partial T$ has minimum energy $\mathcal{E}(S) = \Phi(S) + \int_{T_+} H + \int_{T_-} H$. Here T_+ and T_- are the positive and negative parts of T. Note that any solution must lie in the convex set K, since projection into K can only decrease the energy $\mathcal{E}$.

Theorem. *The absolute curvature problem described above has a solution.*

Proof. The functions Φ and H are nonnegative, so there is an infimum $\mathcal{E}_0$ for $\mathcal{E}(S)$. Suppose T_i is a sequence of currents such that $\mathcal{E}(S_i)$ approaches this infimum ($S_i = S_0 + \partial T_i$). Then by compactness there is a subsequence converging to an integral current T in the flat norm. Then the S_i must converge to $S = S_0 + \partial T$, and the positive and negative parts of T_i converge to the currents T_+ and T_-. By the lower semicontinuity of parametric integrands, we find that $\mathcal{E}(S) \leq \lim \mathcal{E}(S_i)$, and thus $\mathcal{E}(S) = \mathcal{E}_0$, the infimum. □

4. Crystalline approximation theorems

The main theorems of [S] describe how integral currents can be approximated arbitrarily well in a certain kind of cell complex. In fact, the complex can be chosen even without knowledge of the current, if we know just certain bounds on its mass. This allows us to approximate the solution to a minimization problem by the solution in this complex. Here we summarize these results, and additionally check that this approximation to a current has nearly the same energy $\mathcal{E}$ as the original current, so that the algorithm described in section 2 can be used to find good approximations to the minimum-energy surfaces for absolute curvature problems (like those in the definition of the flat curvature flow of [ATW]). In [S] only surface energies with no bulk terms were considered.

4.1. Hypergrids and theorems for surface energy

The cell complexes we use are hypergrids, which are sometimes called multigrids, but are not to be confused with the multigrids used in numerical solution of PDEs. They are projections of cubical grids from higher dimensions into $\mathbf{R}^n$.

Definition. An (N, δ)-*hypergrid* in $\mathbf{R}^n$ is a cell complex obtained by slicing space in N different directions, each time with parallel planes at spacing δ.

The idea here is that if the N directions are spread evenly around the sphere S^{n-1}, then a flat hyperplane can be approximated well using facets in the n nearest directions, in a tent-like surface. Furthermore, if δ is much smaller than any radius of curvature of some smooth hypersurface, we can in some sense follow the curves of this surface, and approximate it well in an (N, δ)-hypergrid.

However, any current can be smoothed so that over most of its area it has small curvatures; the smooth parts can be approximated well in our hypergrid, and the remainder is so small that even a bad approximation there doesn't hurt us. This outlines the proof of the following crystalline approximation theorem.

Theorem [S]. *Any current S can be approximated well in some hypergrid. In particular, given ϵ and bounds on the masses of S and its boundary, we can choose a hypergrid C in which there is a chain $\hat{S}$ approximating S well, in the sense that $\Phi(\hat{S}) \leq (1+\epsilon)\Phi(S)$ and $\mathcal{F}(S,\hat{S}) \leq \epsilon M(S)$.* □

Note that the proof sketched above and given in [S] leads to explicit bounds on the size of the hypergrid needed; if we don't care about the bounds, we can invoke compactness to prove the theorem quite quickly.

From this theorem it follows that given a boundary cycle, we can use our flow-network algorithm to find the surface spanning it with least area (or least energy Φ). Given the desired accuracy, we pick a hypergrid in which the true surface will have a good approximation. Then the algorithm finds the best surface in the hypergrid, which must have close to the true minimum energy.

4.2. Results for the absolute curvature problem

The results described above can easily be extended to apply to the absolute curvature problem. In fact, we merely need to note that in a compact region (like the convex hull of the initial surface S_0), the bulk energy function $H(x)$ is bounded above by a constant H_0. Remember also that Φ is bounded below by Φ_0. Thus we will see that the approximation $\hat{S}$ constructed above has energy close to the energy $\mathcal{E}(S)$ of S.

Theorem. *Any current S can be approximated well in some hypergrid for the absolute curvature problem. In Given a desired accuracy $\epsilon > 0$ and bounds on the masses of a current S and its boundary, we can choose a hypergrid C in which any such S has a good approximation for the absolute curvature problem. By this we mean a chaing $\hat{S}$ such that $\mathcal{E}(\hat{S}) \leq (1+\epsilon)\mathcal{E}(S)$ and $\mathcal{F}(S,\hat{S}) \leq \epsilon M(S)$.*

Proof. Use the approximating $\hat{S}$ from the previous theorem, with a smaller ϵ'. Then

$$\mathcal{E}(\hat{S}) = \Phi(\hat{S}) + \int_{\hat{T}} H \leq (1+\epsilon')\Phi(S) + \int_T H + H_0\mathcal{F}(S,\hat{S})$$
$$\leq (1+\epsilon')\mathcal{E}(S) + \frac{\epsilon' H_0}{\Phi_0}\Phi(S) \leq \left(1+\epsilon'+\epsilon' H_0/\Phi_0\right)\mathcal{E}(S).$$

We see that taking $\epsilon' < \epsilon/(1+H_0/\Phi_0)$ is sufficient to get the desired result. □

This theorem means that for any absolute curvature problem, and any desired accuracy, we can choose a hypergrid in which to look for a minimum energy surface. The surface found by the algorithm of section 2 will find the global minimum in this grid, which cannot have more than $(1+\epsilon)$ times the energy of the true minimum surface.

References

[ATW] Almgren, F.J., J.E. Taylor, L. Wang, Curvature driven flows: a variational approach, SIAM J. Control and Optimization, SICON 31-2, March 1993, 51 pages, to appear

[GTT] Goldberg, A.V., É. Tardos, R.E. Tarjan, Network flow algorithms, Princeton University Department of Computer Science Technical Report CS-TR-216-89, 1989

[G] Gulliver, R., Existence of surfaces with prescribed mean curvature vector, Math Z., 131 (1973), 117–140

[M] Morgan, F., Geometric Measure Theory: A Beginner's Guide. Academic Press, 1988

[S] Sullivan, J.M., A crystalline approximation theorem for hypersurfaces, Princeton University Ph.D. thesis, 1990; also available as Research Report GCG 24 from the Geometry Center, 1300 South Second Street #500, Minneapolis, MN 55454

Boundary regularity for flows of nonparametric surfaces driven by mean curvature

Nina N. Uraltseva

I. Introduction

Let Ω be a bounded domain in $\boldsymbol{R}^n, n \geq 2$, with C^∞ boundary $\partial\Omega$. We consider the following problem describing evolution of nonparametric surfaces $S(t) = (x, u(x,t))$, $x \in \Omega$, with normal speed equal to the mean curvature:

$$u_t = \sqrt{1+|Du|^2}\; H[u] \; in \; \Omega \times [0,\infty) \tag{1}$$

$$u = \psi \; on \; \partial\Omega \times [0,\infty) \tag{2}$$

$$u = u_o \; in \; \Omega \times \{0\}. \tag{3}$$

Here $Du = (u_{x_1}, ..., u_{x_n})$ is the gradient of u, $H[u] = div \frac{Du}{\sqrt{1+|Du|^2}}$ is the mean curvature of $S(t)$, ψ and u_o are given smooth functions of space variables $x = (x_1, ..., x_n)$, u_t is the time-derivative of u.

We denote by

$$\partial_+\Omega = \{x \in \partial\Omega : \; H_{\partial\Omega}(x) \geq 0\}$$

$$\partial_-\Omega = \{x \in \partial\Omega : \; H_{\partial\Omega}(x) < 0\},$$

where $H_{\partial\Omega}(x)$ is the mean curvature of $\partial\Omega$ at x with respect to the inner normal vector $\boldsymbol{n}(x)$ on $\partial\Omega$. The case when $\partial_-\Omega = \emptyset$ was studied by G. Huisken in 1989 who proved the classical solvability of the problem and the convergence of $u(\cdot,t)$ to the corresponding solution of minimal surface equation when $t \to +\infty$.

V. Oliker and the author considered the problem (1)-(3) for general domains such that $\partial_-\Omega \neq \emptyset$. In [OU 1] the case of homogeneous boundary condition $\psi \equiv 0$ has been studied. It was proven that for any $u_o \in C_o^\infty(\Omega)$ there exists some generalized solution which is smooth in $\Omega \times [0,\infty)$ as well as on $\bar{\Omega} \times [T,\infty)$ for some T depending on initial data. Besides, the asymptotics of the solution when $t \to +\infty$ was found.

This solution, however, may develop singularities on $\partial\Omega$ and not attain prescribed boundary data in finite time $t < T$. Most of these results were extended to the case of nonhomogeneous boundary data spanning a minimal surface [OU 2]. Solutions constructed in [OU 1,2] belong to $W_1^1(\Omega)$ for any time $t \geq 0$ and satisfy (1), (3) but, in general, the boundary condition (2) is satisfied only in the following weak sense:

$$\frac{Du\cdot n}{\sqrt{1+|Du|^2}} \in sign\ (u-\psi) \quad a.e.\ on\ \partial\Omega \times (0,T),$$
$$u = \psi \qquad on\ \partial\Omega \times [T,\infty)$$

These results rely on local in x and t estimates of $max\ |Du^\alpha|$ for solutions of regularized problems

$$(4) \qquad u_t^\alpha = \sqrt{1+|Du^\alpha|^2}\ (H[u^\alpha] + \alpha\Delta u^\alpha)\ in\ \Omega \times [0,\infty)$$

$$(5) \qquad u^\alpha = \psi\ on\ \partial\Omega \times [0,\infty)$$

$$(6) \qquad u^\alpha = u_o\ in\ \Omega \times \{0\}.$$

Here $\Delta = \sum_{i=1}^n \frac{\partial^2}{\partial x_i^2}$, $\alpha \in (0,1]$. It may be pointed out, however, that the operators H and Δ behave in quite different ways as $|Du^\alpha| \to \infty$, so the standard methods of local gradient estimates can not be applied to (4). In Lemma 4.1 of [OU 1] the technique of "parabolic" integral inequalities due to De Giorgi has been developed to obtain the uniform with respect to $\alpha \in (0,1]$ local estimates of $max\ |Du^\alpha|$. More elementary estimates

$$(7) \qquad sup_{\Omega\times(0,\infty)}\ |u^\alpha| \leq M_o;\ sup_{\Omega\times(0,\infty)}\ |u_t^\alpha| \leq M_1;$$

$$(8) \qquad sup_{t\in(0,\infty)} \int_\Omega \left(\sqrt{1+|Du^\alpha|^2} + \alpha|Du^\alpha|^2\right) dx \leq M_2$$

enable us to apply Lemma 4.1 to solutions of (4)-(6) (see [OU 1]).

Recently, the author found out that it is possible to establish local estimates for $|Du^\alpha|$ without using the parabolicity of (4). Instead, one can apply some refined results concerning the elliptic equations

$$(9) \qquad H[u^\alpha] + \alpha\Delta u^\alpha = f^\alpha(x, u^\alpha, Du^\alpha)\ in\ \Omega.$$

The uniform in α local C^1-estimates for solutions $u^\alpha(x)$ of such equations were obtained by R. Temam [T] under the following structure conditions on function f^α and its derivatives:

$$|f^\alpha(x,z,p)| \le const, \tag{10}$$

$$-(1+|p|)\, f^\alpha_z + (1+|p|)^2 |f_p| + (1+|p|)^{-1} |p_k\, f^\alpha_{x_k}| \le const. \tag{11}$$

It should be noted that, in general, C^1-estimates are not valid without a hypothesis like (11) even for the case $\alpha = 0$, $f^\alpha = f(x)$. But it turned out that if instead of (10), one assumes that $f^\alpha(x,z,p)$ decreases as $|p|^{-1}$ when $|p| \to \infty$ then one doesn't need any structure conditions on the derivatives of f^α. This remark together with (7) lead to the uniform in α interior gradient estimates for solutions $u^\alpha(x,t)$ of the parabolic problem (4)-(6).

The goal of the present paper is to investigate the behavior of generalized solutions of (1)-(3) near $\partial_-\Omega$ for finite time $t \in (0,T)$. The corresponding elliptic problem was studied by L. Simon [S 1, 2]. In [S 1] a class of equations of minimal surface type, including the equation

$$H[u] = f(x,u,Du) \text{ in } \Omega, \tag{12}$$

with $f(x,z,p)$ satisfying structure conditions (10), (11), was considered. It was proved that a solution u is Holder continuous on $\bar\Omega$ provided the boundary trace of u is Lipschitz continuous on $\partial\Omega$. In [S 2] the tangential gradient of u in a neighborhood of any point $x^o \in \partial_-\Omega$ is estimated for a generalized solution u of the Dirichlet problem for minimal surface equation, that is, for the equation (12) with $f \equiv 0$. In particular, this estimate implies the Lipschitz continuity of the boundary trace of u on compact sets contained in $\partial_-\Omega$.

Here we extend L. Simon's results to the case when f in the right hand side of (12) decreases as $|p|^{-1}$ when $|p| \to \infty$ and no restrictions on the derivatives of f are assumed (Theorem A below). Combining these results with estimates (7), we obtain analogous results for generalized solutions of the parabolic problem (1)-(3). These results are formulated in Theorem B at the end of the paper. As a corollary of Theorem B for domains Ω with $\partial_+\Omega \cap \overline{\partial_-\Omega} = \emptyset$ one can state that the generalized solutions constructed in [OU 1,2] are Holder continuous in $\bar\Omega \times [0,\infty)$ and Lipschitz continuous on $\partial\Omega \times [0,\infty)$.

Many of the calculations given in [S1, 2] remain valid in our case as well, almost without any change. For that reason we restrict ourself here to those places in the proofs where additional explanations and arguments are needed. One should read this paper along with the papers [S1, 2] by L. Simon.

The author wishes to thank G. Anzellotti and A. Visintin for the invitation to visit the University of Trento where the results of this paper were obtained.

2. Elliptic Problem

Throughout this section $u^\alpha = u^\alpha(x)$ are smooth solutions of

$$H[u^\alpha] + \alpha\Delta u^\alpha = f^\alpha(x, Du^\alpha) \text{ in } \Omega \tag{13}$$

$$u^\alpha = \psi \text{ on } \partial\Omega, \ \alpha \in (0,1]. \tag{14}$$

We assume that the functions $f^\alpha(x,p)$ and $\psi(x)$ satisfy the following conditions:

(H-1): $f^\alpha : \bar{\Omega} \times \boldsymbol{R}^n \to \boldsymbol{R}$ are smooth functions such that the inequality holds

$$|f^\alpha(x,p)| \ \leq \ K(1+|p|)^{-1} \tag{15}$$

with a positive constant K. There exists a function $f \in C^1(\Omega \times \boldsymbol{R}^n)$ such that $f^\alpha \to f, f^\alpha_{p_i} \to f_{p_i}, f^\alpha_{x_i} \to f_{x_i}$, when $\alpha \to 0$, uniformly on compact sets in $\Omega \times \boldsymbol{R}^n$

(H-2): $\partial\Omega \in C^4$ and

$$sup_{\partial\Omega}|\psi| \ \leq \ L_o; \ sup_{x,\bar{x}\in\partial\Omega} \frac{|\psi(x) - \psi(\bar{x})|}{|x - \bar{x}|} \ \leq \ L,$$

for some positive constants L_o and L. We denote by c_i, c various constants independent on α and determined by n, K, Ω, L_o and L. The dependence on other quantities will be specified.

Proposition 1. The following estimates hold:

$$sup_\Omega \ |u^\alpha| \ \leq \ c_o, \tag{16}$$

$$\int_\Omega \left(\sqrt{1+|Du^\alpha|^2} \ + \ \alpha|Du^\alpha|^2\right) \, dx \leq c_1, \tag{17}$$

$$sup_{\Omega'} \ |Du^\alpha| \ \leq \ c(\Omega'), \ \Omega \subset\subset \Omega. \tag{18}$$

The constant $c(\Omega')$ depends on the distance from Ω' to $\partial\Omega$.

Proof. The estimate (16) may be obtained by A. D. Aleksandrov's method (see [GT]).

Let $g : \boldsymbol{R} \to \boldsymbol{R}$ be an arbitrary continuous positive function, and Ω_+ be the upper contact set for $\hat{u}^\alpha = max \ \{u^\alpha; sup \ \psi_{\partial\Omega}\}$, that is,

$$\Omega_+ = \{x \in \Omega : \ (y-x) \cdot D\hat{u}^\alpha(x) \ + \ \hat{u}^\alpha(x) \ \geq \ \hat{u}^\alpha(y) \ \forall y \in \Omega\}$$

We write the equation (13) in a form

$$F_{ij}^{\alpha}(D^{\alpha}u)\, u_{x_i x_j}^{\alpha} = f^{\alpha}(x, Du^{\alpha})$$

with

$$F_{ij}^{\alpha}(p) = (1+|p|^2)^{-1/2}\left[\delta_i^j - (1+|p|^2)^{-1} p_i p_j\right] + \alpha\, \delta_i^j$$

Hence, we get (see [GT]), setting $h := sup_{\Omega}\, u^{\alpha} - sup_{\partial\Omega}\psi > 0$,

(19) $$\int_{B_{h/\mathcal{D}}} g(p)dp \le n^{-n} \int_{\Omega_+} g(Du^{\alpha}) \left[det(F_{ij}^{\alpha}(Du^{\alpha}))\right]^{-1} |f^{\alpha}(x, Du^{\alpha})|^n\, dx$$

where $B_{h/\mathcal{D}} = \{p \in \boldsymbol{R}^n : |p| < h/\mathcal{D}\}$, $\mathcal{D} = diam\ \Omega$. It is easy to see that $det\,(F_{ij}^{\alpha}(p)) \ge (1+|p|^2)^{\frac{(n+2)}{2}}$.

Thus, taking into account (15) and choosing $g(p) = (1+|p|^2)^{-1}$ we obtain

$$\int_{B_{h/\mathcal{D}}} (1+|p|^2)^{-1}\, dp \le (K/n)^n \int_{\Omega_+} dx \le (K/n)^n\ meas\ \Omega.$$

This implies the estimate $h \le c\mathcal{D}$ and so

$$sup_{\Omega} u^{\alpha} \le L_o + c\mathcal{D}.$$

The same estimate is, evidently, true for $sup_{\Omega}(-u^{\alpha})$, so we get (16). In order to obtain (17), we observe that

$$\int_{\Omega}\left[\frac{u_{x_i}^{\alpha}\, \eta_{x_i}}{\sqrt{1+|Du^{\alpha}|^2}} + \alpha\, u_{x_i}^{\alpha}\, \eta_{x_i} + f^{\alpha}(x, Du^{\alpha})\, \eta\right] dx = 0,\ \forall \eta \in H_0^1(\Omega).$$

Letting here $\eta = u^{\alpha} - \psi$ and using (15), (16), and (H-2), we easily arrive at (17).

As it was already mentioned, the estimate (18) can be obtained by a modification of the arguments of Temam [T], where such an estimate is established under conditions (10) and (11).

The Proposition 1 together with (H-2) imply

Proposition 2. There exist a function $u \in C^2(\Omega)$ such that $u^{\alpha_k} \to u$ in $C_{loc}^2(\Omega)$ for some sequence $\alpha_k \to 0$, and

(20) $$H[u] = f(x, Du)\ in\ \Omega,$$

(21) $$sup_{\Omega}\, |u| \le c_o;\ sup_{\Omega'}\, |Du| \le c(\Omega')\ \forall \Omega' \subset\subset \Omega,$$

$$(22) \qquad \int_\Omega \sqrt{1+|Du|^2}\,dx \;\le\; c_1$$

Below we omit subscript k and write u^α instead of u^{α_k}.

Proposition 3. (Compare with (1.8) in [S 2]). There exists a limit

$$(23) \qquad lim_{\alpha\to 0}\, \alpha \int_\Omega |Du^\alpha|^2\,dx \;=\; 0.$$

Proof. Let $Lip_\psi(\Omega)$ be a set of Lipschitz functions in $\bar\Omega$ coinciding with ψ on $\partial\Omega$, and

$$(24) \qquad g^\alpha(x) \;=\; f^\alpha(x, Du^\alpha(x)),\; g(x) \;=\; f(x, Du(x)).$$

Proposition 2 and hypothesis (H-1) imply that $g, g^\alpha \in L_\infty(\Omega)$ and $g^\alpha \to g$ for $\alpha \to 0$, uniformly on compact sets in Ω as well as in $L_q(\Omega)$ $\forall q < \infty$.

For $\varphi \in Lip_\psi(\Omega)$ consider the functionals

$$I(\varphi) \;=\; \int_\Omega \left(\sqrt{1+|D\varphi|^2} + g\varphi\right) dx,$$

$$I_\alpha(\varphi) \;=\; \int_\Omega \left(\sqrt{1+|D\varphi|^2} + g^\alpha\varphi\right) dx,$$

$$J_\alpha(\varphi) \;=\; I_\alpha(\varphi) \;+\; \frac{\alpha}{2}\int_\Omega |D\varphi|^2\,dx.$$

All of them are convex and because of (13), (14), u^α is a minimum point of J_α. So we have for any $\varphi \in Lip_\psi(\Omega)$,

$$\overline{lim}_{\alpha\to 0}\, J_\alpha(u^\alpha) \;\le\; lim_{\alpha\to 0}\, J_\alpha(\varphi) \;=\; I(\varphi).$$

On the other hand, we also have the inequality

$$inf_{\phi\in\, Lip_\psi(\Omega)}\, I(\varphi) \;\le\; \underline{\lim}_{\alpha\to 0}\, I(u^\alpha).$$

Hence one can conclude that

$$\overline{lim}_{\alpha\to 0}\, J_\alpha(u^\alpha) \;\le\; \underline{\lim}_{\alpha\to 0}\, I(u^\alpha) \;=\; \underline{\lim}_{\alpha\to 0}\, I_\alpha(u^\alpha).$$

It evidently implies (23). □

Now we recall some notations from [S 2].

$$v \;=\; \sqrt{1+|Du|^2},\; v_\alpha \;=\; \sqrt{1+|Du^\alpha|^2},$$

$$(25) \qquad \nu_i \;=\; v^{-1}\, u_{x_i},\; \nu_i^\alpha \;=\; v_\alpha^{-1}\, u_{x_i}^\alpha,\; i = 1, \ldots, n,$$

$$d(x) = dist\{x; \partial\Omega\}, \ \Omega_t = \{x \in \Omega : d(x) < t\},$$

$$B_\rho(x^o) = \{x \in \mathbf{R}^n : |x - x^o| < \rho\}.$$

For some point $x^o \in \partial_-\Omega$ and small δ we introduce a C^2 coordinate system $y = (y_1, ..., y_n)$ in $\bar\Omega \cap B_\delta(x^o)$ such that

$$y_n = d(x), \ e^{nl}(y) = e^{ln}(y) = 0 \text{ for } l = 1, ..., n-1, \tag{26}$$

where $(e^{ij})(y) = J(x)\, J'(x)$, $J = \left(\frac{\partial y_i}{\partial x_j}\right)$ is a nondegenerate Jacobian matrix of the map $y = y(x)$. Denote by

$$G_\rho = y(\bar\Omega \cap B_\rho(x^o)) \ \rho \in (0, \delta].$$

Following [S 2], we write a "wave" over a function of x-variables if it is calculated in y-variables, that is,

$$\tilde u(y) = u(x), \ \tilde u^\alpha(y) = u^\alpha(x), \ \tilde f(y, \tilde u_y) = f(x, Du)$$

and so on. Then by definition of $e^{ij}(y)$ we obtain

$$\tilde v = \sqrt{1 + e^{ij}\, \tilde u_{y_i}\, \tilde u_{y_j}}; \ \tilde v_\alpha = \sqrt{1 + e^{ij}\, \tilde u^\alpha_{y_i}\, \tilde u^\alpha_{y_j}}$$

We also use the notation:

$$\tilde\nu_i = \tilde v^{-1}\, e^{ij}\, \tilde u_{y_j}, \ \tilde\nu_i^\alpha = \tilde v_\alpha^{-1}\, e^{ij}\, \tilde u^\alpha_{y_j},$$

$$g^{ij} = e^{ij} - \tilde\nu_i\, \tilde\nu_j, \ g^{ij}_\alpha = e^{ij} - \tilde\nu_i^\alpha\, \tilde\nu_j^\alpha,$$

$$z^2 = 1 + \sum_{k,l=1}^{n-1} e^{kl}\, \tilde u_{y_k}\, \tilde u_{y_l} = \tilde v^2 - \tilde u^2_{y_n};$$

$$z_\alpha^2 = 1 + \sum_{k,l=1}^{n-1} e^{kl}\, \tilde u^\alpha_{y_k}\, \tilde u^\alpha_{y_l} = \tilde v_\alpha^2 - |\tilde u^\alpha_{y_n}|^2,$$

$$\chi = \frac{z^2}{v^2}, \ \chi_\alpha = \frac{z_\alpha^2}{v_\alpha^2}, \ \tau = \sqrt{1 + \sum_{l=1}^{n-1} \tilde u^2_{y_l}}, \ \tau^\alpha = \sqrt{1 + \sum_{l=1}^{n-1} |\tilde u^\alpha_{y_l}|^2}$$

Using this notation, one can rewrite the equation (13) in the form

$$\frac{d}{dx_i}\left(\nu_i^\alpha + \alpha\, u^\alpha_{x_i}\right) = f^\alpha, \tag{27}$$

or, in y-variables,

$$\frac{d}{dy_i}\left\{\mu\left(\tilde{\nu}_i^\alpha + \alpha\, e^{ij}\,\tilde{u}_{y_j}^\alpha\right)\right\} = \mu\,\tilde{f}^\alpha \tag{28}$$

where $\mu = det\left(\frac{\partial x_i}{\partial y_j}\right) = [det\,(e^{ij})]^{-1/2}$.

It is not difficult to deduce from (27) the relations

(29)

$$\frac{d}{dx_j}\left[v_\alpha\left(\delta_i^j - \nu_i^\alpha\,\nu_j^\alpha\right) + \frac{\alpha}{2}\,v_\alpha^2\left(\delta_i^j - 2\,\nu_i^\alpha\,v_j^\alpha\right)\right] = -\,v_\alpha\,\nu_i^\alpha\,f^\alpha\,,\quad i = 1,...,n.$$

In the case where $f^\alpha \equiv 0$ the relations (28) and (29) coincide respectively with (1.13) and (1.10) in [S 2]. The last one was a starting point in [S 2] in the proof of Lemma 1. In our case, for non-zero f^α, it is necessary to add one more term on the right hand side of the inequality in Lemma 1. This term is present because the right hand side in (29) does not vanish. It is equal to

$$\int_{G_\delta\cap\{y:y_n\,>\,t\}} \mu\,\tilde{v}_\alpha\,\tilde{\nu}_i^\alpha\,\tilde{f}^\alpha\,h\,d_{y_i}\,dy$$

Because of (15) it may be bounded by $c_3K\int_{G_\delta\cap\{y_n\,>\,t\}}|h|\,|\tilde{u}_{y_n}^\alpha|\,\tilde{v}_\alpha^{-1}\,dy$, with a constant c_3 depending only on the map $x \to y$.

In the same way as in [S 2], we obtain from Lemma 1 two important corollaries.

Corollary 1. Let $u = lim_{\alpha\to 0}\,u^\alpha$. Then for any non-negative function $h \in C^1(G_\delta)$ vanishing on $\partial G_\delta - \{y_n = 0\}$ the inequality

$$\int_{G_\delta\cap\{y_n\,>\,t\}} \Delta\,d\,h\tilde{v}\,dy + \int_{G_\delta\cap\{y_n\,=\,t\}} h\,\chi\,\tilde{v}\,dy' \le \tag{30}$$

$$\le c_2\int_{G_\delta\cap\{y_n\,>\,t\}}\left(h\,\chi\,\tilde{v} + \sqrt{\chi}\,\sqrt{g^{ij}\,h_{y_i}\,h_{y_j}}\,\tilde{v}\right)dy + c_3\,K\int_{G_\delta\cap\{y_n\,>\,t\}} h\,\frac{|\tilde{u}_{y_n}|}{\tilde{v}}dy$$

holds for almost each $t > 0$. Here $dy' = dy_1...dy_{n-1}$. It should be noted that the Proposition 3 is essential for the Proof of (30).

In comparison to the corresponding expression in Corollary 1 in [S 2] we have in (30) an additional term – the last integral on the right hand side.

We estimate this term now. At first, as in [S 2], we choose δ so small that

$$\Delta d \ge \kappa > 0 \text{ in } B_\delta(x^0) \cap \Omega\,. \tag{31}$$

Let

$$G'_{\delta,t} = \{y \in G_\delta : y_n > t : |\tilde{u}_{y_n}| > \frac{\kappa}{2c_3K}\, \tilde{v}^2\},$$

$$G''_{\delta,t} = \{y \in G_\delta : y_n > t;\ |\tilde{u}_{y_n}| \le \frac{\kappa}{2c_3K}\, \tilde{v}^2\}.$$

Then because of (31), we have

$$c_3\, K \int_{G''_{\delta,t}} h\, \frac{|\tilde{u}_{y_n}|}{\tilde{v}}\, dy \le \frac{1}{2} \int_{G_\delta \cap \{y_n > t\}} \Delta\, d\, h\, \tilde{v}\, d\, y.$$

On the other hand, if $y \in G'_{\delta,t}$ then

$$z^2 + \tilde{u}^2_{y_n} = \tilde{v}^2 \le 2c_3\, K\, \kappa^{-1}|\tilde{u}_{y_n}| \le \tilde{u}^2_{y_n} + c_3^2\, K^2\, \kappa^{-2},$$

that is, $z^2 \le c_3^2\, K^2\, \kappa^{-2}$, and therefore

$$\tau \le c_4 \text{ on } G'_{\delta,t}$$

for some c_4 depending on κ. Thus, replacing h by h^2 in (30), we obtain the inequality

$$\frac{\kappa}{2} \int_{G_\delta \cap \{y_n > t\}} h^2\, \tilde{v}\, dy + \int_{G_\delta \cap \{y_n = t\}} h^2\, \chi \tilde{v}\, dy' \le \tag{32}$$

$$\le c_2 \int_{G_\delta \cap \{y_n > t\}} \left(h^2\, \chi + h\, \sqrt{\chi\, g^{ij}\, h_{y_i}\, h_{y_j}}\right) \tilde{v} dy + c_3 K \int_{G_\delta \cap \{y_n > t; \tau \le c_4\}} h^2 \tilde{v}^{-1} |\tilde{u}_{y_n}|\, dy$$

The inequality (32) as well as (1.22) in [S 2] will be applied to functions h with supports on a set where τ is sufficiently large. In particular, if $h = 0$ while $\tau \le c_4$, then the last integral on the right hand side of (32) vanishes and (30) coincides with (1.22) in [S 2].

The Corollary 2 from [S 2] holds in our case with the corresponding constant depending on K as well.

The proof of the Theorem 1 carries over without significant changes but it is necessary to add

$$- \sum_{l=1}^{n-1} \frac{\partial}{\partial y_l} \left(\mu\, \tilde{f}^\alpha\, \tilde{u}^\alpha_{y_l}\right) \tag{33}$$

to the right hand side of (2.4). This theorem was used in [S 2] in the proof of Corollary 3. The statement of this Corollary is valid also in our case. However, its proof needs additional arguments because of the term (33). Namely, we have to estimate the following integral

(34) $$J = 2 \int_{G_\delta} \mu \tilde{f}^\alpha \sum_{l=1}^{n-1} \tilde{u}^\alpha_{y_l} \left[\gamma\gamma' \tau^\alpha_{y_l} (\tau^\alpha)^{-1} \zeta^2 + \gamma^2 \zeta \zeta_{y_l}\right] dy$$

where $\gamma = \gamma(\log \tau^\alpha)$, $\gamma \in C^1(\boldsymbol{R})$, $\gamma \geq 0$, $\zeta \in C^1(G_\delta)$ and $\gamma^2 (\log \tau^\alpha) \zeta^2$ has a compact support in G_δ.

First we establish the following

Proposition 4. The inequality

(35) $$\sum_{k,l=1}^{n-1} e^{kl} \xi_k \xi_l \leq z_\alpha^2 g^{ij} \xi_i \xi_j$$

holds for any $\xi = (\xi_1, ..., \xi_n) \in \boldsymbol{R}^n$.

Proof. It is convenient to use the following inner product and norm in $\boldsymbol{R}^n$

$$< \xi, \eta > = e^{ij} \xi_i \eta_j, \ ||\xi|| = < \xi, \xi >^{1/2}.$$

Then by definition of g_α^{ij} we may write

(36) $$g_\alpha^{ij} \xi_i \xi_j = ||\xi||^2 - (\tilde{v}^\alpha)^{-2} < \xi, \tilde{u}^\alpha_y > ^2,$$

where $\tilde{u}^\alpha_y = (\tilde{u}_{y_1}, ..., \tilde{u}^\alpha_{y_n})$. Let us denote by $\xi' = (\xi_1, ..., \xi_{n-1}, 0)$ and respectively $\tilde{u}^\alpha_{y'} = (\tilde{u}^\alpha_{y_1}, ..., \tilde{u}^\alpha_{y_{n-1}}, 0)$. It follows from (26) that

(37) $$||\xi||^2 = ||\xi'||^2 + \xi_n^2,$$

$$< \xi, \tilde{u}^\alpha_y >^2 = \left[< \xi', \tilde{u}^\alpha_{y'} > + \xi_n \tilde{u}^\alpha_{y_n}\right]^2 \leq$$

$$\leq q < \xi', \tilde{u}^\alpha_{y'} >^2 + \frac{q}{q-1} \left(\xi_n \tilde{u}^\alpha_{y_n}\right)^2 \ \forall q > 1$$

For $q = \tilde{v}_\alpha^2 z_\alpha^{-2}$, $\frac{q}{q-1} = \tilde{v}_\alpha^2 \left(\tilde{u}^\alpha_{y_n}\right)^{-2}$ it gives

(38) $$< \xi, \tilde{u}^\alpha_y >^2 \leq \tilde{v}_\alpha^2 \left[z_\alpha^{-2} ||\xi'||^2 ||\tilde{u}^\alpha_{y'}||^2 + \xi_n^2\right]$$

Here we have also used the Schwartz inequality. From (36)-(38) we obtain the inequality

$$g_\alpha^{ij} \xi_i \xi_j \geq ||\xi'||^2 \left(1 - z_\alpha^{-2} ||\tilde{u}^\alpha_{y'}||^2\right) = z_\alpha^{-2} ||\xi'||^2$$

which coincides with (35).

As a consequence of (35), we have for any $h \in C^1(G_\delta)$

$$|h_{y_l}(y)| \leq c\,\tau_\alpha \sqrt{g_\alpha^{ij}\, h_{y_i}\, h_{y_j}}\; l = 1, ..., n-1,\; y \in G_\delta.$$

Applying this inequality one can easily estimate the integral (34) over terms similar to those from the proof of Corollary 3 in [S 2].

Corollary 3 together with the inequality (1.22) was the basis of the proof of Theorem 2 in [S 2]. That theorem states that for an arbitrary function $h \in C^1(\bar{G}_\delta - \{y : y_n = 0\})$, vanishing on $\partial G_\delta - \{y : y_n = 0\}$, Sobolev inequality holds. It also holds in our case if, in addition, $h(y)$ vanishes when $\tau(y) \leq c_4$. Indeed, for such functions h the inequality (32) above is just the same as (1.22) in [S 2].

The main result of [S 2], Theorem 3, contains an estimate of the tangential gradient of u in $G_{\delta/4}$. To obtain such an estimate, Sobolev inequality was applied to the function $h(y) = \gamma(w(y))\,\bar{\zeta}(y)$, where $w = log\,\tau$, $\gamma \in C^1(\boldsymbol{R})$, $\gamma(t) = 0$ for $t < \lambda + \frac{\theta}{2}$, $\gamma(t) = 1$ for $t > \lambda + \theta$. It was assumed that $\zeta(x)$ are cut-off functions for sequence of balls of radii $\rho \in (\delta/2, \delta)$, with θ being positive numbers tending to zero, and λ a fixed, sufficiently large, positive constant. Now we have to require, additionally, that $\lambda \geq log\,c_4$. Then $h(y) = \gamma(log\,\tau(y))\,\zeta(y) = 0$ while $\tau(y) < c_4$. As it was proven above, it is possible to apply Sobolev inequality to such h and therefore one can estimate $sup_{G_{\delta/4}}\,\tau$ by the same arguments as in [S 2].

These tangential gradient estimates enable us to prove the Holder continuity of u in $B_{\delta/8}(x^o) \cap \bar{\Omega}$. It was mentioned in the Introduction that such a result has been proven in [S 1] for a class of equations including [12] provided structure conditions of type (10), (11) are fulfilled. A careful examination of all the proofs in [S 1] shows that assumptions of type (10), (11) there may be replaced by the condition (15) on f. Such an examination is rather routine and we do not present it here.

Now, we can state the following result for the elliptic problem.

Theorem A. Let u^α be smooth solutions of (13), (14) and

$$u^{\alpha_k} \to u \text{ in } C^2_{loc}(\Omega) \text{ for } \alpha_k \to 0.$$

If (H-1), (H-2) hold then, u satisfies (20)-(22). If, moreover, $x^o \in \partial_-\Omega$ and the inequality (31) holds for some positive κ and δ then

$$|u(x) - u(\bar{x})| \leq c|x - \bar{x}|^\beta,\; x, \bar{x} \in \bar{\Omega} \cap B_{\delta/8}(x^o)$$

and

$$|Du(x)|^2 - (Du(x)\cdot n(x))^2 \leq c,\; x \in B_{\delta/4}(x^o)$$

Here the constants c and $\beta \in (0,1)$ depend only on $n, \delta, \kappa, L_0, L, K$ and on $\partial\Omega$.

3. Parabolic Problem

In [OU 1,2] a generalized solution of (1)-(3) was defined as a limit

(39) $$u = lim_{\alpha_k \to 0} u^{\alpha_k}$$

where u^α were the solutions of regularized problems (4)-(6). It was proven that the convergence in (39) is uniform on compact sets of $\Omega \times [0, \infty)$ as well as on compact sets of $\bar{\Omega} \times [T, \infty)$ for some large $T > 0$. Moreover, the same convergence was proven for derivatives of u^{α_k}. Now for any given $t > 0$ one can consider $u^{\alpha_k}(\cdot, t)$ as a solution of (13), (14) with

$$f^\alpha(x,p) = (1 + |p|^2)^{-1/2} u_t^\alpha(x,t), \ \alpha = \alpha_k.$$

The hypothesis (H-1) is evidently true for such functions f^α with K in (15) coinciding with M_1 from (7) and $f(x,p) = (1+|p|^2)^{-1/2} u_t(x,t)$. Applying Theorem A to $u(\cdot, t)$ we come to the following

Theorem B. Let $u(x,t)$ be a generalized solution of (1)–(3), considered in [OU 1,2]. If (31) holds for some $x^o \in \partial_-\Omega$ and positive κ, δ, then the estimates

$$|u(x,t) - u(\bar{x},t)| \le c|x - \bar{x}|^\beta, \ x, \bar{x} \in \bar{\Omega} \times B_{\delta/8}(x^o),$$

$$|Du(x,t)|^2 - (Du(x,t) \cdot n(x))^2 \le c, \ x \in \bar{\Omega} \times B_{\delta/4}(x^o)$$

are valid for any $t \in [0, \infty)$. The constants c and $\beta \in (0,1)$ depend only on n, κ, δ, $C^2(\bar{\Omega})$-norm of u_o, on constants L_o, L from (H-1) and on $\partial\Omega$.

Remark. The dependence on C^2-norm of u_o is because of the constant M_1 from (7).

REFERENCES

[GT] Gilbarg, D. and Trudinger, N.S., *Elliptic Partial Differential Equations of Second Order*, Springer-Verlag, p. 511, 1983.

[OU 1] Oliker, V. I. and Uraltseva, N. N., *Evolution of Nonparametric Surfaces with Speed Depending on Curvature, II: The Mean Curvature Case.*, Commun. on Pure and Appl. Math. XLVI, 97-135, 1993.

[OU 2] Oliker, V. I. and Uraltseva, N. N., *Evolution of nonparametric surfaces with speed depending on curvature, III: Some remarks on mean curvature and anisotropic flows, in: Degenerate D-iffusions, Wei-Ming Ni, L. A. Peletier, and J.-L. Vazques, eds.*, IMA Volumes in Mathematics and its Applications, Vol. 47, Springer-Verlag, New York, to appear.

[S 1] Simon, L., *Global estimates of Holder continuity for a class of divergence-form elliptic equations*, Arch. for Rational Mech. and Analysis, 56, 253-272, 1974.

[S 2] Simon, L., *Boundary regularity for solutions of the non-parametric least area problem*, Annals of Math., 103, 429-455, 1976.

[T] Temam, R., *Solutions généralisées de certaines équations du type hypersurface minimales*, Arch. Rational Mech. Anal. 44, 121-156, 1971.

ST. PETERSBURG STATE UNIVERSITY, RUSSIA, AND EMORY UNIVERSITY, ATLANTA

Some questions of De Giorgi about mean curvature flow of triply-periodic surfaces

Brian White

ABSTRACT. This paper analyzes the long-term behavior of certain non-trivial triply periodic surfaces under the mean-curvature flow.

Let u_t be the solution of the level-set mean curvature flow equation $\frac{d}{dt}u_t = (\delta_{ij} - |Du_t|^{-2}D_i u_t D_j u_t)D_{ij}u_t$ with initial values

$$u_0(x, y, z) = \sin x + \sin y + \sin z$$

In particular, this means that for each s, the level set ${u_t}^{-1}(s)$ moves (as t increases) with velocity at each point equal to the mean curvature at that point (provided the level set is a smooth surface). (See [CGG] or [ES] for basic properties of the level-set mean curvature flow.) De Giorgi has asked whether the following hold for the particular initial function u_0 given above:

(1) For $s \neq 0$, ${u_t}^{-1}(s)$ converges to $\emptyset$ as $t \to \infty$.
(2) For $s \neq 0$, ${u_t}^{-1}(s)$ disappears within a finite time $t = t(s)$.
(3) The disappearing times $t(s)$ are bounded above. That is, there is a time $T < \infty$ such ${u_t}^{-1}(s)$ for all $s \neq 0$ and $t \geq T$.

These questions are related to one of the main open questions about level set flow: can an initially smooth surface (which of course must stay smooth for a short time) later "fatten" (develop non-empty interior)? Note that (3) is equivalent to $u_T \equiv 0$. Since (as is proved below) the 0-level set is initially smooth, this would give an example of such fattening.

In this note we show that assertions 1 and 2 are true, but 3 is false. This does not of course settle the fattening question in general. Nonetheless, the analysis below is interesting because it is very unusual to be able to describe in such detail the long-term behavior of the mean-curvature flow for such nontrivial initial data as the u_0 above. Also, some of the techniques below may be useful in more general situations. (For example, the proof below implies that for *any* triply-periodic initial function u_0, if one of the level sets flows as $t \to \infty$ to a smooth connected surface, then each of the other level sets vanishes in a finite time.)

Proposition 1. *The 0-level sets $u_t^{-1}(0)$ are smooth for all $t \geq 0$ and converge smoothly as $t \to \infty$ to a triply-periodic minimal surface.*

Proof. First note that the 0-level set is initially smooth everywhere. To see this, observe that

$$\nabla u_0(x, y, z) = (\cos x, \cos y, \cos z)$$

This gradient cannot vanish except where x, y, and z are odd multiples of $\frac{\pi}{2}$, which implies that $\sin x$, $\sin y$, and $\sin z$ are each ± 1, so that $u_0(x, y, z)$ is an odd number. In particular, ∇u_0 cannot vanish where u_0 does, so the 0-level set $u_t^{-1}(0)$ is initially smooth. Consequently it will stay smooth at least for small t.

Consider the portion S of $u_0^{-1}(0)$ in the region

$$0 \leq x \leq \pi/2, \quad 0 \leq y \leq \pi/2, \quad -\pi \leq z \leq 0$$

One easily checks that the boundary of S consists of four line segments P_0P_1, P_1P_2, P_2P_3, and P_3P_0, where

$$\begin{array}{ll} P_0 = (0,0,0) & P_1 = (\frac{\pi}{2}, 0, -\frac{\pi}{2}) \\ P_2 = (0,0,-\pi) & P_3 = (0, \frac{\pi}{2}, -\frac{\pi}{2}) \end{array}$$

Furthermore, for each of these four segments, the surface $u_0^{-1}(0)$ contains the entire line of which the segment is a part. Let $\phi : R^3 \to R^3$ be rotation through π about one of those lines. It is also easy to check that

$$u_0 \circ \phi = -u_0$$

(One way to see this is as follows. If R_1 and R_2 are the reflections in the $y = 0$ and $x = -z$ planes, respectively, then R_1 followed by R_2 is the rotation ϕ by π about the line P_0P_1. Thus

$$\begin{aligned} u_0 \circ \phi(x, y, z) &= u_0(R_2(R_1(x, y, z))) = u_0(R_2(x, -y, z)) \\ &= u_0(-z, -y, -x) = -u_0(x, y, z) \end{aligned}$$

The proof for the other lines is similar.)

It follows that $u_0^{-1}(0)$ is invariant under the group $\mathcal{G}$ of symmetries generated by these rotations. Indeed,

$$u_0^{-1}(0) = \bigcup_{g \in \mathcal{G}} g(S)$$

I claim that S is the graph over a convex quadrilateral Q in the $x = -y$ plane of a smooth function f_0:

$$S = \{(x, y, z) : \tfrac{x+y}{\sqrt{2}} = f(\tfrac{x-y}{\sqrt{2}}, z)\}$$

This is because the inner product of the normal ∇u_0 to S and the normal $(1, 1, 0)$ to the $x = -y$ plane is always positive. This can be seen as follows:

$$\nabla u_0 \cdot (1, 1, 0) = (\cos x, \cos y, \cos z) \cdot (1, 1, 0) = \cos x + \cos y$$

Since x and y are both between 0 and $\pi/2$ on S, this is positive unless $x = y = \pi/2$. But $u_0(x, y, z) = 0$ precludes x and y from both being $\pi/2$.

By standard parabolic PDE, there is a smooth solution

$$F : [0, \infty) \times Q \to R$$

to the nonparametric equation for mean curvature flow with initial values $F(0, \cdot) = f_0$ and with dirichlet boundary conditions, and $F(t, \cdot)$ converges smoothly to a solution of the minimal surface equation as $t \to \infty$. (Of course the convexity of the domain Q is essential.) Let S_t be the graph of $F(t, \cdot)$, and let

$$M_t = \bigcup_{g \in \mathcal{G}} g(S^t)$$

Then M_t is a smooth family of surfaces that flow by mean curvature. (If this is not clear, observe that M_t certainly flows by mean curvature except possibly along the lines of rotational symmetry, which we are holding fixed by imposing dirichlet boundary conditions on $F(\cdot, \cdot)$. But any surface with a line of rotational symmetry has zero mean curvature along that line, so M_t in fact flows with velocity everywhere equal to the mean curvature.) Also, $M_0 = {u_0}^{-1}(0)$. Thus $M_t = {u_t}^{-1}(0)$ for all t, because the classical mean-curvature flow and the level-set version of mean curvature flow coincide whenever the classical flow makes sense: that is, as long as the surface remains smooth. (See theorem 6.1 of [ES].) □

Proposition 2. *Let K be any triply periodic closed set disjoint from $M_0 = {u_0}^{-1}(0)$. Let K_t be the result of letting K evolve for time t as a level-set under the level set mean-curvature flow. Then there is a $T > 0$ such that $K_T = \emptyset$.*

Lemma. *Let $\bar{U}$ be a compact region in R^3 or in a flat 3-torus such that the mean curvature of the boundary of U is everywhere $\geq H > 0$. Let $K \subset U$ be a compact set and K_t be the result of letting K evolve for time t as a level set under the level set mean curvature flow. Then*

$$\tfrac{d}{dt}_* \operatorname{dist}(K_t, \partial U) \geq H$$

for all t until K_t disappears, where

$$\operatorname{dist}(A, B) = \inf\{|a - b| : a \in A, b \in B\}$$

and $\frac{d}{dt}_* f(t) = \liminf_{h \to 0^+} h^{-1}(f(t + h) - f(t))$. *In particular,* $K_t = \emptyset$ *when*

$$t > H^{-1} \max_{p \in U} \operatorname{dist}(p, \partial U)$$

Proof of lemma. Let V_t be the level set flow with $V_0 = \partial U$. Then V_t is smooth for small t so

$$\tfrac{d}{dt} \operatorname{dist}(V_t, V_0) \geq H \qquad (t = 0) \tag{1}$$

Also, $\text{dist}(K_t, V_t) \geq \text{dist}(K_0, V_0)$ (because the distance between disjoint sets increases under the level-set mean curvature flow; see theorem 7.3 of [ES]). Thus

$$\begin{aligned} \text{dist}(K_t, V_0) &\geq \text{dist}(K_t, V_t) + \text{dist}(V_t, V_0) \\ &\geq \text{dist}(K_0, V_0) + \text{dist}(V_t, V_0) \end{aligned} \tag{2}$$

The result follows immediately from (1) and (2). □

Proof of proposition 2. Let $M = \lim_{t\to\infty} M_t$ where $M_t = {u_t}^{-1}(0)$. Since M is a complete smooth non-planar minimal surface, it is unstable. Thus the first eigenfunction w of the Jacobi (or second variation) operator has a negative eigenvalue. Of course this eigenfunction does not change sign. (This is true for the first eigenfunction of any second-order self-adjoint elliptic operator on a connected oriented manifold with or without boundary. See for example [CH, p. 451] or [GT,§8.12]). Thus for sufficiently small s, the surface

$$\hat{M}(s) = \{x + sw(x) \cdot \nu(x) : x \in M\}$$

is strictly mean convex. That is, the mean curvature vector of $\hat{M}(s)$ is nowhere 0 and points away from M. (Here $\nu(x)$ is the unit normal to M at x.)

Since K_0 is disjoint from M_0, the distance from K_t to M_t increases with t. Since M_t converges to M smoothly, this means for some large τ, K_τ is disjoint from M. Therefore it is disjoint from $\hat{M}(s)$ and $\hat{M}(-s)$ for sufficiently small s. Let U be the region consisting of those points that are separated from M by $\hat{M}(s) \cup \hat{M}(-s)$. Then U has strictly mean convex boundary and K_τ lies in U, so proposition 2 follows immediately from the lemma. □

References

[CGG] Y. G. Chen, Y. Giga, and S. Goto, *Uniqueness and existence of viscosity solutions of generalized mean curvature flow equations*, J. Diff. Geom. **33** (1991), 749–786.

[CH] R. Courant and D. Hilbert, *Methods of Mathematical Physics Vol. I*, Wiley Interscience.

[ES] L. C. Evans and J. Spruck, *Motion of level sets by mean curvature I*, J. Diff. Geom. **33** (1991), 635–681.

[GT] D. Gilbarg and N. Trudinger, *Elliptic Partial Differential Equations of Second Order*, 2nd edition, Springer, 1983.

List of Contributors

J.F. Blowey, Department of Mathematical Sciences, University of Durham, Durham, DH1 3LE, England

Guy Bouchitté, Department of Mathematics, University of Toulon et du Var, BP 132, F-83957 La Garde Cedex, France

Gunduz Caginalp, Department of Mathematics, University of Pittsburgh, Pittsburgh, PA 15260, USA

David L. Chopp, Department of Mathematics, University of California, Los Angeles, California 90024, USA

Ennio De Giorgi, Scuola Normale Superiore, Piazza dei Cavalieri 7, 56126 Pisa, Italy

Charles M. Elliott, Mathematics Division, Sussex University, Brighton, BN1 9QH, England

Maurizio Falcone, Dipartimento di Matematica, Universitá La Sapienza, P. Aldo Moro 2, 00185 Roma, Italy

Morton E. Gurtin, Department of Mathematics, Carnegie Mellon University, Pittsburgh, PA 15213-3890, USA

Danielle Hilhorst, Laboratoire d'Analyse Numérique, CNRS et Université Paris Sud, F-91405 Orsay Cedex, France

Elisabeth Logak, Département de Mathématiques et d'Informatique, Ecole Normale Supérieure, 45 rue d'Ulm, F-75230 Paris Cedex 05, France

C. Lopez, CEREMADE, Université Paris Dauphine, Place du marechal de Lattre de Tassigny, F-75775 Paris Cedex, France

Jean Michel Morel, CEREMADE, Université Paris Dauphine, Place du marechal de Lattre de Tassigny, F-75775 Paris Cedex, France

Yasumasa Nishiura, Faculty of Integrated Arts and Sciences, Hiroshima University, Kagami-yama 1-7-1, Higashi-Hiroshima City, Hiroshima 724, Japan

Ricardo H. Nochetto, Department of Mathematics and Institute for Physical Science and Technology, University of Maryland, College Park, MD 20742, USA

Maurizio Paolini, Dipartimento di Matematica, Università di Milano, Via C. Saldini 50, 20133 Milano, Italy, and Istituto di Analisi Numerica del CNR, Corso Carlo Alberto 5, 27100 Pavia, Italy

Paolo Podio Guidugli, Dipartimento di Ingegneria Civile, Universitá Tor Vergata, Via della Ricerca Scientifica, 00173 Roma, Italy

Sergio Rovida, Istituto di Analisi Numerica del CNR, Corso Carlo Alberto 5, 27100 Pavia, Italy

P. Seppecher, Department of Mathematics, University of Toulon et du Var, BP 132, F-83957 La Garde Cedex, France

James A. Sethian, Department of Mathematics, University of California, Berkeley, California 94720, USA

John M. Sullivan, The Geometry Center, 1300 South Second Street, Minneapolis, MN 55454, USA

Nina N. Uraltseva, Steklov Institute of Mathematics, St. Petersburg Branch (POMI), Fontanka 27, St. Petersburg 191011, Russia

Claudio Verdi, Dipartimento di Matematica, Università di Milano, Via C. Saldini 50, 20133 Milano, Italy, and Istituto di Analisi Numerica del CNR, Corso Carlo Alberto 5, 27100 Pavia, Italy

Giorgio Vergara Caffarelli, Dipartimento di Metodi e Modelli Matematici per le Scienze Applicate, Università La Sapienza, Via A. Scarpa 10, 00161 Roma, Italy

Brian White, Mathematics Department, Stanford University, Stanford, CA 94305, USA

W. Xie, Department of Mathematics, University of Pittsburgh, Pittsburgh, PA 15260, USA

List of Participants

Frederick J. Almgren (Princeton University, USA)
Robert Almgren (The University of Chicago, USA)
Gabriele Anzellotti (Università di Trento, Italy)
Elisabetta Barozzi (Università di Trento, Italy)
Ana Cristina Barroso (Carnegie Mellon University, Pittsburgh, USA)
Giovanni Bellettini (Università di Bologna, Italy)
Marino Belloni (Università di Pisa, Italy)
Danilo Benarros (Università di Trento, Italy)
Luca Bonaventura (Università di Trento, Italy)
Guy Bouchitté (University of Toulon et du Var, France)
Andrea Braides (Università di Brescia, Italy)
Giuseppe Buttazzo (Università di Pisa, Italy)
Gunduz Caginalp (University of Pittsburgh, USA)
Francine Catté (Université de Paris-Dauphine, France)
Paolo Cermelli (Università di Torino, Italy)
Claire Chan (Stanford University, USA)
Vito Antonio Cimmelli (Università della Basilicata, Italy)
Alain Damlamian (Ecole Polytechnique Palaiseau, France)
Ennio De Giorgi (Scuola Normale Superiore, Pisa, Italy)
Silvano Delladio (Università di Trento, Italy)
Francesco Dell'Isola (Università "La Sapienza", Roma, Italy)
Antonio Di Carlo (Università "La Sapienza", Roma, Italy)
Gerhard Dziuk (Albert Ludwigs Universität Freiburg, Germany)
Klaus Ecker (The University of Melbourne, Parkville, Australia)
Charles M. Elliott (Sussex University, Brighton, England)
Michele Emmer (Università di Viterbo, Italy)
Maurizio Falcone (Università "La Sapienza", Roma, Italy)
Robert Finn (Stanford University, USA)
Irene Fonseca (Carnegie Mellon University, Pittsburgh, USA)
Lorenzo Freddi (Università di Pisa, Italy)
Mariano Giaquinta (Università di Firenze, Italy)
Eduardo H. Gonzalez (Università di Padova, Italy)
Robert Gulliver (University of Minnesota, Minneapolis, USA)
Morton E. Gurtin (Carnegie Mellon University, Pittsburgh, USA)
Stefan Hildebrandt (Universität Bonn, Germany)
Danielle Hilhorst (Université Paris Sud, Orsay, France)
Ulrich Hornung (Universität München, Germany)
Gerhard Huisken (Universität Tübingen, Germany)
John E. Hutchinson (Australian National University, Canberra, Australia)

Thomas Ilmanen (Institute for Advanced Study, Princeton, USA)
Ryo Kobayashi (Ryukoku University, Seta, Japan)
Witold Kosinski (Università "La Sapienza", Roma, Italy)
Elisabeth Logak (Ecole Normale Supérieure, Paris, France)
Stephan Luckhaus (Universität Bonn, Germany)
Enrico Magenes (Università di Pavia, Italy)
Halil Mete Soner (Carnegie Mellon University, Pittsburgh, USA)
Mario Miranda (Università di Trento, Italy)
Hans D. Mittelmann (Arizona State University, Tempe, USA)
Giuseppe Modica (Università di Firenze, Italy)
Jean Michel Morel (Université Paris Dauphine, France)
Paola Nardinocchi (Università "La Sapienza", Roma, Italy)
Ricardo H. Nochetto (University of Maryland, College Park, USA)
Vladimir Oliker (Emory University, Atlanta, USA)
Enza Orlandi (Università "La Sapienza", Roma, Italy)
Elisabetta Ossanna (Università di Trento, Italy)
Diego Pallara (Univesità di Lecce, Italy)
Maurizio Paolini (Università di Milano, Italy)
Colette Picard (Université de Paris-Sud, Orsay, France)
Paolo Podio Guidugli (Università "Tor Vergata", Roma, Italy)
Errico Presutti (Università "Tor Vergata", Roma, Italy)
Klaus Räwer (Universität Bonn, Germany)
Andrew Roosen (Rutgers University, New Brunswick, USA)
Friedrich Sauvigny (Universität Bonn, Germany)
Michelle Schatzman (Université Lyon I, France)
Giuseppe Scianna (Università di Siena, Italy)
P. Seppecher (University of Toulon et du Var, France)
Raul Serapioni (Politecnico di Milano, Italy)
James A. Sethian (University of California, Berkeley, USA)
Antonio Siconolfi (Università "La Sapienza", Roma, Italy)
Sergio Solimini (Università di Lecce, Italy)
Jiri Soucek (Czech Academy of Sciences, Praha, Czech Republic)
Panagiotis E. Souganidis (University of Wisconsin, Madison, USA)
Joel Spruck (University of Massachusetts, Amherst, USA)
Peter Sternberg (Indiana University, Bloomington, USA)
John M. Sullivan (The Geometry Center, Minneapolis, USA)
Nicola Taddia (Università di Pisa, Italy)
Italo Tamanini (Università di Trento, Italy)
Luc Tartar (Carnegie Mellon University, Pittsburgh, USA)
Jean Taylor (Rutgers University, New Brunswick, USA)
Alessandro Tiero (Università "Tor Vergata", Roma, Italy)
Vincenzo Tortorelli (Scuola Normale Superiore, Pisa, Italy)
Neil S. Trudinger (Australian National University, Canberra, Australia)
Nina N. Uraltseva (Steklov Institute of Mathematics, St. Petersburg, Russia)

Claudio Verdi (Università di Milano, Italy)
Vincenzo Vespri (Università di Pavia, Italy)
Augusto Visintin (Università di Trento, Italy)
Brian White (Stanford University, USA)

Bei Fragen zur Produktsicherheit wenden Sie sich bitte an:
If you have any questions regarding product safety,
please contact:

Walter de Gruyter GmbH
Genthiner Straße 13
10785 Berlin
productsafety@degruyterbrill.com